#4027

# THE MOTHS OF AMERICA NORTH OF MEXICO

# The Moths of America North of Mexico

INCLUDING GREENLAND

FASCICLE 6.1

# GELECHIOIDEA

## COSMOPTERIGIDAE

RONALD W. HODGES

SYSTEMATIC ENTOMOLOGY LABORATORY, SCIENCE AND EDUCATION ADMINISTRATION
UNITED STATES DEPARTMENT OF AGRICULTURE
BELTSVILLE, MARYLAND

COLOR PHOTOGRAPHS BY
RICHARD B. AND TATIANA DOMINICK AND CHARLES R. EDWARDS
THE WEDGE ENTOMOLOGICAL LABORATORY

LINE DRAWINGS BY
ELAINE R. HODGES
ANNE R. RICHARDSON
RONALD W. HODGES
LINDA A. HEATH

1978
*LONDON*

E. W. CLASSEY LIMITED AND
THE WEDGE ENTOMOLOGICAL RESEARCH FOUNDATION

This work is to be cited as
Hodges, Ronald W., *in* Dominick, R. B., *et al.*, 1978,
*The Moths of America North of Mexico*,
Fasc. 6.1, *Gelechioidea* (in part): *Cosmopterigidae*

DISTRIBUTORS

E. W. CLASSEY LTD
Park Road, Faringdon, Oxfordshire SN7 7DR, England

B. H. BLACKWELL LTD
Broad Street, Oxford, England

ENTOMOLOGICAL REPRINT SPECIALISTS
P.O. Box 77224, Dockweiler Station, Los Angeles, California 90007, U.S.A.

ANTIQUARIAAT JUNK
(Dr. R. Schierenberg & Sons N.V.), Postbox 5, Lochem, Holland



ISBN 0 900848 31 6: COMPLETE WORK   ISBN 0 86096 001 3: FASCICLE 6.1
LIBRARY OF CONGRESS CATALOG CARD NUMBER 78–149292

The text has been set in Monotype Baskerville and printed
on Unicorn Wove specially made for The Curwen Press
by Grosvenor Chater. The color plates have been reproduced in
four colors and printed by offset litho on Couché matt coated cartridge.

Design, color reproduction, printing and binding
by The Curwen Press, London, England

## CONTENTS

*TO MY WIFE*

# AUTHOR'S ACKNOWLEDGEMENTS

It is my pleasure to thank the following persons for loan of specimens to study and photograph, for loan of type-specimens and information about type-specimens, for information on life histories and host plants, for critical review of the manuscript and for numerous other favors: A. Blanchard, J. D. Bradley, J. F. G. Clarke, I. F. B. Common, C. V. Covell, Jr., J. P. Donahue, R. L. Fischer, M. O. Glenn, H. J. Hannemann, C. P. Kimball, G. E. Lewis, C. R. Patrick, L. L. Pechuman, J. A. Powell, K. Sattler, D. Smith, G. T. Strickland, L. G. Swash and B. Wright. To my fellow editors for their abundant help and constructive criticism, to Sally B. Adams for technical assistance during the preparation of this part, to Elaine R. Hodges, Anne R. Richardson and Linda A. Heath for the numerous and excellent line drawings. Finally to R. B. and T. Dominick and C. R. Edwards for the photographic work go special thanks.

# THE MOTHS OF AMERICA NORTH OF MEXICO

SUPERFAMILY

## GELECHIOIDEA

The Gelechioidea are among the largest and most important lepidopteran superfamilies and were formerly included in the Tineina. They are ditrysian Lepidoptera related to Yponomeutoidea and Copromorphoidea. They occur in all major zoogeographic regions with various groups exceptionally well developed in each, e.g., Holarctic Region: Coleophoridae, Elachistidae, Depressariinae, Gelechiinae, Scythridae, Momphidae; Palearctic Region: Symmocinae, Pleurotini; Oriental Region: Lecithocerinae; Australian Region: Xyloryctinae, Oecophorini, Stathmopodini, Cosmopteriginae, Gelechiidae; Oceania: Cosmopteriginae, Autostichinae; southern Ethiopian Region: Gelechiidae; and New World: Blastobasinae, Ethmiinae, Gelechiidae, Stenomatinae, Chrysopeleiinae. The Neotropical and Ethiopian Regions have not been adequately sampled or the groups in these regions studied to permit many generalizations.

Gelechioids usually are very small to medium-sized moths, with the head smooth scaled, the labial palpus upturned and often sickle shaped, the tongue scaled on the anterior surface basally, the maxillary palpus one to four segmented and folded over the base of the tongue, the chaetosema absent, the forewing usually with M absent in the cell and the tympanic organs absent. They are separated from the following five ditrysian microlepidopterous superfamilies as noted. The Cossoidea have a naked tongue, when present, and the forewing has M well developed and forked in the cell. The Tortricoidea have a rough-scaled head, porrect labial palpus, naked tongue, chaetosema usually present and the forewing with M present or absent in the cell. The Tineoidea have a rough-scaled head normally, naked tongue, maxillary palpus one to five segmented, drooping, porrect or ascending, not folded over base of tongue, and forewing with M (usually?) present in the cell. The Yponomeutoidea have the tongue usually naked, maxillary palpus various, rarely folded over base of tongue, and labial palpus various, sometimes upturned, third segment rarely acute. The Copromorphoidea have a naked tongue.

The Gelechioidea have the following characters: Head: smooth scaled; tongue usually developed, scaled on anterior surface, at least at base; labial palpus three segmented, usually upturned, rarely porrect with third segment slightly upturned, third segment normally acute; maxillary palpus one to four segmented, folded over base of tongue; ocelli present or absent; antenna simple or ciliate, rarely pectinate or bipectinate, pecten present or absent; ocelli present or absent; chaetosemata absent. Forewing narrowly lanceolate with apex acute to broad, pterostigma present or absent, cell open or closed, 10 to 12 veins present, 1A present or absent, 2A and 3A forked basally; retinaculum subcostal in male, usually subcubital in female; sometimes with secondary wing coupling mechanism of very small, closely spaced setae in a basal band near posterior margin. Hindwing linear to broad, fringe short to over five times width of wing; cell closed or open; six to nine veins present; frenulum a single bristle in male, usually three bristles in female; cubital pecten sometimes present; base of wing sometimes with hair pencils. Wings rarely reduced or non-functional, then linear or short lobes. Legs: epiphysis present; tibial spurs 0, 2, 4; hindtibia usually with numerous long scales on dorsal surface. Abdomen: frequently with strong dorsal spines; often with secondary hair pencils from lateral or ventral surface in male; segments seven and eight sometimes modified in male. Male genitalia: simple and symmetrical to highly specialized, by reduction or modification, and asymmetrical; aedoeagus sometimes ankylosed, rarely unsclerotized; valvae sometimes reduced and replaced functionally by modifications of preceding segments. Female: ovipositor usually adapted for external placement of eggs, sometimes for piercing; ostium bursae rarely on dorsal surface of abdomen. Egg: flat type, normally laid singly. Larva: three prothoracic prespiracular setae; abdominal setae L1 and L2 approximate, rarely widely separated; crochets in a circle or ellipse, or rarely two transverse rows. Pupa: obtect, frontoclypeal suture present or absent, without dorsal abdominal spines, maxillary palpus usually visible, labial palpus and prothoracic femur seldom exposed, usually segments five to seven movable in male, five and six in female.

The adults usually are inconspicuous night-flying moths in temperate regions. Some are brightly colored day fliers, particularly in the tropics. Most are attracted to light, but a significant number are not, and these include many of the Depressariinae of the Oecophoridae. Most species feed as adults and some are long lived, overwintering in this stage. Very few species appear to be attracted to baits of rotting fruits or other materials. Usually the moths are solitary, but some hibernating species congregate in very large numbers. Most species are very small to small, but some Xyloryctinae have a wing length in excess of 30 mm. Flight of many species is weak.

Eggs usually are laid singly on the host plant, sometimes inserted into flowers or crevices on the host. Larvae usually live in a shelter: leaf, stem or needle mines; stem or root galls; in flower buds, rolled or folded leaves; externally but with a portable case; in seeds; scavenging in fungi, dead wood, fungal galls; or a stem chamber but feeding on leaves. Pupation usually occurs in the larval abode with or without a cocoon, sometimes in a cell at the surface of the ground or just below the surface of the ground. The pupa does not emerge from the cocoon, gall, mine etc. at eclosion of the adult. In temperate regions most species are univoltine.

Very little is known of the immature stages. Most reports in the literature are from rearings of field-collected larvae or pupae. Extremely few species have been reared through a complete life cycle.

Economically important species are the pink bollworm, *Pectinophora gossypiella* (Saunders); the potato tuberworm, *Phthorimaea operculella* (Zeller); the Angoumois grain moth, *Sitotroga cerealella* (Olivier); the juniper webworm, *Dichomeris marginella* (Fabricius); the obscure-barred coast groundling, *Scrobipalpa ocellatella* (Boyd); lodgepole needleminer, *Coleotechnites milleri* (Busck); *Ithome concolorella* (Chambers); sweet-clover root borer, *Walshia miscecolorella* (Chambers); larch casebearer, *Coleophora laricella* (Hübner); elm casebearer, *C. ulmifoliella* McDunnough; the white-shouldered house moth, *Endrosis sarcitrella* (Linnaeus); and the brown house moth, *Hofmannophila pseudospretella* (Stainton).

## CLASSIFICATION

Families of microlepidoptera have been defined by vestiture, wing shape and venation, presence or absence of pterostigma, posture and shape of labial palpus, length and structure of antenna, presence or absence of tongue, ocelli and chaetosemata, presence or absence of spines on the dorsal surface of the abdomen, foreleg with or without epiphysis, presence or absence of whorls of spines at the apexes of tarsal segments and genital characters. Using these characters and some from the pupae, workers have proposed about 25 family names and several infrafamilial names in the Gelechioidea.

A survey of classifications of the Gelechioidea since its proposal by Fracker (1915) (Table 1) shows the development of the superfamily concept. Fracker's criteria were larval characters even though he did not give a set to define the superfamily. He (1915: 48) stated "The grouping and the division into families is principally a compilation from Walsingham and from published and unpublished work of August Busck. At the same time we believe that nothing in the larval structure precludes this arrangement and that it is a more natural classification than any published more than a decade ago [Dyar, 1903]." The treatment of Mosher is based on pupal characters, that of Forbes on larval, pupal and adult characters, that of McDunnough on current practice, that of Common on adult, pupal and larval characters and that of Bradley (although not stated) on adult characters. Fracker, Mosher, Forbes and McDunnough dealt with the North American, north of Mexico, fauna; Common with the world

TABLE I
CLASSIFICATIONS OF THE GELECHIOIDEA

| FRACKER (1915) | MOSHER (1916) | FORBES (1923) | MCDUNNOUGH (1939) | COMMON (1970) | BRADLEY (1972) |
|---|---|---|---|---|---|
| Ethmiidae | Lavernidae | Oecophoridae | Cosmopterygidae | Coleophoridae | Coleophoridae |
| Stenomidae | Scythridae | Xylorictidae | Epermeniidae | Agonoxenidae | Elachistidae |
| Hemerophilidae | Gelechiidae | (Stenomidae) | Gelechiidae | Elachistidae | Oecophoridae |
| Gelechiidae | Chrysopeleiidae | Gelechiidae | (incl. *Symmoca*) | Scythridae | Oecophorinae |
| Oecophoridae | Oecophoridae | (incl. *Symmoca*) | Oecophoridae | Stathmopodidae | Depressariinae |
| Blastobasidae | Stenomidae | Blastobasidae | Blastobasidae | Oecophoridae | Ethmiidae |
| Cosmopterygidae | Cosmopterygidae | Lavernidae | Stenomidae | Ethmiidae | Gelechiidae |
| | Elachistidae | (Momphidae, | Ethmiidae | Timyridae | Gelechiinae |
| | | Cosmopterygidae) | | Cosmopterigidae | Symmocinae |
| | | | | Cosmopteriginae | Blastobasidae |
| | | | | Momphinae | Stathmopodidae |
| | | | | Walshiinae | Momphidae |
| | | | | Metachandidae | Batrachedrinae |
| | | | | Anomologidae | Momphinae |
| | | | | Pterolonchidae | Cosmopteriginae |
| | | | | Blastobasidae | Blastodacninae |
| | | | | Xyloryctidae | Chrysopeleiinae |
| | | | | Stenomidae | Scythrididae |
| | | | | Gelechiidae | |
| | | | | (incl. *Symmoca*) | |
| | | | | Physoptilidae | |
| | | | | Strepsimanidae | |

TABLE II
PROPOSED CLASSIFICATION OF THE GELECHIOIDEA
(Colored figures of representative North American species indicated in parentheses)

**Oecophoridae**
- Depressariinae
  - Depressariini (1:19, 21–23)
  - Amphisbatini (1:24)
- Ethmiinae (1:10, 12–16)
- Peleopodinae
- Autostichinae
- Xyloryctinae
- Stenomatinae (1:8, 9, 11)
- Oecophorinae
  - Oecophorini (1:25–27; 2:6)
  - Stathmopodini
  - Pleurotini (1:28)
- Chimabachinae
- Deuterogoniinae
- Hypertrophinae

**Elachistidae**
- Coelopoetinae (2:7)
- Elachistinae (2:8–13)

**Pterolonchidae**

**Coleophoridae**
- Coleophorinae (1:2; 2:2, 3)
- Batrachedrinae (1:31, 32)

**Agonoxenidae**
- Agonoxeninae
- Blastodacninae
  - Blastodacnini (1:33, 34)
  - Parametriotini (2:15)

**Blastobasidae**
- Blastobasinae
  - Blastobasini (1:4–7)
  - Pigritiini (2:4, 5)
- Symmocinae (1:3, 35, 36)

**Momphidae** (1:29, 30; 2:14)

**Scythridae** (1:1; 2:1)

**Cosmopterigidae**
- Cosmopteriginae (1:17, 18, 20; 2:20)
- Antequerinae (4:44; 5:14)
- Chrysopeleiinae (1:62–64; 2:19)

**Gelechiidae**
- Anomologinae (1:37, 38, 40; 2:16, 18)
- Gelechiinae (1:39, 41–51, 54; 2:17)
- Anacampsinae (1:52, 53)
- Dichomeridinae (1:55–57)
- Chelariinae (1:58, 60, 61)
- Lecithocerinae (1:59)
- Physoptilinae

fauna with definitions of taxa occurring in Australia; and Bradley with the British fauna.

Most of these family group names are associated with distinctive groups of genera. However, when genital characters are considered in conjunction with other characters, the relationships and relative rank of many of the names must be reassessed. Some genera, closely associated on genital characters, are differently associated on venation. I have been concerned with this problem for several years and have reached tentative conclusions on the relative ranking of the groups and the relationships among some families but have not reached an equal degree of refinement for all. Table 11 shows my tentative classification of the Gelechioidea.

Ancestral gelechioids probably had relatively broad wings with a well-developed discal cell in fore- and hindwing; forewing with complete venation with 1A present; hindwing with nine veins, $R_s$ and $M_1$ separate; dorsal surface of the abdomen unspined; male and female abdomen with all segments unmodified; male genitalia symmetrical; valva a broad lobe; aedoeagus free; juxta a simple plate; gnathos a semicircular band with small spicules on dorsomedial surface; uncus a sclerotized lobe; female genitalia conservative, signum single, a series of inwardly directed, clustered projections; larva a concealed feeder on living plant tissue.

Specializations in modern species are numerous, and in some instances the male genitalia are among the most highly modified within the order. Narrowing of the fore- and hindwings has been associated with reduction in number of veins and usually with stalking of $R_4$ and $R_5$ and sometimes $R_4$, $R_5$, $M_1$ and $M_2$ in the forewing; an unformed cell and the loss of 1A. Characters of the hindwing are $R_s$ and $M_1$ stalked; loss of the base of M in the cell; $M_2$, $M_3$, $Cu_1$ and $Cu_2$ arising from a common stalk; and gradual reduction and loss of anal veins. Some very narrow-winged moths have only one or two strong veins; the other veins are markedly reduced. The abdomen sometimes has stout tergal setae and sternal scale tufts and rarely pleural scale tufts; in males modification of the eighth and sometimes seventh segment to form plates or bands that rarely replace the function of the valvae; in females a reduction in number of abdominal segments; migration of the ostium bursae to the dorsal surface; sclerotized plates (sometimes highly sculptured) around the ostium bursae; ductus bursae and corpus bursae variously sclerotized; development of zero to three, variously shaped signa; apophyses massive and heavily sclerotized; male genitalia with valvae divided into three or four discrete parts; aedoeagus ankylosed, often with saccus or juxta; manica heavily sclerotized; juxta with lateral lobes; transtilla developed; gnathos spinose or setose medially, sometimes with two arms, each terminating in a spined lobe; uncus and socii variously developed; uncus and/or gnathos absent. The larvae are concealed feeders, usually leaf rollers, tiers or folders. Some mine leaves or stems, are gall makers, are scavengers on plant or animal tissue, and some are case makers that mine leaves from the case.

Some explanations and comments on my classification are as follows: The Oecophoridae are as given in Fascicle 6.2 but with the addition of the Ethmiinae, Autostichinae, Xyloryctinae, Stenomatinae, Stathmopodini and Hypertrophinae. Although the male genital structure of *Ethmia* Hübner is somewhat different from that of other oecophorids, it is the gnathos, when present, that is particularly different. Venation, habitus, female genitalia and pupa ally *Ethmia* with the Depressariinae.

Xyloryctinae are a compact group associated with Depressariinae and Autostichinae. They are variable in forewing venation with 1A present or absent and in hindwing venation with $R_s$ and $M_1$ stalked, approximated or distant basally. Male and female genital characters are generally consistent. As the family was conceived by Meyrick, it contained an anomalous array of genera, some of which are referable to the Autostichinae, Depressariinae, Stenomatinae and Lecithocerinae. I place these genera as follows. Xyloryctinae: *Acompsogma* Meyrick, *Amorbaea* Meyrick, *Arignota* Turner, *Athrypsiastis* Meyrick, *Callicopis* Meyrick, *Capnolocha* Meyrick, *Chironeura* Diakonoff, *Cilicitis* Meyrick, *Clepsigenes* Meyrick, *Cryptophasa* McLeay, *Letogenes* Meyrick, *Linoclostis* Meyrick, *Metathrinca* Meyrick, *Niphorycta* Meyrick, *Pansepta* Meyrick, *Paralecta* Turner, *Phthonerodes* Meyrick, *Protrachyntis* Meyrick, *Scieropepla* Meyrick, *Stachyneura* Diakonoff, *Thymiatris* Meyrick, *Trypherantis* Meyrick, *Uzucha* Walker, *Xylodriadella* Fletcher and *Xylorycta* Meyrick. *Prothamnodes* Meyrick probably is not a xyloryctine but is retained in this subfamily until it can be placed correctly. Depressariinae: *Acria* Stephens, *Deloryctis* Meyrick and *Epichostis* Meyrick. Autostichinae: *Lasiodictis* Meyrick, *Nephantis* Meyrick, *Procometis* Meyrick, *Ptochorytis* Meyrick and questionably *Gemerodes* Meyrick. Lecithocerinae: *Aeolanthes* Meyrick, *Antolaea* Meyrick, *Epimactis* Meyrick, *Myriopleura* Meyrick, *Odites* Walsingham, *Rhizosthenes* Meyrick and *Trichernis* Meyrick. Gelechiidae without subfamily assignment: *Amphitrias* Meyrick and *Antithyra* Meyrick.

Autostichinae are a group of Asian-Oceanic genera and species that have been placed mainly in the Gelechiidae and in part in the Xyloryctidae. They have the forewing with 1A usually present and the hindwing rounded or slightly produced apically and $R_s$ and $M_1$ stalked or approximate basally; male genitalia symmetrical, gnathos a slender band and slightly expanded medially, uncus broad basally and sometimes apically; female not distinctive. Included genera are *Apethistis* Meyrick, *Autosticha* Meyrick, *Demiophila* Meyrick, *Encrasima* Meyrick, *Epicoenia* Meyrick, *Ischnodoris* Meyrick, *Nephantis* Meyrick, *Pachnistis* Meyrick, *Paradoris* Meyrick, *Procometis* Meyrick, *Protobathra* Meyrick, *Semnolocha* Meyrick, *Stochastica* Meyrick, *Stoeberhinus* Butler and *Thyrocopa* Meyrick.

Stenomatinae are a numerous, relatively homogeneous, mainly neotropical group most closely related to Xyloryctinae and Autostichinae. Vein 1A usually is present in the forewing, and $R_s$ and $M_1$ usually are stalked in the hindwing. Male genitalia consistently have forked or palmate setae arising from the valvae, female genitalia are not consistently different from those of related oecophorid subfamilies.

Stathmopodini are very closely allied with Oecophorini on characters of the male and female genitalia, head and larval feeding habits. They differ in the generally more narrow wings, the forewing usually with $M_1$ absent and the hindwing with the cell open and $R_s$ and $M_1$ approximate or stalked basally.

Elachistidae are closely related to Oecophoridae and are questionably separate. They may be polyphyletic with *Onceroptila* Braun, *Stephensia* Stainton, *Hemiprosopa* Braun, *Elachista* Treitschke, *Cosmiotes* Clemens, *Dicranoctetes* Braun and *Perittia* Stainton derived from the Depressariinae and *Coelopoeta* Walsingham derived from the Oecophorini. Wing venation, habitus and larval feeding habits will separate them from the Oecophoridae. I propose the name Coelopoetinae, NEW SUBFAMILY, type-genus *Coelopoeta* Walsingham, for this one genus. Braun (1948: 8) gives characters separating Coelopoetinae from Elachistinae.

Pterolonchidae are somewhat anomalous. They appear to be closely allied with Pleurotini in some respects but differ in lacking the tongue, the forewing having $R_4$ and $R_5$ fused and both male and female genitalia being unlike those of Pleurotini.

Coleophoridae have two subfamilies, Coleophorinae and Batrachedrinae, and are most closely related to the Agonoxenidae. I have transferred the Batrachedrinae from the Momphidae because of differences in the male and female genitalia and venation. The forewing with $Cu_2$ usually arising at a right angle from the cell and the larvae of some species of *Batrachedra* Herrich-Schäffer making cases as do those of *Coleophora* Zeller associate the Batrachedrinae with Coleophorinae.

Agonoxenidae with two subfamilies, Agonoxeninae and Blastodacninae, are a small family based on venational, genital, larval and pupal characters. Agonoxeninae have one genus, *Agonoxena* Meyrick. Blastodacninae have two tribes: Blastodacnini with *Blastodacna* Wocke, *Glyphipteryx* Curtis, *Cladobrostis* Meyrick, *Glaucacna* Forbes, *Heinemannia* Wocke, *Homeoprepes* Clarke, *Microcolona* Meyrick, *Nanodacna* Clarke, *Pammeces* Zeller, *Spuleria* Hofmann and *Zaratha* Walker; Parametriotini with *Aetia* Chambers (= *Tetanocentria* Rebel, 1902, NEW SYNONYMY; = *Platybathra* Meyrick, 1912, NEW SYNONYMY) and *Parametriotes* Kuznetsov. Agonoxenidae are related to the Coleophoridae and may belong in the same family.

Blastobasidae are similar in many characters to Coleophoridae and Agonoxenidae, but the male genitalia differ. The Blastobasinae have a pterostigma in the forewing; Symmocinae do not. I have transferred Symmocinae to Blastobasidae mainly because *Gerdana caritella* Busck has characters that bridge the gap between the two groups: 1A is developed at the margin of the forewing, $R_s$ and $M_1$ are separate basally in the hindwing (blastobasid characters), and the male genitalia are distinctly symmocid. I consider Holcopogonidae to be a junior synonym of Symmocinae, NEW SYNONYMY.

Momphidae are somewhat puzzling. With the removal of the Batrachedrinae only the genus *Mompha* Hübner remains. It is similar to Blastobasidae but differs in lacking a pterostigma; the male genitalia lack the gnathos; the female genitalia have the bursa with paired, somewhat hook-shaped signa; and the larvae feed on living plant tissue.

Scythridae are the most highly evolved gelechioid family considering the reduction of veins, modification of male genitalia (reduced, asymmetrical, parts fused, lost) and specialization of the eighth abdominal segment in males. Larvae

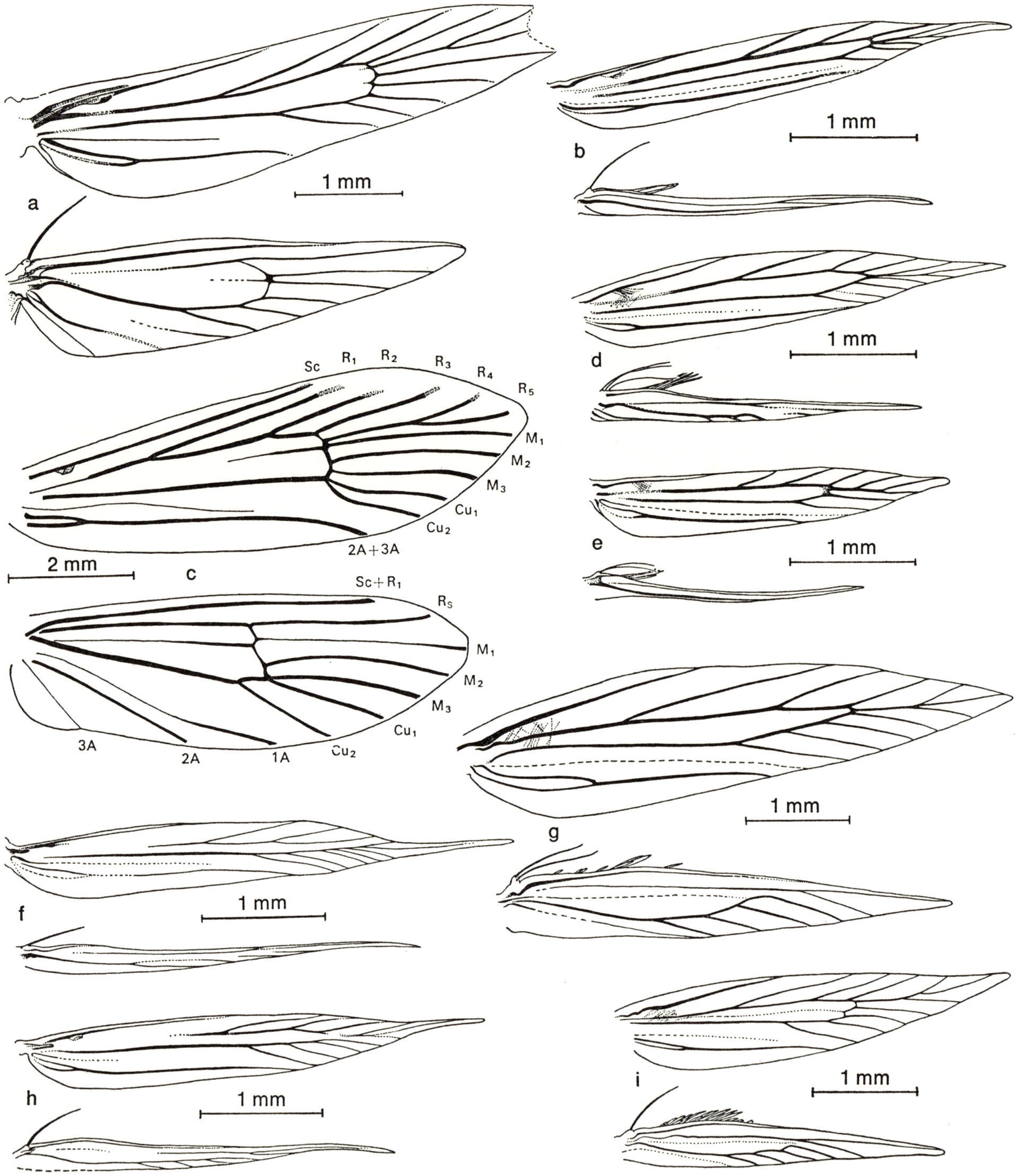


FIGURE 1: VENATION OF COSMOPTERIGIDAE SPECIES

*a.* Fore- and hindwing, *Euclemensia bassettella.*
*b.* Fore- and hindwing, *Tanygona lignicolorella.*
*c.* Fore- and hindwing, *Antequera acertella.*
*d.* Fore- and hindwing, *Eralea albalineella.*
*e.* Fore- and hindwing, *Melanocinclis lineigera.*
*f.* Fore- and hindwing, *Cosmopterix zieglerella.*
*g.* Fore- and hindwing, *Stagmatophora sexnotella.*
*h.* Fore- and hindwing, *Pebops ipomoeae.*
*i.* Fore- and hindwing, *Stagmatophora heydeniella.*

feed on living plant tissue and are leaf miners and/or skeletonizers. Mackay (1972: 23) suggested that the Scythridae are a highly specialized group of the Blastobasidae because larvae of species of each have a submental pit. However, Weisman (personal communication) says that a submental pit is present in the following gelechioids: *Martyringa* Busck, *Endrosis* Hübner some Autostichinae and Xyloryctinae (Oecophoridae); some *Batrachedra* Herrich-Schäffer and *Homaledra* Busck (Coleophoridae); Blastobasinae (Blastobasidae); and some Gelechiinae (Gelechiidae). Presence of a submental pit in the larva, although suggestive of relationship, is not a definite family character.

Cosmopterigidae have three subfamilies with similar venation but different male genitalia and habitus. In Antequerinae the aedoeagus is free, and the paired lobes of the uncus (gnathos?) are symmetrical. In Cosmopteriginae the aedoeagus is ankylosed by a heavily sclerotized manica with the diaphragma; the uncus is absent; and the brachia of the gnathos are asymmetrical. In Chrysopeleiinae the aedoeagus is ankylosed, usually by a heavily sclerotized, ventral band to the saccus; the uncus may be present; and the gnathos is absent.

Gelechiidae are a large, diverse family with several subfamilies. I have placed *Anomologa demens* Meyrick (type-genus of Anomologidae) in the Gelechiidae. On characters other than the reduced tongue and somewhat aberrant labial palpus *Anomologa* Meyrick is referable to the Gelechiidae. *Euryctista hobohmi* Janse from South Africa, in the Apatetrinae, is a very similar moth with somewhat reduced tongue, normal labial palpi and male genitalia. For this reason I treat *Anomologa* as a gelechiid and Apatetrinae as a junior synonym of Anomologinae, NEW SYNONYMY. Lecithocerinae differ from other Gelechiidae mainly in the uncus-gnathos structure of the male genitalia. This difference is no greater than the differences in male genitalia among the other subfamilies; thus, I treat them as a subfamily of Gelechiidae. The two species of *Physoptila* Meyrick have 1A present in the forewing, an oecophorid character. In the hindwing vein $M_1$ is fused with $R_s$ or $M_2$. Either is an anomalous condition. The shape of the hindwing and aspect of the male genitalia are similar to Gelechiidae. Tentatively I treat the Physoptilinae as a subfamily of Gelechiidae.

I have studied the type-specimens of *Strepsimanes scieropis* Meyrick and find it to be a noctuoid. Strepsimanidae Meyrick, 1930 is transferred to the superfamily Noctuoidea Latreille, 1809.

FAMILY

## Cosmopterigidae

Type-genus: *Cosmopterix* Hübner, 1825.

Cosmopterygidae Heinemann, [1876] 1877, *in* Heinemann and Wocke, *Die Schmetterlinge Deutschlands und der Schweiz*, **2** (2): 403.

NOTE—Heinemann spelled the family name Cosmopterygidae based on an emended spelling *Cosmopteryx*, which is etymologically correct and which has been used by almost all authors. The family name, based on the arbitrary combination of letters *Cosmopterix*, is correctly spelled Cosmoptericidae. Proponents of usage argue for the spelling Cosmopterigidae, and it is used in this Fascicle.

The Cosmopterigidae, with more than 1200 described species, occur in all zoogeographic areas and are prominent in Oceania. Walsingham (1907) described approximately 250 species from the Hawaiian Islands, and of these 177 species belong in *Hyposmocoma* Butler. Twenty-six genera and 180 species occur in North America north of Mexico. *Cosmopterix* is cosmopolitan; *Limnaecia* and *Stagmatophora* would be cosmopolitan except for their absence in South America; *Pyroderces* is pantropical to warm temperate; *Euclemensia* and *Sorhagenia* are holarctic; *Ithome*, *Pristen*, *Stilbosis*, *Triclonella* and *Walshia* are nearctic and neotropical; and *Afeda*, *Anoncia*, *Antequera*, *Chrysopeleia*, *Eralea*, *Melanocinclis*, *Nepotula*, *Obithome*, *Pebops*, *Periploca*, *Pristen*, *Siskiwitia*, *Synploca*, *Tanygona* and *Teladoma* are endemic to North America.

Cosmopterigid adults have narrow to very narrow wings. Those with narrow wings have the apices rounded; those with very narrow wings have the apices acuminate. They may be reliably distinguished from species of other gelechioid families only by characters of the male genitalia. The male genitalia have the aedoeagus usually ankylosed, the vinculum reduced, the uncus or gnathos absent and the gnathos as two usually asymmetrical lobes that lack apical setae. Most very narrow-winged species can be recognized by the absence of vein 1A in the forewing. Broader winged species can be confused with Agonoxenidae, Blastobasidae, Gelechiidae and Oecophoridae. They differ from Blastobasidae and Oecophoridae by lacking vein 1A in the forewing, from Agonoxenidae by lacking either uncus or gnathos and from Gelechiidae by the apex of the hindwing being rounded rather than produced to a slender point.

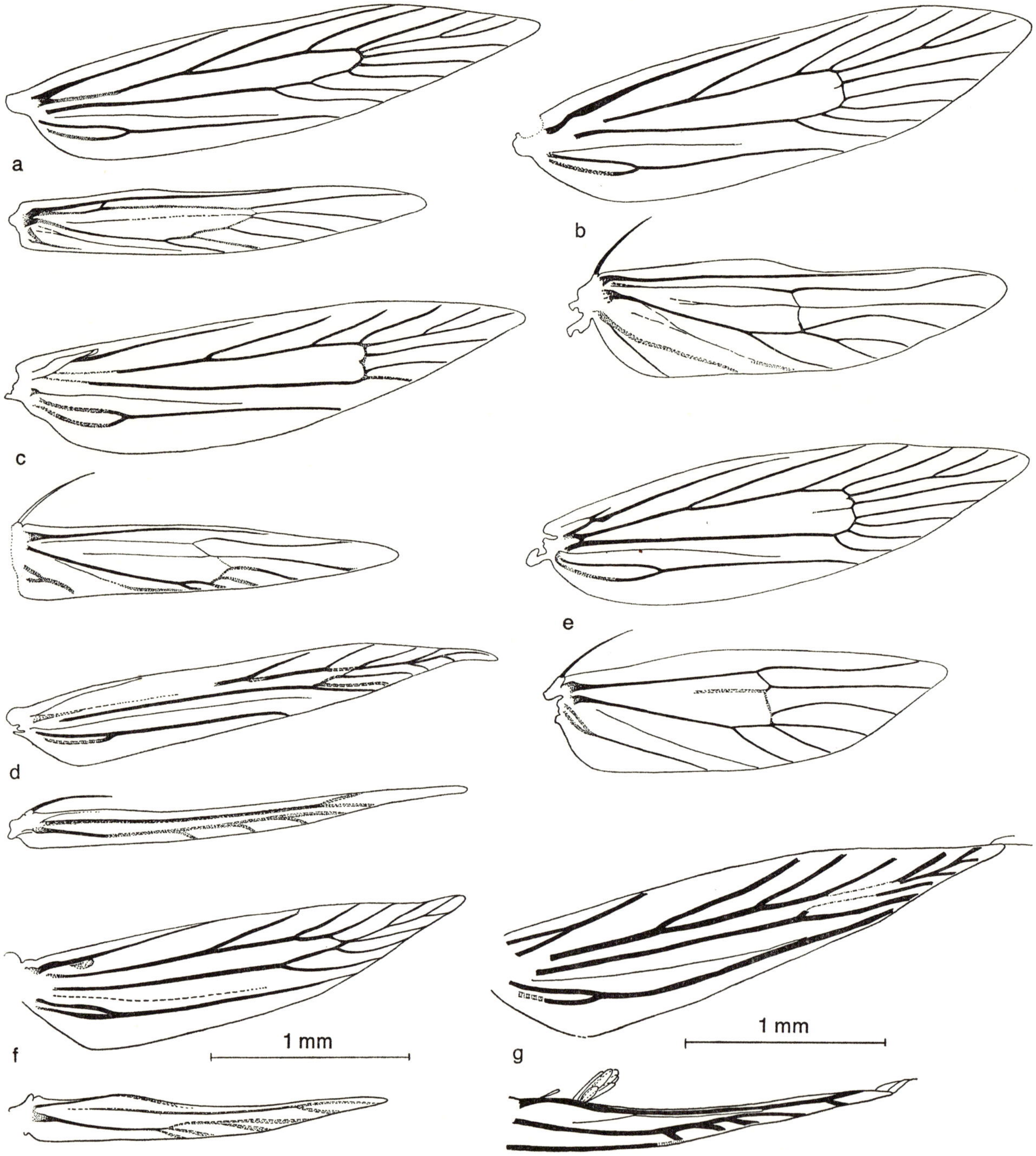


FIGURE 2: VENATION OF COSMOPTERIGIDAE SPECIES

*a*. Fore- and hindwing, *Limnaecia phragmitella*.
*b*. Fore- and hindwing, *Triclonella pergandeella*.
*c*. Fore- and hindwing, *Teladoma helianthi*.
*d*. Fore- and hindwing, *Pyroderces rileyi*.
*e*. Fore- and hindwing, *Anoncia conia*.
*f*. Fore- and hindwing, *Siskiwitia alticolans*.
*g*. Fore- and hindwing, *Periploca orichalcella*.

Not to scale: *a*, *b*, *c*, *d*, *e*.

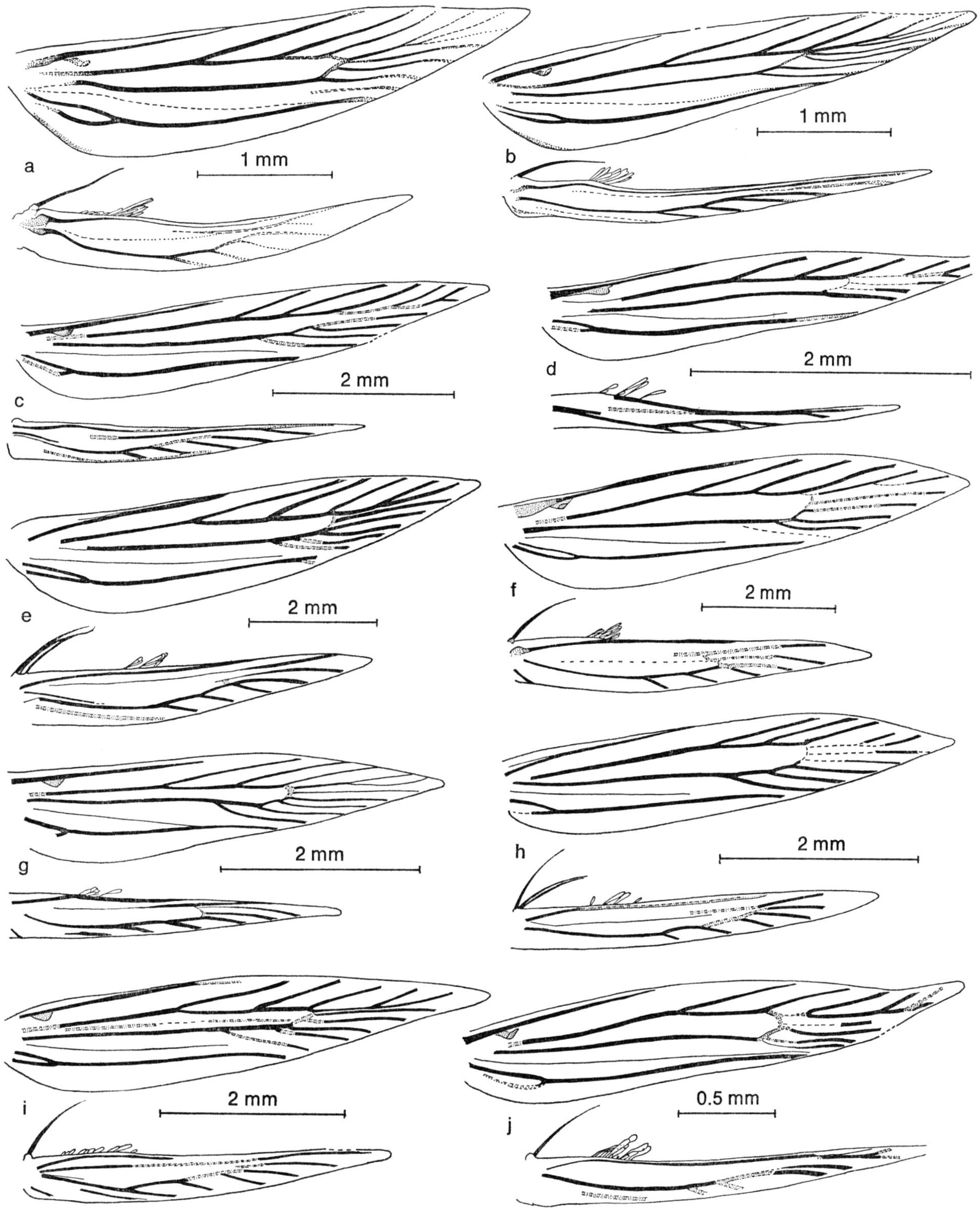


FIGURE 3: VENATION OF COSMOPTERIGIDAE SPECIES

*a*. Fore- and hindwing, *Pristen corusca*.
*b*. Fore- and hindwing, *Synploca gumia*.
*c*. Fore- and hindwing, *Stilbosis nubila*.
*d*. Fore- and hindwing, *Chrysopeleia purpuriella*.
*e*. Fore- and hindwing, *Walshia amorphella*.
*f*. Fore- and hindwing, *Nepotula secura*.
*g*. Fore- and hindwing, *Afeda biloba*.
*h*. Fore- and hindwing, *Perimede erransella*.
*i*. Fore- and hindwing, *Sorhagenia nimbosa*.
*j*. Fore- and hindwing, *Ithome concolorella*.

Major characters of the Cosmopterigidae are as follows: labial palpus recurved, very slender and sickle shaped to somewhat stout; antenna $\frac{1}{2}$ to $\frac{3}{4}$ length of forewing; pecten present or absent and of few hairlike setae or numerous, somewhat broader scales; ocelli present or absent. Forewing broadly lanceolate to very slender, apex rounded to acuminate, raised scales present or absent; cell usually closed; 10 to 12 veins present; 1A absent; $R_4$ and $R_5$ stalked, sometimes $R_4$, $R_5$, $M_1$ and $M_2$ stalked; 2A and 3A forked basally. Hindwing lanceolate to linear, apex rounded to acute, cell open or closed, usually open in narrow-winged forms; sometimes with a tuft of stout scales on costal margin near base in narrow-winged forms; six to eight veins present; $R_s$ and $M_1$ stalked in narrow-winged forms, separate and somewhat parallel in broader winged forms. Abdomen without stout setae on dorsal surface; male seventh and eighth segments sometimes modified as heavily sclerotized plates, often with lobes from intersegmental membrane between eighth segment and genital capsule. Male genitalia symmetrical or asymmetrical; aedoeagus usually ankylosed, cornuti present or absent; vinculum and tegumen often highly reduced; gnathos present and asymmetrical or absent; uncus present or absent. Female with apophyses short to long, sometimes massive; ostium bursae on seventh segment; corpus bursae a single or double lobe; signa sometimes absent, usually present and paired. Larval and pupal characters are given in the subfamily treatments.

Adult habits and behavior are very poorly known. Most species have been collected at light, but some species of *Cosmopterix* are both diurnal and nocturnal, and *Ithome concolorella* is crepuscular and nocturnal. Larvae feed internally as leaf miners, stem or root gall makers, seed feeders, within flower buds or scavengers on plant materials. Ovoviviparity probably occurs in *Stilbosis* and *Synploca* (Hodges, 1964).

Morphological trends in males begin with the primitive genera *Antequera* and *Euclemensia* in which the valvae are simple lobes; the aedoeagus is free; the juxta is well developed and slightly asymmetrical; the vinculum is well developed; and the lobes of the gnathos are symmetrical. This type of genitalia may have been derived from Oecophoridae in which the lobes of the gnathos are free, the uncus is absent, and the anal tube is dorsad of its normal position. In these two genera the forewing lacks vein 1A. Subsequent development in the Cosmopteriginae led to reduction of the vinculum, fusion of the juxta with the aedoeagus, and reduction of the left lobe of the gnathos. Derivation of the terminal, lobed structures from the gnathos or uncus is problematical. In most broad-winged species the lobes may be elements of the uncus with a few setae at the base representing the socius area. Simultaneously, the anal tube has migrated dorsally between these lobes, and it touches the caudal margin of the tegumen. In narrow-winged forms the two lobes appear to be gnathos, particularly in *Cosmopterix* species that have a striate pattern on the dorsal surface of the right lobe, a character of the Oecophorinae. Juxtal lobes occur in many genera. These may be free distally and parallel with the valvae or fused basally and parallel with the aedoeagus. In some specialized genera the aedoeagus is tripartite, consisting of a distal bulbous lobe, a basal cylinder that extends into the distal lobe and an incomplete flange at the base of the distal lobe. Modifications of the seventh and eighth abdominal segments and the intersegmental membrane just before the genital capsule occur in many advanced genera.

In the Chrysopeleiinae modification of the male genitalia and associated abdominal segments is in different directions. The valva can be a relatively simple lobe as in *Periploca*, a lobe with well-defined fracture zones as in *Walshia*, highly complex with three well-defined lobes as in *Sorhagenia*, reduced and replaced in function by the eighth abdominal sternum as in *Chrysopeleia* or a pair of lateral lobes from the intersegmental membrane as in *Obithome*. The aedoeagus is reduced in *Sorhagenia*. Modification of the female is that for supposed piercing oviposition. The apophyses are massive, and the lateral margins of the ovipositor are heavily sclerotized and slightly serrate in *Ithome* and *Obithome*.

Meyrick (1915) proposed Diplosaridae for species which have a broader hindwing with regularly arched costal margin and which lack the tuft of stout scales at the base of the costa. In 1922 he proposed Scaeosophides for a group of Oecophoridae which differ in having the male antenna moderately to minutely ciliated, the forewing with vein $R_5$ running to the costal margin, and the hindwing with a moderately broad apex and a hyaline patch in the cell causing a modification in venation. Clarke (1955: 21) synonymized Scaeosophidae with Cosmopterigidae, and Diakonoff (1967: 160) treated this group as a subfamily of the Cosmopterigidae. Genital characters

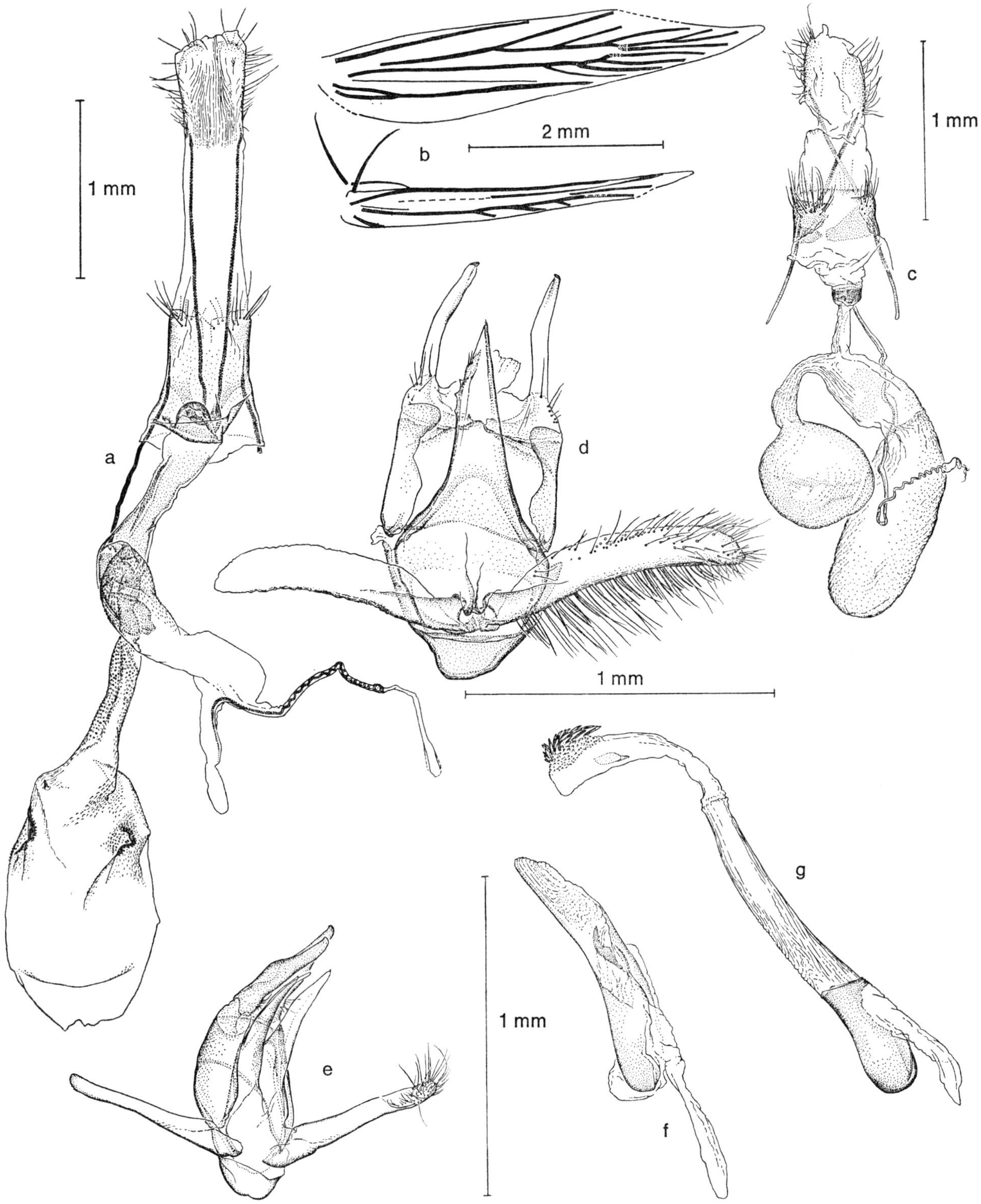


FIGURE 4: VENATION AND GENITAL FEATURES OF COSMOPTERIGIDAE SPECIES

*a*. Female genitalia, *Antequera acertella*.
*b*. Fore- and hindwing, *Obithome punctiferella*.
*c*. Female genitalia, *Euclemensia bassettella*.
*d*. Male genitalia (aedoeagus omitted), *Antequera acertella*.
*e*. Male genitalia (aedoeagus omitted), *Euclemensia bassettella*.
*f*. Aedoeagus, *Euclemensia bassettella*.
*g*. Aedoeagus, *Antequera acertella*.

of species assigned to the Diplosaridae and Scaeosophinae are as for Cosmopteriginae, so I treat the three names as synonymous.

Classification of the genera is based on labial palpi, antennae, ocelli, wing venation and scaling, genitalia and eighth segment of the male abdomen. Larvae and pupae of too few species are known to make comparative statements based on their characters.

The following arrangement of subfamilies and genera is used in this work.

FAMILY **COSMOPTERIGIDAE**

SUBFAMILY **Antequerinae**

| | | |
|---|---|---|
| *Antequera* | *Euclemensia* | |

SUBFAMILY **Cosmopteriginae**

| | | |
|---|---|---|
| *Cosmopterix* | *Melanocinclis* | *Teladoma* |
| *Pebops* | *Stagmatophora* | *Triclonella* |
| *Tanygona* | *Pyroderces* | *Anoncia* |
| *Eralea* | *Limnaecia* | |

SUBFAMILY **Chrysopeleiinae**

| | | |
|---|---|---|
| *Periploca* | *Chrysopeleia* | *Sorhagenia* |
| *Siskiwitia* | *Walshia* | *Ithome* |
| *Pristen* | *Nepotula* | *Obithome* |
| *Synploca* | *Afeda* | |
| *Stilbosis* | *Perimede* | |

## KEY TO SUBFAMILIES OF COSMOPTERIGIDAE

1. Male with nearly symmetrical genitalia, aedoeagus not ankylosed, lobes of juxta symmetrical to slightly asymmetrical, valva simple lobe, vinculum well developed, lobes of gnathos symmetrical; female with ostium bursae in membrane between seventh and eighth abdominal sterna; antenna without pecten.......... Antequerinae this page

— Male with symmetrical or asymmetrical genitalia, aedoeagus ankylosed, vinculum and tegumen usually reduced, gnathos absent or asymmetrical, right lobe larger than left lobe; female with ostium bursae on seventh abdominal segment, often surrounded or preceded by heavily sclerotized lamellae; antenna with or without pecten...... 2

2. Male genitalia asymmetrical, gnathos two separate lobes, uncus absent, aedoeagus ankylosed with diaphragma; labial palpus often very slender, smooth scaled, sickle shaped; male abdomen often appearing blunt apically; forewing often brightly colored with well-defined pattern, without patches of raised scales........ Cosmopteriginae p. 19

— Male genitalia symmetrical or asymmetrical, gnathos absent, uncus present or absent, aedoeagus ankylosed with saccus; labial palpus never extremely slender, often stout; male abdomen tapering to apex; forewing usually dark gray, shining black or brown, often nearly unicolorous or shades of gray, often with patches of raised scales...................... Chrysopeleiinae p. 98

SUBFAMILY

## **Antequerinae** Hodges, NEW SUBFAMILY

Antequerinae Hodges.
Type-genus: *Antequera* Clarke, 1941.

The Antequerinae are a small group with two genera and four species occurring in Great Britain and North America.

Characters of the subfamily are as follows: forewing lanceolate, apex rounded, cell closed, 12 veins present. Hindwing lanceolate, base of costal margin lacking stout scales, cell closed, eight veins present, $R_s$ and $M_1$ separate. Male with seventh and eighth abdominal segments essentially unmodified; genitalia symmetrical or slightly asymmetrical, aedoeagus free, lobes of gnathos or uncus symmetrical, vinculum well developed. Female with ostium bursae between seventh and eighth abdominal sterna, lacking sclerotized lamellae around ostium bursae, signa present and paired or absent.

Larva: characterization of the last instar larva is based on *Euclemensia bassettella* and *schwarziella.* Characters are drawn from a chart made by Heinrich and Capps of the U.S. Department of Agriculture, and terminology is that proposed by Hinton (1946). Head smooth, dark, reticulated; not flattened; submental pit lacking; ocelli six; adfrontal ridges extending nearly to vertex; seta P2 posterolateral of P1; L1 not near hind margin; A1, A2 and A3 not in a line. Body with granulose skin. Thorax with prolegs short, length much less than width; prothorax with cervical sclerite mainly dark, without color pattern, prespiracular wart with two setae, D2 higher than XD1, D2 approximately equal to XD2 and SD2, SV1 and SV2 on vi; meso- and metathorax with SV1 on vi. Abdominal segments one to six with D2 approximate to D1 and lower than D1, L1 approximate to

L2 but not on same pinaculum, L1 higher than L2; segments one, eight and nine with SV1 (A1 with SV2 in *bassettella*); segments three to seven with SV1 and SV2 (A3–6 with SV3 in *bassettella*); segment eight with SD1 dorsoposterad of normal spiracle, L3 caudoventrad of L1 and L2, seta D2 not much closer than on A7, paired D2 not on same pinaculum, L1 and L2 not on same pinaculum; segment nine with D2 not much closer than on A8 and not on single pinaculum, D1 not directly in front of D2, D1 and SD1 not on a single pinaculum, D1 not directly in front of SD1, D1, D2 and SD1 in a line, D1 closer to D2 than to SD1; anal segment with L3 present, L1 and L2 absent. Crochets uniordinal and uniserial in a complete circle. Pupa unknown.

Larvae of *Euclemensia* are internal parasitoids on kermes scale insects.

## KEY TO GENERA OF ANTEQUERINAE

1. Head very smooth scaled; forewing broadest at $\frac{1}{4}$ length, with orange or yellow scales ........ *Euclemensia* p. 18
— Head somewhat rough scaled; forewing broadest at $\frac{3}{4}$ length, without bright-colored scales ........ *Antequera* this page

GENUS
## *Antequera* Clarke

*Antequera* Clarke, 1941, *Proc. U.S. Natl. Mus.*, **90**: 270, pl. 4, figs. 27, 28; pl. 9, figs. 63, 63a, 63b; pl. 18, fig. 111.
Type-species: *Semioscopis acertella* Busck, 1913. Original designation.

*Antequera* is a monobasic genus whose single species occurs in southern California and Baja California. Genital characters ally *Antequera* most closely with *Euclemensia*. Adults of *Antequera* lack ocelli, are slightly rough scaled, have non-metallic scales, have very broadly lanceolate fore- and hindwings, have veins $Cu_1$ and $Cu_2$ connate from the end of the cell in the forewing and $M_3$ and $Cu_1$ connate from the end of the cell in the hindwing. Adults of *Euclemensia* have ocelli, are extremely smooth scaled, have metallic scales, have lanceolate fore- and hindwings, have the forewing with veins $Cu_1$ and $Cu_2$ distant and arising before the end of the cell and $M_3$ and $Cu_1$ separate in the hindwing.

Characters of the genus are as follows: tongue well developed, scaled on anterior surface nearly to apex; labial palpus slightly rough scaled on first and second segments, second segment longer than third segment; frons, vertex and occiput slightly rough scaled; ocelli absent; antenna with sensory cilia about equal to depth of segments in male, very short in female, pecten absent. Hindtibia with first pair of spurs at $\frac{3}{4}$ length of segment. Forewing broadly lanceolate; 12 veins present; $R_1$ arising before $\frac{1}{2}$ length of cell; $R_4$ and $R_5$ stalked; $M_1$, $M_2$, $M_3$ and $Cu_1$ separate; $Cu_1$ and $Cu_2$ connate from end of cell, $Cu_2$ slightly downcurved then running parallel to margin. Hindwing broadly lanceolate; costal margin slightly convex; eight veins present; $M_2$ arising closer to $M_3$ than to $M_1$; $M_3$ and $Cu_1$ connate basally, $Cu_2$ arising before end of cell. Abdomen without stout setae dorsally, sterna heavily sclerotized, eighth segment nearly unmodified. Male genitalia nearly symmetrical; valvae symmetrical, slender lobes; diaphragma around aedoeagus heavily sclerotized, extending posterad as long, asymmetrical cones; aedoeagus nearly straight, basal $\frac{1}{3}$ more heavily sclerotized than distal $\frac{2}{3}$, cornuti present in vesica; brachia symmetrical, broad based, slender on distal $\frac{3}{4}$. Female with eighth abdominal sternum lightly sclerotized laterally, a row of setae on caudal margin; ostium bursae on anterior margin of eighth abdominal sternum or in membrane between seventh and eighth abdominal sterna; basal $\frac{1}{2}$ of ductus bursae heavily sclerotized, rest slightly sclerotized with spiculose walls; corpus bursae with paired signa, each a sclerotized plate with numerous, inwardly projecting points; base of ductus seminalis broad, arising almost as a pouch from $\frac{3}{5}$ length of heavily sclerotized part of ductus bursae; apophyses heavily sclerotized, relatively slender, apophyses anteriores shorter than apophyses posteriores, diverging from base.

The immature stages are unknown.

### *Antequera acertella* (Busck)
PL. 4, FIGS. 43, 44. TEXT FIGS. 1 *c*; 4 *a*, *d*, *g* (McD. 8462).

*Semioscopis acertella* Busck, 1913, *Jour. Ent. and Zool., Pomona College*, **5**: 100.
Type-locality: San Diego, California. [USNM]

*Borkhausenia exstimulata* Meyrick, 1928, *Exotic Microlepidoptera*, **3**: 468. NEW SYNONYMY.
Type-locality: Mexico, Lower California. [BMNH]

*Acertella* differs from other cosmopterigids by the combination of forewing color pattern, lack of antennal pecten and hindleg with first pair of tibial spurs at $\frac{3}{4}$ length of tibia and genitalia. Wing length varies from 7.5 to 9.0 mm. Forewing color pattern varies in intensity and hue. Some specimens have yellowish-brown forewings; most have gray forewings.

The immature stages are unknown.

*Acertella* has been collected from Ventura County, California south into Baja California. Most specimens have been collected from 24 October to 15 December. A specimen from Ventura County was collected on 6 May.

GENUS

## *Euclemensia* Grote

*Hamadryas* Clemens, 1864, *Proc. Ent. Soc. Philadelphia*, **2**: 423.
Type-species: *Hamadryas bassettella* Clemens, 1864. Monotypy.

NOTE—*Hamadryas* Clemens, 1864, is a homonym of *Hamadryas* Hübner, 1806, in the Insecta.

*Euclemensia* Grote, 1878, *Can. Ent.*, **10**: 69.
Type-species: *Hamadryas bassettella* Clemens, 1864. Monotypy as type-species of *Hamadryas* Clemens, 1864.

NOTE—*Euclemensia* Grote is a replacement name for *Hamadryas* Clemens, 1864.

*Euclemensia* is a small genus of three species. Two occur in North America; one occurs in Great Britain. *Euclemensia* is most closely related to *Antequera* but can be separated by the presence of metallic scales on the body and wings. Differences are given under *Antequera* (page 17).

Characters of the genus are as follows: tongue short, scaled basally; labial palpus recurved, extending to or beyond vertex, smooth scaled, second and third segments about equal in length; ocelli present; antenna $\frac{3}{5}$ to $\frac{3}{4}$ length of forewing, pecten absent, male densely setose on ventral surface, female with restricted sensory areas; frons, vertex and occiput extremely smooth scaled. Forewing lanceolate, apex acute; 12 veins present; $R_1$ from $\frac{1}{2}$ length of cell; $R_4$ and $R_5$ stalked, $R_5$ to costal margin; $M_1$, $M_2$, $M_3$ and $Cu_1$ nearly equidistant; $Cu_2$ arising at $\frac{2}{3}$ length of cell. Hindwing lanceolate, costal margin convex to $\frac{1}{2}$ length then slightly concave; eight veins present; $R_s$ and $M_1$ somewhat distant basally; $Cu_1$ and $Cu_2$ distant basally, $Cu_2$ from $\frac{2}{3}$ length of cell. Male with abdominal sterna more heavily sclerotized than terga; eighth segment slightly modified in shape; genitalia nearly symmetrical; valvae simple lobes of nearly uniform width; diaphragma heavily sclerotized around aedoeagus, developed as pair of lateral lobes extending beyond aedoeagus; aedoeagus a stout cylinder, basal $\frac{1}{3}$ to $\frac{2}{5}$ more heavily sclerotized than distal part, two large cornuti in vesica; brachia symmetrical, tapering to acute apices. Female with ostium bursae in membrane anterad of eighth abdominal sternum; base of ductus bursae heavily sclerotized as short ring, rest membranous; corpus bursae a pair of unequal pouches, signa absent; apophyses anteriores short, $1\frac{1}{2}$ to 2 times length of eighth abdominal segment; apophyses posteriores longer than apophyses anteriores.

Larvae of the North American species are parasitoids of armored scales, *Kermes* Latreille. Pupation occurs within the body of the scale.

### KEY TO SPECIES OF *EUCLEMENSIA*

1. Forewing with orange areas contiguous, without orange scales on posterior margin at $\frac{2}{3}$ length, East Coast west to Missouri and eastern Texas . . . . . . . . . . . . *bassettella* this page
— Forewing with orange areas discontiguous, with a patch of orange scales on posterior margin at $\frac{2}{3}$ length, southern Arizona . . . . . . . . . . . . *schwarziella* p. 19

### *Euclemensia bassettella* (Clemens)

PL. 6, FIG. 14. TEXT FIGS. 1 *a*; 4 *c*, *e*, *f* (McD. 8806).

*Hamadryas bassettella* Clemens, 1864, *Proc. Ent. Soc. Philadelphia*, **2**: 423.
Type-locality: Waterbury, Connecticut. [ANSP]

*Bassettella* is similar to *schwarziella*, but the two species can be separated as indicated in the key. Wing length varies from 4.0 to 6.2 mm. Moths vary from yellow orange to red orange on the forewing, and worn specimens tend to be pale. Variation in size reflects the size of the host: small moths emerging from small hosts.

The larva is an internal parasitoid on *Kermes* species (Homoptera, Kermesidae). The prolegs are rudimentary and have a circle of uniordinal hooks. Pupation occurs in the host, and the moth emerges through a circular slit cut by the last instar larva according to Forbes (1923: 357).

*Bassettella* has been collected from New Hampshire, southern Ontario and central Illinois south to central Florida and eastern Texas. Moths have been collected from 8 May to 10 September.

*Euclemensia schwarziella* Busck
PL. 6, FIG. 15 (McD. 8807).

*Euclemensia schwarziella* Busck, 1900 [1901], *Jour. New York Ent. Soc.*, **8**: 239, pl. 9, fig. 3.
Type-locality: Santa Rita Mts., Arizona. [USNM]
NOTE—The lectotype, ♀, present designation, bears the following labels: 1. "Sta. Rita Mts. Ariz. on *Kermes* sp. on Oak. iss. Jun. 20. 98."; 2. "AB 2."; 3. "AB. *Euclemensia* n. sp."; 4. "Type No. 5356 U.S.N.M."; 5. "*Euclemensia schwarziella* Type! Busck"; 6. "LECTOTYPE *schwarziella* Bsk. By R. W. Hodges 1975."

*Schwarziella* can be separated from *bassettella* as indicated in the key. Additionally, the male has two stout cornuti in the vesica; the male of *bassettella* has one cornutus. Wing length varies from 5.0 to 5.3 mm. The forewing pattern varies in degree of separation of the orange marks.

The type series was reared from a *Kermes* scale on oak.

*Schwarziella* has been collected in the Santa Rita Mountains and at Nogales, Arizona. Adults have emerged from 20 June to 8 July.

SUBFAMILY
## **Cosmopteriginae** Heinemann

Cosmopterygidae Heinemann, [1876] 1877, *in* Heinemann and Wocke, *Die Schmetterlinge Deutschlands und der Schweiz*, **2** (2): 403.
Type-genus: *Cosmopterix* Hübner, 1825.

Diplosaridae Meyrick, 1915, *Exotic Microlepidoptera*, **1**: 339.
Type-genus: *Diplosara* Meyrick, 1883.

Scaeosophides Meyrick, 1922, *in* Wytsman, *Genera Insectorum*, **180**: 48.
Type-genus: *Scaeosopha* Meyrick, 1914.

The Cosmopteriginae are the largest of the three subfamilies and comprise 74 genera and more than 1100 species. These occur on all continents except Antarctica. Eleven genera and 91 species are found in America north of Mexico.

Major characters of the Cosmopteriginae are as follows: head smooth scaled; tongue scaled basally; labial palpus slender to very slender and sickle shaped, normally extending beyond vertex; maxillary palpus folded over base of tongue; antenna $\frac{2}{3}$ to equal length of forewing, simple, ventral surface usually with setose sensory areas, pecten present or absent and of few hairlike setae or numerous, somewhat broader scales; ocelli present or absent. Forewing lanceolate, apex rounded to acuminate; 10 to 12 veins present, radials always present; $R_4$ and $R_5$ stalked, sometimes with $M_1$ and rarely with $M_2$; cell usually closed, very narrow in narrow-winged forms. Hindwing lanceolate to linear, apex rounded to acuminate; cell open or closed, usually open in narrow-winged forms; a tuft of stout scales on costal margin near base in narrow-winged forms; $R_s$ and $M_1$ stalked or approximate in narrow-winged forms, separate and somewhat parallel in broader winged forms. Abdomen without stout setae on dorsal surface, male seventh and eighth segments sometimes modified as heavily sclerotized plates. Male genitalia asymmetrical; valvae strongly asymmetrical to nearly symmetrical, sometimes with lobes from base or from base and diaphragm; aedoeagus ankylosed with anellus; anellus often with a pair of ventral or ventrolateral lobes; cornuti present or absent; vinculum usually poorly developed, often undeveloped in saccal area; transtilla absent; free lobes of gnathos asymmetrical; uncus absent; intersegmental area before genitalia often with variously developed lobes. Female genitalia with apophyses short to long; well-defined lamellae vaginales often projecting from seventh sternum as a tube; corpus bursae a single or double lobe; signa sometimes absent, usually present and paired; ostium bursae usually on seventh abdominal sternum, but migrating to dorsal surface in at least one genus; apophyses anteriores usually separate basally, but sometimes connected basally and asymmetrical.

Larva: characterization of the last instar larva is based on *Anoncia conia*, *A. diveni*, *A. sphacelina*, *Cosmopterix clandestinella*, *Limnaecia phragmitella*, *Pebops ipomoeae*, *Pyroderces rileyi*, *P. badia*, *Stagmatophora iridella* and *Triclonella pergandeella*. Head smooth; flattened or not; submental pit lacking; ocelli six; front extending $\frac{3}{4}$ or more distance to vertex; adfrontal ridge extending to vertex; seta P2 laterad, posterad or lateroposterad of P1; L1 below level of P1 (approximately equal to level of P1 in *A. diveni*); A3 closer to L1 than to A2 (approximately equidistant in *Pebops ipomoeae*); A1, A2, A3 and L1 not in line (nearly in line in *A. conia* and *A. sphacelina*). Thorax: cervical sclerite without secondary setae; prespiracular wart with three setae; TI with D2 higher than XD1 (D2 approximately level with XD1 in *C. clandestinella*, lower than XD1 in *P. ipomoeae* and *T. pergandeella*); vi with two setae, SV1 and SV2 on TI; coxae not touching on TI; coxae less than two widths apart on TIII (two widths apart in *L. phragmitella*. Abdomen: A1–A6 with D2 remote from and lower than D1; L1 approximate to L2 and on same pinaculum (on separate pinacula in *T. pergandeella*); A8 with spiracle normal (versus dorsad); D2 not much closer than on D7, paired D2 not on a pinaculum; L3 ventrocaudad of L1 and L2; A9 with D2 closer than on A8, paired D2 not on a single pinaculum, L3 ventrocaudad of L1 and L2; A9 with D2 closer than on A8, paired D2 not on single pinaculum, D1, D2 and SD1 in a line (not in a line in *T. pergandeella*), SD1 not closer to L1 and L2 than to D1 and D2, L1, L2 and L3 present (L3 absent in *A. conia*). SV group: A1 with SV1 and SV2 (and SV3 in *Pyroderces badia* and *P. rileyi*); A2–A6 with SV1, SV2 and SV3 (many setae in *A. conia* and *A. sphacelina*); A7 with SV1 and SV2; A8 and A9 with SV1. A10 lacking anal fork; prolegs with secondary setae (*A. diveni*, *A. sphacelina*, *L. phragmitella*, *P. badia*, *P. rileyi* and *T. pergandeella*) or without secondary setae (*A. conia*, *C. clandestinella*, *S. iridella* and *Pebops ipomoeae*). Crochets more than six (absent in *S. iridella*), uniserial, uniordinal (*C. clandestinella*, *Pyroderces badia*, *P. rileyi* and *Pebops ipomoeae*) or biordinal (*Anoncia*, *L. phragmitella* and *T. pergandeella*), in complete circle (penellipse in *A. conia*, *A. diveni* and *A. sphacelina*), weaker outwardly and continuous on anal legs.

Pupa (based on *A. conia*, *A. diveni*, *C. scirpicola*, *Pyroderces rileyi* and *T. pergandeella*): frontoclypeal suture incomplete; mandible and eyepiece defined; appendages firmly appressed to body, extending to posterior margin of fifth abdominal segment in *Anoncia* and *Triclonella*, to the posterior margin of sixth abdominal segment in *Cosmopterix* and *Pyroderces*; abdomen moveable between segments six and seven and perhaps seven and eight in *Cosmopterix* and *Pyroderces* and between four and five, five and six, and six and seven in *Anoncia* and *Triclonella*; cremaster present; entire body slender (width 3.4: length 16) in *C. scirpicola* or stout (width 5.6: length 12.4) in *Anoncia* and *Triclonella*.

Larvae usually are leaf miners, sometimes stem miners, seed feeders or scavengers on plant tissue.

## KEY TO GENERA OF COSMOPTERIGINAE

1. Hindwing with cell usually closed, crossveins closing cell directed anterobasally from cubitus, veins $R_s$ and $M_1$ separate and generally parallel . . . . . 2

— Hindwing with cell usually open, crossveins directed anterodistally from cubitus, veins $R_s$ and $M_1$ stalked or sometimes connate, diverging to wing margin . . . . . 4

2. Labial palpus slender, very smooth scaled, extending well beyond vertex; ocelli present; hindwing with $M_3$ and $Cu_1$ connate or stalked . . . . . *Triclonella* p. 52

— Labial palpus somewhat stout, slightly rough scaled, not reaching vertex; ocelli usually absent, when present, hindwing with $M_3$ and $Cu_1$ separate and nearly parallel . . . . . 3

3. Hindwing with $M_3$ and $Cu_1$ connate or stalked, rarely separate; forewing with $Cu_1$ and $Cu_2$ slightly downcurved from cell then parallel with $M_3$; ocelli absent; male aedoeagus without dorsal projection from midregion . . . . . *Anoncia* p. 55

— Hindwing with $M_3$ and $Cu_1$ separate and parallel; forewing usually without $Cu_1$ and $Cu_2$, when present at right angle with cell to margin; ocelli present or absent; male aedoeagus with a dorsal projection from midregion . . . . . *Teladoma* p. 49

4. Forewing with spots of metallic scales (shining silver, bluish silver or gold) . . . . . 5

— Forewing without metallic scales . . . . . 7

5. Forewing with pattern of transverse bands of metallic and colored scales from $\frac{1}{3}$ to $\frac{2}{3}$ wing length and a unicolorous basal brown area usually with four silver or bluish-white lines . . . . . 6

— Forewing with pattern of triangular white marks on costal margin and metallic silver spots on, or near, posterior margin, basal $\frac{1}{4}$ without longitudinal lines . . . . . *Stagmatophora* p. 44

6. Male with tripartite aedoeagus (text figure 5 *c*), right brachium much more heavily sclerotized apically than basally; forewing

with anteromedial and posteromedial blue-white lines, when present, usually continuous . . . . . . . . . . . . . . . . . . . . . . *Cosmopterix* this page

— Male with undivided aedoeagus (text figure 7 *d*), right brachium uniformly sclerotized; forewing with anteromedial and posteromedial blue-white lines interrupted by dark-brown scales . . . . . . . . . . . . . . . . . . . . . *Pebops* p. 38

7. Hindwing with veins $M_2$, $M_3$, $Cu_1$ and $Cu_2$ well developed . . . . . . . . . . . . . . . . . . . . . . . . . . . 8

— Hindwing with veins $M_2$, $M_3$, $Cu_1$ and $Cu_2$ absent or scarcely defined (not tubular veins) . . . . . . . . . . . . . . . . . . . . . . . . . . . . . . . . 9

8. Forewing with veins $R_4$ and $R_5$ stalked, $M_1$ separate; hindwing with veins $R_s$ and $M_1$ separate . . . . . . . . . . . . . . . . . . . . . . . . . . *Limnaecia* p. 48

— Forewing with veins $R_4$, $R_5$ and $M_1$ stalked; hindwing with veins $R_s$ and $M_1$ stalked . . . . . . . . . . . . . . . . . . . . . . . . . . . . . . . . . . . . *Pyroderces* p. 46

9. Antennal pecten a slender seta; male with valva nearly uniformly slender, bent at nearly right angle at $\frac{2}{3}$ length (text figure 7 *e*) . . . . . . . . . . . . . . . . . . . . . . . . . . . . . *Tanygona* p. 39

— Antennal pecten six to eight slender setae; male with valva broadly lobate, often expanded distally (text figure 7 *f*) . . . . . . . . . . . 10

10. Forewing with $M_2$ and $Cu_1$ present; hindwing with $M_2$, $M_3$, $Cu_1$ and $Cu_2$ evanescent; female with paired signa . . . . . . . . . . . . . . . . *Eralea* p. 40

— Forewing without $M_2$ and $Cu_1$; hindwing with only Sc, $R_s$, $Cu_1$ and 2A developed; female without signa . . . . . . . . . . . . . . *Melanocinclis* p. 41

GENUS

## *Cosmopterix* Hübner

*Cosmopterix* Hübner [1825], *Verzeichniss Bekannter Schmettlinge* [sic], 424.

Type-species: *Tinea zieglerella* Hübner, 1805–1810. Designated by the International Commission on Zoological Nomenclature under the plenary powers, 1969, *Bull. Zool. Nomenclature*, **25**: 150.

NOTE—The emended spelling, *Cosmopteryx*, was proposed by Zeller, 1839, *Isis von Oken*, **32**: 210.

*Cosmopterix* is a large genus with more than 140 species in the world. Thirty species occur in North America north of Mexico. Most species are eastern, particularly southeastern; some occur in the Southwest. Fifteen species have been reported from Florida; none from British Columbia, Alberta, Montana, Idaho and Washington. Whether the apparent absence of the genus from these states reflects distribution or inadequate collecting remains to be determined.

*Cosmopterix* is close to *Pebops* and *Tanygona* but may be separated from them by the complete venation in the forewing, complete venation in the hindwing, male genitalia with apex of right brachium very heavily sclerotized and set off by degree of sclerotization from the basal part, very short tegumen essentially a heavily sclerotized rim, tripartite aedoeagus consisting of a basal and distal unit with a flange partially surrounding the juncture between them, juxtal lobes free and valva broad basally. *Pebops* has the right brachium uniformly sclerotized, a relatively long tegumen without a distinct rim, a cylindrical aedoeagus without distinct subunits, the juxtal lobes closely associated with the aedoeagus and the valva slender basally. *Tanygona* lacks vein $Cu_2$ in the forewing, and veins $M_2$, $M_3$, $Cu_1$ and $Cu_2$ are not distinguishable, if present, in the hindwing.

Characters of *Cosmopterix* are as follows: labial palpus smooth scaled, third segment slightly longer than second segment and at an angle with second segment; head smooth scaled; ocelli absent; basal segment of antenna with pecten a single basal seta. Forewing narrowly lanceolate with anterior and posterior margins moving together at $\frac{2}{3}$ to $\frac{3}{4}$ wing length, apex extremely slender; 12 veins present; $R_4$, $R_5$ and $M_1$ stalked, $M_1$ arising before or after $R_4$; $M_2$ approximate to connate with $R_{4+5}+M_1$; $Cu_2$ arising basad of $R_2$; 1A absent; 2A and 3A forked basally; ventral surface with series of setae that interlock with stout scales on anterior margin of hindwing at $\frac{1}{5}$ wing length. Hindwing linear, tapering gradually to apex, eight veins usually present but weakly developed; $M_1$ stalked with $R_s$; cell open; anal veins very weak. Male genitalia: asymmetrical; valvae symmetrical in our species, broad basally; juxta a pair of lobes, often broadened apically; aedoeagus ankylosed, tripartite, a bulbous distal part, a more slender and basally broadened basal part and a flange partially encircling the insertion of the basal part into the distal part; tegumen a heavily sclerotized band, relatively short; gnathos asymmetrical, right brachium very heavily sclerotized and often with shallow indentations distally, left brachium short, uniformly sclerotized. Female with posterior margin of seventh sternum often modified; lamella antevaginalis usually snowshoe shaped with ostium bursae near apex; ductus bursae very slender, often with pair of sclerotized patches just before the beginning of the corpus bursae; ductus seminalis arising from a lobe at base of ductus bursae; one or two signa usually present; apophyses anteriores and posteriores slender, apophyses anteriores often connected dorsally by a lightly sclerotized band. Abdomen with scales on dorsal surface having extremely dentate posterior margins, contrasting with nearly smooth-margined scales on other surfaces. Hindtibia has long scales on dorsal surface.

*Cosmopterix* larvae are leaf miners on Cyperaceae, Gramineae, Leguminosae and Compositae in our area. Pupation occurs in the larval mine. Many of the species have more than one generation per year, and in the South some apparently breed throughout the year.

Superficially, all *Cosmopterix* species are similar in appearance. The second and third segments of the labial palpus are linearly banded with white or bluish-white lines. The dorsal surface of the head and thorax has a series of longitudinal lines that most extensively consists of a medial line, a pair of lines above the eyes and continued on the inner margins of the tegulae and a pair of lines on the outer margins of the tegulae. Any or all of these lines may be absent. The base of the forewing usually has four lines, one starting from the base near the costal margin and running distad and posterad, the costomedial line anterad of the fold, the posteromedial line posterad of the fold and one usually on the posterior margin of the wing starting from the base and extending toward the transverse fascia. A fifth line may occur on the costal margin from the transverse fascia running basad. These lines may be broad, fused or absent. A transverse band of yellow or orange scales bounded basally by a band or two spots of silver scales and distally by a pair of silver spots is at $\frac{2}{3}$ the wing length. And, a white line is on the posterior margin of the wing beyond the transverse fascia. The dorsal surface of the hindwing and ventral surface of both wings are immaculate except for some white scales on the costal margin of the forewing at the end of the fascia. The scales have metallic reflections of blue, purple, gold, bronze etc. causing the colors to vary and appear different at each angle of light reflection.

Variation may be present in most characters discussed. For this reason illustrations of the genitalia are included as an additional guide for species determination.

## KEY TO SPECIES OF *COSMOPTERIX*
(*C. chalybaeella* not included)

1. Forewing with basal lines .................... 2
— Forewing without basal lines ........... *sinelinea* p. 24

2. Forewing with basal lines very broad, confluent .................................. 3
— Forewing with basal lines narrow, separate..... 4

3. Antenna with apical two segments white preceded by seven bronze segments and one white segment.......................... *nitens* p. 24
— Antenna with apical four segments white, preceded by five lead-colored segments and five white segments.................. *molybdina* p. 25

4. Thorax without lateral white or silver-blue lines.................................... 5
— Thorax with lateral white lines.............. 11

5. Forewing with a basal line just before posterior margin..................... *bendidia* p. 26
— Forewing with basal line on posterior margin................................. 6

6. Thorax with white lines dorsally; forewing with basal bluish-white lines narrow, anteromedial line basad of posteromedial line.................................... 7
— Thorax without white lines dorsally; forewing with basal bluish-white lines short and broad, posteromedial line basad of anteromedial line...................... *clandestinella* p. 27

7. Forewing with anteromedial line running from base almost to transverse fascia.......... 8
— Forewing with anteromedial line short........ 9

8. Antenna with apical $\frac{1}{2}$ segment gray brown .................................. *opulenta* p. 31
— Antenna with apical four segments white . . ............................. *chisosensis* p. 31

9. Antenna with apex white; forewing with distal silver spots bordering transverse fascia separate, subcostal spot offset distally . . . 10
— Antenna with apex brown; forewing with distal silver spots bordering transverse fascia fused, not offset; dorsal surface of abdomen dark purplish . . . . . . . . . . . . . . . . . . . . . *pulchrimella* p. 25

10. Antenna with apical eight segments white, rest brown . . . . . . . . . . . . . . . . . . . . . . . . . . . *bacata* p. 35
— Antenna with apical three segments white preceded by five brown segments, four white segments then brown ones . . . . *gemmiferella* p. 34

11. Forewing with anteromedial line starting at base . . . . . . . . . . . . . . . . . . . . . . . . . . . . . . . . . . . 12
— Forewing with anteromedial line starting beyond base . . . . . . . . . . . . . . . . . . . . . . . . . . . . 16

12. Forewing with transverse fascia surrounding basal silver spots, extending basally to $\frac{1}{3}$ wing length . . . . . . . . . . . . . . . . . . . . *quadrilineella* p. 32
— Forewing with transverse fascia not surrounding basal silver spots, starting at $\frac{1}{2}$ wing length . . . . . . . . . . . . . . . . . . . . . . . . . . . . 13

13. Forewing with transverse fascia extending between distal silver spots as a narrow line . . . . . . . . . . . . . . . . . . . . . . . . *floridanella* (in part) p. 37
— Forewing with transverse fascia extending between distal silver spots broadly and becoming wider beyond them . . . . . . . . . . . . . . 14

14. Antenna with apical segment gray or partly gray preceded by three white segments, six gray segments and one white segment . . . . . . . . 15
— Antenna with apical four segments white preceded by three brown segments . . . . *minutella* p. 33

15. Antenna with apex of last segment gray preceded by $2\frac{1}{2}$ white segments, one brown segment and one white segment . . . . . . . *dicacula* p. 30
— Antenna with apical segment gray preceded by three white segments and six gray segments . . . . . . . . . . . . . . . . . . . . . . . . . . . . *dapifera* p. 29

16. Forewing pale, yellowish gray or light brown . . . . . . . . . . . . . . . . . . . . . . . . . . . . . . . . . 17
— Forewing dark, yellowish brown or brown . . . . 22

17. Antenna with apical three or four segments white . . . . . . . . . . . . . . . . . . . . . . . . . . . . . . . . . . 18
— Antenna with apex gray brown or brown . . . . . 19

18. Antenna with apical three or four segments white preceded by five brown segments, one white segment, one brown segment and five white segments . . . . . . . . . . . . . . . . . . *facunda* p. 38
— Antenna with apical four segments white preceded by one brown segment, one white segment, three brown segments, one white segment and one brown segment . . . . . . . . . . . . . . . . . . . . . . . . . . . . . . . . . . *floridanella* (in part) p. 37

19. Forewing with costal scales beyond transverse fascia concolorous with base of wing, antenna with apical two segments dark brown . . . . . . . . . . . . . . . . . . . . . . . . . . . *delicatella* p. 29
— Forewing with costal scales beyond transverse fascia yellow brown, not concolorous with base of wing; antenna with apical four or more segments gray brown or brown . . . . . . 20

20. Antenna with apical four segments brown preceded by one white segment . . . . . . . . . . . . . 21
— Antenna with apical nine segments gray brown preceded by one white segment, one gray-brown segment and one white segment . . . . . . . . . . . . . . . . . . . . . . . . . . . . . . . . . . *ebriola* p. 37

21. Hindtibia with short zone of white scales at base; male genitalia with distal $\frac{1}{2}$ of left brachium tapering, convex-up to apex, maximum width of aedoeagus $\frac{1}{2}$ length of aedoeagus . . . . . . . . . . . . . . . . . . . . . . . . *fernaldella* p. 37
— Hindtibia with row of white scales extending from ventral margin at base to dorsal margin at $\frac{1}{2}$ length; male genitalia with distal $\frac{1}{2}$ of left brachium tapering concave-up to apex, maximum width of aedoeagus slightly more than $\frac{2}{3}$ length of aedoeagus . . . . . . . . . . . . . . . . . . . . . . . . . . . . . . . . . . . *scirpicola* p. 36

22. Forewing with small patch of shining blue-white scales on posterior margin at midpoint between transverse fascia and apex, a tuft of white scales on costal margin beyond transverse fascia . . . . . . . . . . . . . . . . . . . *montisella* p. 27

— Forewing with a bluish-white line on posterior margin to apex (sometimes incomplete), tuft of white scales on costal margin beyond transverse fascia present or absent....23

23. Antenna with apex brown..................24
— Antenna with apex white..................26

24. Antenna with apical segment partly brown ...................................*magophila* p. 28
— Antenna with apical two segments brown.....25

25. Antenna with apical two segments brown preceded by two white segments, forewing with transverse fascia bordered basally by solid silver band...................*attenuatella* p. 26
— Antenna with apical three segments brown preceded by two white segments, forewing with transverse fascia bordered basally with silver band nearly divided into two spots.. ...................................*damnosa* p. 35

26. Antenna with apical three or four segments white preceded by four or five brown segments, four or five white segments then brown segments....................*clemensella* p. 35
— Antennal pattern different, usually second white area interrupted by a brown segment...27

27. Forewing with anteromedial line longer than posteromedial line....................29
— Forewing with anteromedial line shorter than posteromedial line or equal to it........28

28. Metathoracic tarsus with a white line on dorsal surface of segments two through five, forewing with transverse fascia bordered basally by two silver spots............*lespedezae* p. 30
— Metathoracic tarsus without a white line on dorsal surface, forewing with transverse fascia bordered basally by a silver band.... *abdita* p. 33

29. Forewing with distal silver-white spots bordered by black basally...............*inopis* p. 33
— Forewing without black scales bordering distal silver-white spots basally.........*gracilens* p. 28

*Cosmopterix nitens* Walsingham

PL. 2, FIG. 20. TEXT FIGS. 5 *a*; 19 *a* (McD. 7687).

*Cosmopteryx* [sic] *nitens* Walsingham, 1889, *Insect Life*, **1**: 289.
Type-locality: Southwest Texas. [BMNH]

*Nitens* is nearest *sinelinea* in general appearance and male and female genitalia but may be separated by the fused basal lines on the forewing, the ground color of the forewing anterad of the transverse fascia unicolorous, and the fringe on the posterior margin of the forewing distad of the fascia concolorous with the wing. The forewing of *sinelinea* lacks the basal lines; the part of the wing posterad of the fold is metallic gray to yellowish gray and contrasts with the more yellow-gray scales on the anterior part of the wing; and the fringe on the posterior margin distad of the transverse fascia is yellowish gray and contrasts with the wing. Wing length varies from 3.9 to 4.5 mm. The anterior part of the fascia may be mainly gray or pale yellow and concolorous with the posterior part.

The food plant is unknown.

*Nitens* occurs from coastal South Carolina to southwestern Texas. Adults are multivoltine, probably throughout the range, and are present from March and April until October. They can be extremely abundant at light.

*Cosmopterix sinelinea* Hodges, NEW SPECIES

PL. 6, FIG. 1. TEXT FIGS. 5 *b*; 19 *f*.

*Cosmopterix sinelinea* Hodges.
Type-locality: The Wedge, McClellanville, South Carolina. [USNM]

Upper surface as figured. Head: tongue and maxillary palpus yellowish gray; labial palpus mainly dark brown, a pale yellowish-gray to white line on ventral surface of second segment and anterior surface of third segment, mainly pale yellowish gray on dorsal surface of second segment and posterior surface of third segment; antenna dark gray brown, ventral surface of scape pale yellowish gray, a slightly curved yellowish-gray line on anterior margin of scape, apical two segments of antenna pale yellowish gray preceded by seven brown segments, one pale yellowish-gray segment then brown segments to scape; frons shining silver with metallic-blue reflections; vertex shining yellowish gray brown with silver or purple metallic reflections, lacking a white line above eye. Thorax same color as for vertex and occiput, lacking pale-metallic lines. Forewing mainly dark yellowish gray with yellow to silver-gold metallic reflections; area posterad of fold more nearly gray than area anterad of fold; yellow fascia from $\frac{1}{2}$ to $\frac{3}{4}$ wing length bordered basally by a band of metallic-silver scales, divided medially by band of black scales; distal $\frac{1}{4}$ of wing mainly dark yellowish gray; a row of silver scales on posterior margin from $\frac{4}{5}$ wing length to apex; fringe on posterior margin from fascia nearly to apex pale yellowish gray, contrasting with darker yellowish gray of wing. Hindwing mainly dark yellowish gray, fringe slightly paler. Foreleg: coxa and femur mainly dark yellowish gray to gray brown with metallic reflections, apex of femur pale yellowish gray; tibia and tarsus dark yellowish gray to brown, epiphysis pale yellowish gray, pale-yellow line on anterior surface interrupted on tarsal segments three and four. Midleg: coxa and femur pale yellowish gray with shining purple reflections, apex of coxa pale yellow to yellowish gray; tibia and tarsus darker yellowish gray, a few pale scales at middle and apex of tibia and on fourth and fifth tarsal segments. Hindleg: coxa and femur as for midleg; tibia and tarsus dark brown, an angled, silver-blue band at base, middle and apex, tibial spurs dark gray brown with pale yellowish-gray scales apically; tarsus mainly dark yellowish gray, apices of first, second, fourth and fifth segments pale yellowish gray. Abdomen: dorsal surface dark, greasy yellowish brown to gray brown, lateral margins and apices of some segments metallic silver with metallic-blue and purplish reflections; ventral surface mainly pale yellow, apices of segments slightly paler. Wing length: 4.3–4.8 mm. Male genitalia: as in text figure 5 *b*. Female genitalia: as in text figure 19 *f*.

The immature stages are unknown.

TYPES: Holotype: ♂. The Wedge, McClellanville, South Carolina; 26 Aug. 1971; R. W. Hodges; USNM genitalia slide 3772. USNM. Paratypes: 8 ♂♂, 16 ♀♀. Same data as for holotype. Paratypes to WPC; USNM.

*Sinelinea* is distinguished from other species in the genus by the absence of basal lines on the forewing. It is a derivative of *nitens* and has very similar male and female genitalia. Additional differences are greater wing length, contrasting shades of color on the basal $\frac{1}{2}$ of the forewing, the pale, posterior fringe on the distal part of the forewing and the off-white pale areas on the antenna.

*Sinelinea* is variable in wing length and the transverse fascia of the forewing which may be pale yellow on the anterior and posterior margins, yellowish gray on the anterior margin and pale yellow on the posterior margin or dark yellowish gray throughout.

*Sinelinea* is known only from the type-locality. It can be incredibly abundant. The short type series was taken from tens of thousands of specimens in a walk-in light trap and at other lights at The Wedge.

### *Cosmopterix molybdina* Hodges

PL. 2, FIG. 21. TEXT FIGS. 5 *c*; 19 *d*.

*Cosmopterix molybdina* Hodges, 1962, *Ent. Americana*, **42** (n.s.): 19, figs. 28, 89, 136.
Type-locality: Bar Harbor, Maine. [CU]

*Molybdina* may be recognized by the shining lead-colored pro- and mesothorax above, by the fused and shining lead-colored basal lines on the forewing and by the distal part of the antenna having the color combination of four white segments preceded by five brown segments, four white segments, then brown segments. Wing length varies from 4.0 to 4.4 mm. The transverse fascia on the forewing is yellow in eastern specimens and orange in western specimens and may be dusted with black scales. *Molybdina* is closely related to *nitens*, *sinelinea* and *pulchrimella* on genital characters, and the male of *molybdina* is particularly similar to that of *pulchrimella*.

The larva has been reared as a leaf miner on morning glory, *Ipomoea* species, at Salem, Oregon and moon vine, probably another species of *Ipomoea*, at Portland, Oregon.

*Molybdina* has been collected at Bar Harbor, Maine in late June and mid-July, near Berkeley, California in mid-April and in western Oregon. The Oregon specimens were collected as larvae in mid- to late June, and one series of adults emerged the following January. The geographic distribution suggests that *molybdina* has been introduced into North America.

### *Cosmopterix pulchrimella* Chambers

PL. 2, FIGS. 22, 23. TEXT FIGS. 5 *d*; 19 *b* (McD. 7674).

*Cosmopteryx* [sic] *pulchrimella* Chambers, 1875, *Cincinnati Quart. Jour. Sci.*, **2**: 231.
Type-locality: Covington, Kentucky. [MCZ]

*Pulchrimella* is defined by the bluish-white line above each eye, by the pro- and mesothorax having a medial bluish-white line and lacking lateral lines, by the medial lines near the base of the forewing being short with the anteromedial line more basad than the posteromedial line and by the distal part of the antenna with the color combination of $2\frac{1}{2}$ brown segments preceded by $1\frac{1}{2}$ white segments, five brown segments, one white segment, one brown segment, two segments white with brown and then brown segments. Wing length varies from 3.0 to 4.0 mm. The transverse fascia on the forewing varies from pale orange to darker gray orange, and it may be dusted with brown scales on the posterior $\frac{1}{2}$.

Larvae have been reared from Pennsylvania pellitory, *Parietaria pensylvanica* Muhl., and Canada clearweed, *Pilea pumila* (L.) Gray. According to Busck (1906: 709) the larva is a leaf miner.

*Pulchrimella* has a wide geographic range from Martha's Vineyard, Massachusetts and southeast Wyoming south to southern Florida and southern Arizona. It is multiple brooded, and in Florida occurs as an adult throughout the year.

### *Cosmopterix bendidia* Hodges

PL. 2, FIG. 24. TEXT FIGS. 5 *e*; 19 *e*.

*Cosmopterix bendidia* Hodges, 1962, *Ent. Americana*, **42** (n.s.): 22, figs. 25, 93, 138.
Type-locality: Madera Canyon, 4880′, Santa Rita Mts., Santa Cruz Co., Arizona. [CU]

*Bendidia* may be recognized by the bluish-white line above each eye and the absence of white lines on the thorax, by the distal part of the antenna with the color combination of four white segments preceded by five brown segments, one white segment, one brown segment, two mainly white segments then brown segments, with slender anteromedial and posteromedial bluish-white lines on the basal $\frac{1}{2}$ of the forewing and by the basal bluish-white line slightly anterad of the posterior margin of the forewing. Wing length varies from 3.2 to 4.5 mm. The bluish-white lines on the basal $\frac{1}{2}$ of the forewing vary in length. The yellow-orange fascia may be bounded by the outer silver spots, or the yellow-orange scales may be present beyond the silver spots and extend to the blue-white line on the posterior margin of the distal $\frac{1}{4}$ of the wing. This line may be complete or a short dash at $\frac{4}{5}$ the wing length and a line at the apex. The apex of the antenna usually is white but may be pale gray.

The larva is a leaf miner on morning glory, *Ipomoea* species.

*Bendidia* occurs from Martha's Vineyard, Massachusetts and northwest Arkansas south to southern Florida, southern Texas and southern Arizona. It is probably multiple brooded throughout much of its range. Specimens have been collected in late May and July through September. Reared adults emerged from mid-February through early March at Hyattsville, Maryland.

### *Cosmopterix attenuatella* (Walker)

PL. 2, FIGS. 25, 26. TEXT FIGS. 5 *g*; 19 *c* (McD. 7677).

*Gelechia attenuatella* Walker, 1864, *List of the Specimens of Lepidopterous Insects in the Collection of the British Museum*, **30**: 1019.
Type-locality: Jamaica. [BMNH]

*Cosmopteryx* [sic] *flavofasciata* Wollaston, 1879, *Ann. Mag. Nat. Hist.*, (5) **3**: 438.
Type-locality: St. Helena. [BMNH]

*Cosmopteryx* [sic] *mimetis* Meyrick, 1897, *Proc. Linn. Soc. New South Wales*, **22**: 339.
Type-locality: Newcastle and Sydney, New South Wales. [BMNH]

*Cosmopteryx* [sic] *antillia* Forbes, 1931, *Jour. Dept. Agric. Porto Rico*, **4**: 356, pl. 42, fig. 6.
Type-locality: Coamo Sprgs., Puerto Rico. [CU]

*Cosmopteryx* [sic] *superba* Gozmány, 1960, *Annales Historico-Naturales Musei Nationalis Hungarici*, **52**: 419, fig. 4A.
Type-locality: Sohag, Egypt. [Hungarian Natural History Museum]

*Attenuatella* has three white to bluish-white lines on the head and thorax. The slender and short medial lines near the base of the forewing are unequal in length with the posteromedial line longer and more distad than the anteromedial line. The orange transverse fascia on the forewing is bordered basally by a silver band, and it extends between the two silver spots distally. The apical color pattern of the antenna is two brown segments preceded by two white segments, five brown segments, one white segment, one brown segment, one white segment then brown segments. The dorsal surface of the abdomen has prominent orange scales. A characteristic feature of the male genitalia is the slender aedoeagus. Wing length varies from 3.0 to 4.7 mm. The transverse fascia on the forewing is light orange or orange red, and the ground color of the forewing is dark brown. The bluish-white line on the posterior margin of the forewing beyond the fascia may be complete or a short dash followed by a longer line to the apex. The posteromedial line at the base of the forewing usually is longer and more distad than the anteromedial line but may be the same length and start at the same distance from the base of the wing.

Fletcher (1933: 1) summarized information on the life history. The larva is a leaf miner on flatsedges, *Cyperus* species. It feeds near the leaf tip, working toward the apex. Pupation occurs in the mine.

*Attenuatella* has been collected in Florida throughout the year, in southern Louisiana in April and August and in extreme southern Texas. *Attenuatella* is a tropical species that just reaches the southern part of our area. Although the complete range is not known, *attenuatella* probably is pantropical.

### *Cosmopterix clandestinella* Busck

PL. 2, FIG. 27. TEXT FIGS. 5 *i*; 19 *g* (McD. 7688).

*Cosmopteryx* [sic] *clandestinella* Busck, 1906, *Proc. U.S. Natl. Mus.*, **30**: 712.
Type-locality: [Twining City], District of Columbia. [USNM]

NOTE—Busck did not give a type-locality in the original description, but he did give a USNM type number. The information entered in the USNM type register is "six specimens, type and cotypes". "Twining City, D.C., July 1899."

*Clandestinella* is superficially like *pulchrimella* but may be separated as follows. The apical four segments of the antenna are white preceded by four brown segments, three white segments then brown segments. A white line borders each eye, and the thorax lacks white lines. The blue-white lines at the base of the forewing are reduced with the anterior line a short band; the anteromedial and posteromedial lines are short and very broad with the posteromedial line more basad; and the posterior line is absent. The transverse fascia is orange to pale orange and is contained between the silver basal and distal bands. A few shining pale-blue scales are on the posterior margin of the forewing at $\frac{4}{5}$ the wing length, and a white streak is at the apex. The dorsal surface of the abdomen is dark brown; the ventral surface is shining cream white to off white; and the dorsolateral margins have bluish reflections. The male genitalia are distinctive with the juxtal lobes slender medially, expanding to a triangular distal part. In the female the genital plate is narrow and cylindrical, and the posterior margin of the seventh sternum is a rounded projection medially. Wing length varies from 3.2 to 4.0 mm. The anteromedial and posteromedial lines on the forewing are separate or fused, and the fascia may be heavily marked with brown scales. In females the projection on the posterior margin of the seventh sternum is rounded to nearly straight.

Busck (1906: 713) reared the larva on *Panicum clandestinum* L. He said the following: "The mine is an irregular longitudinal clear blotch-mine with the frass ejected at one end. The larva is light green with short light hairs and with yellow head and thoracic shield. At maturity it assumes a brilliant wine-red color in three longitudinal stripes, and cuts a circular piece out of the epidermis of its mine, which it bends lengthwise and uses for a cocoon exactly like the genus *Cycloplasis* Clemens." Larvae were collected in early June, and adults emerged in late July.

*Clandestinella* occurs from Massachusetts and southern Ohio south to Virginia and North Carolina. Most specimens have been reared. Adults have emerged in late May and early June and again in late July and August.

### *Cosmopterix montisella* Chambers

PL. 2, FIGS. 28–30. TEXT FIGS. 5*f*; 19 *h* (McD. 7679, 7686).

*Cosmopteryx* [sic] *montisella* Chambers, 1875, *Cincinnati Quart. Jour. Sci.*, **2**: 297.
Type-locality: Spanish Bar [mouth of Fall R., Clear Creek, 7800'], Colorado. [MCZ]

*Cosmopteryx* [sic] *unicolorella* Walsingham, 1889, *Insect Life*, **1**: 291.
Type-locality: Siskiyou Co., California. [BMNH]

*Montisella* is dark brown with three blue-white lines on the dorsal surface of the head and thorax. The blue-white lines on the basal part of the forewing are separate, and the line on the posterior margin extends to $\frac{1}{4}$ or $\frac{2}{5}$ the wing length. The transverse fascia is bordered basally by a broad silver band and distally by two silver spots; the anterior spot starts approximately where the posterior spot ends. A small tuft of shining blue-white scales is on the posterior margin at $\frac{4}{5}$ the wing length, and the apex of the forewing has a tuft of white scales. Superficially, *montisella* is similar to *clemensella*, but the two species usually may be separated by nature of the distal line on the forewing: broken in *montisella*, continuous in *clemensella*. In the males the aedoeagus of *montisella* is broadest at the base of the distal part, whereas in *clemensella* the aedoeagus is broadest at ½ of the distal part. In the females the posterior margin of the seventh sternum is emarginate medially in *montisella*, broadly emarginate for nearly the entire width and produced medially in *clemensella*. *Montisella* is highly variable in several characters. Wing length varies from 4.5 to 7.0 mm. The costomedial and posteromedial lines at the base of the forewing may be equal in length, or either may be longer, and either may start nearer the base of the wing. The transverse fascia on the forewing is pale to dark orange or dark brown. When the fascia is dark brown, the costal part usually is paler than the posterior part. The apical part of the antenna may be brown for the terminal 2½ segments preceded by one white segment, 5½ brown segments, one white segment, 1½ brown segments, then alternating ½ segments of white and brown. Or, the terminal 3½ or four segments are white preceded by five brown segments, one white segment, 1½ brown segments then alternating ½ segments of white and brown. Specimens with a dark fascia often have dark antennae. The dorsal surface of the abdomen usually is pale to medium orange, but in specimens with a dark fascia on the forewing the dorsal surface of the abdomen is gray brown or dark brown.

The immature stages and host are unknown.

*Montisella* occurs from Ithaca, New York and Oregon south to New Mexico, Arizona and southern California. Few specimens have been collected in the East, and all have come from the northern tier of states. Single specimens have been collected in Arkansas and Iowa. *Montisella* may be common in Arizona and southern California. It may be single brooded throughout much of the range but have an extended emergence period. Most records are for late June into September. In southern Arizona two broods are indicated, July and August and late September.

### *Cosmopterix magophila* Meyrick

PL. 2, FIG. 31. TEXT FIG. 5*h* (McD. 7690).

*Cosmopteryx* [sic] *magophila* Meyrick, 1919, *Exotic Microlepidoptera*, **2**: 282.
Type-locality: Southern Pines, North Carolina. [BMNH]

*Magophila* is dark brown. The apical segment of the antenna is gray to brown preceded by three white segments, five brown segments, one white segment, one brown segment, two white segments then brown segments with white on the dorsal surface. Three white lines are on the dorsal surface of the head and thorax. Four white lines are on the basal ¼ of the forewing; the costomedial and posteromedial lines are short and nearly equal in length, and the costomedial line starts nearer the base. The yellow fascia is bordered basally by two offset silver spots of which the anterior one is basad of the posterior one and distally by a larger and more basally located posterior spot and a smaller, anterior spot. The yellow passes between the distal spots and is followed by a white line on the posterior margin of the wing. The dorsal surface of the abdomen is yellowish brown medially, yellowish gray laterally and ventrally. The male genitalia are similar to those of *montisella*. Superficially, *magophila* may be confused with *clemensella*, but in *magophila* the apex of the antenna is gray to gray brown, in *clemensella* the apex of the antenna is white. Variation in the few specimens is limited to wing length, 3.3 to 3.8 mm, and intensity of color of the transverse fascia.

The food plant is unknown.

*Magophila* has been collected in the piedmont of North Carolina in mid-May and mid-August and in northwest Arkansas in late August.

### *Cosmopterix gracilens* Hodges

PL. 2, FIG. 32. TEXT FIG. 5*j*.

*Cosmopterix gracilens* Hodges, 1962, *Ent. Americana*, **42** (n.s.): 30, figs. 30, 143.
Type-locality: Palmerlee [Huachuca Mts.], Arizona. [USNM]

The single specimen of *gracilens* has a wing length of 5.3 mm. The color scheme of the distal part of the antenna is four white segments preceded by five brown segments, one white segment, one brown segment, 2½ white segments, 2½ brown segments then six segments of alternating white and brown ½ segments. The dorsal surface of the head, thorax and forewings is brown with dark red-orange reflections. Three bluish-white lines are on the head and thorax. The basal ½ of the forewing has four slender bluish-white lines of which the anteromedial line starts basad of the posteromedial line. The relative lengths of these two lines is different on each wing. The intense orangish-yellow fascia is bordered basally by an oblique silver band formed by the coalescence of two spots. Distally, the yellow mark passes between the large silver spot on the posterior margin and the very small silver spot on the costal margin. A bluish-white line on the posterior margin of the forewing extends from the yellow mark to the apex. The male genitalia are similar to those of *magophila*, but the aedoeagus is larger relative to the entire genital structure, and the apex of the right brachium is slightly more blunt than is that of *magophila*. In the same geographic area *gracilens* can be confused with *inopis*, but the forewings of *inopis* are darker brown with less reddish-orange reflections than for *gracilens*, and the antenna of *inopis* lacks the six segments of alternating white and brown ½ segments. In the male genitalia the valva is quadrate distally in *inopis* and broadly rounded in *gracilens*.

Nothing is known of the immature stages.

*Gracilens* is known only from the type-locality.

### *Cosmopterix dapifera* Hodges

PL. 2, FIGS. 33, 34. TEXT FIGS. 5*p*; 20*a*.

*Cosmopterix dapifera* Hodges, 1962, *Ent. Americana*, **42** (n.s.): 31, figs. 32, 91, 144.
Type-locality: Madera Canyon, 4880′, Santa Rita Mts., Santa Cruz Co., Arizona. [CU]

In *dapifera* the apical ½ segment of the antenna is gray brown to brown preceded by 3½ white segments, 5½ brown segments, ½ white segment then brown segments. The dorsal surface of the head, thorax and forewings is shining yellowish brown with the reflections varying into metallic gold. The head has three bluish-white to white lines, and the thorax has five bluish-white lines. The base of the forewing has four white lines of which the anteromedial line extends from the base nearly to the transverse fascia and the posteromedial line extends from ¼ the wing length to an extension of the orange fascia. The orange fascia is bordered basally by two silver spots; the anterior spot is separated from the costal margin and the posterior spot by yellow scales and is slightly basad of the posterior spot. Distally, the orange fascia extends between two silver spots and becomes broad beyond them extending from the costal margin to the posterior margin of the wing. A white line is on the posterior margin of the forewing from the yellow scales to the apex. The basal, anterior silver spot is followed by black scales. The dorsal surface of the abdomen is brown medially, greasy gray white laterally and ventrally. Wing length varies from 3.5 to 4.5 mm. The transverse fascia varies from pale yellowish orange to intense orange. Eastern specimens of *dapifera* can be confused with *dicacula* and *minutella*, but the apical ½ segment of the antenna of *dapifera* is gray, that of *minutella* is white, and in *dapifera* the third colored area is 5½ brown segments, in *dicacula* this area has one brown segment, one white segment and three brown segments. Additionally, the orange mark in *dicacula* is more extensive anterior to the basal silver spots, and the posterior silver spot is preceded by black scales. No black scales precede the posterior silver spot in *dapifera*. In the male genitalia the aedoeagus of *dapifera* is large relative to the genital capsule, small in *dicacula*. The right brachium of *dapifera* has a small expanded part immediately preceding the apex, whereas in *minutella* the expanded part of the right brachium is ½ the length of the brachium.

The immature stages are unknown.

*Dapifera* occurs from Tennessee and northwest Arkansas south to central Florida and south-central Arizona. Adults have been collected from early May through August.

### *Cosmopterix delicatella* Walsingham

PL. 2, FIGS. 35, 36. TEXT FIG. 5*m* (McD. 7683).

*Cosmopteryx* [sic] *delicatella* Walsingham, 1889, *Insect Life*, **1**: 290.
Type-locality: North Carolina. [BMNH]

*Delicatella* is yellowish brown with shining yellow or gold reflections. The apex of the antenna has two brown segments preceded by two white segments, four brown segments, one white segment, one brown segment, one white segment, ½ brown segment, ½ white segment, four or five brown segments, then alternating ½ segments of white and brown. The dorsal surface of the head and thorax has three white lines. The base of the forewing has four white lines of which the anteromedial line is slightly basad of the posteromedial line and usually is slightly longer than the posteromedial line. The orange fascia has four shining silver spots with the two on the posterior margin of the wing closer together than the two subcostal spots. The basal, subcostal spot is nearer the base of the wing than the basal posterior spot, and it is immediately followed by black-tipped scales. The fascia passes broadly both basally and distally between the pairs of silver spots. Distally, the fascia is followed by a white line that runs to the apex of the wing along the posterior margin. The dorsal surface of the abdomen is greasy yellow brown to gray. *Delicatella* is somewhat similar to *quadrilineella*, *ebriola*, *floridanella* and *scirpicola* but can be separated by the antennal coloration and white marks at the base of the forewing. Wing length varies from 3.3 to 5.0 mm. The forewing varies in relative lengths of the anteromedial and posteromedial lines, development of the orange fascia that may surround the basal, posterior silver spot and the anterior, distal spot and size of the anterior, distal spot.

The immature stages of *delicatella* are unknown.

*Delicatella* occurs on the coastal plain from Martha's Vineyard, Massachusetts to south-central Florida and west to Biloxi, Mississippi. In the North adults have been collected from late May through July, and in Florida they have been collected from late March into early June.

### *Cosmopterix dicacula* Hodges

PL. 2, FIG. 37. TEXT FIGS. 5 *l*; 20 *b*.

*Cosmopterix dicacula* Hodges, 1962, *Ent. Americana*, **42** (n.s.): 34, figs. 34, 146.
Type-locality: Gainesville, Florida. [CU]

*Dicacula* is yellowish brown with shining reflections. The extreme apex of the antenna is gray brown preceded by two white segments, one brown segment, one white segment, three brown segments, one white segment, one brown segment, one white segment, one brown segment, one white segment then brown segments. The dorsal surface of the head and thorax has three white lines. The base of the forewing has four white lines of which the anteromedial line extends from the base of the wing to the transverse fascia, and the posteromedial line is about ½ the length of the anteromedial line and extends to the transverse fascia. The light-orange fascia is bordered basally and distally by a pair of shining silver spots and extends basally and distally between the pair of spots. The anterior, basal spot is basad of the posterior, basal spot and is followed by a row of black scales. The posterior, basal spot is preceded by orange scales and bordered by a row of black scales basally. The distal silver spots are much smaller than the basal ones and are nearly opposite one another. A row of white scales extends from the orange fascia to the apex of the wing along the posterior margin. *Dicacula* is very similar to *delicatella* but can be separated by the anteromedial line extending from the base of the wing and by the basal, posterior silver spot being preceded by black scales. Wing length is 3.8 mm.

The immature stages are unknown.

*Dicacula* is known from a single male collected at Gainesville, Florida on 7 July.

### *Cosmopterix lespedezae* Walsingham

PL. 2, FIG. 38. TEXT FIGS. 5 *k*; 20 *d* (McD. 7677).

*Cosmopteryx* [sic] *lespedezae* Walsingham, 1882, *Proc. Zool. Soc. London*, 1882: 198.
Type-locality: none given [Texas]. [USNM]

*Lespedezae* is dark brown with a pale-yellow to yellowish-orange transverse fascia on the forewing. The apical four segments of the antenna are white preceded by five brown segments, one white segment, one brown segment, three white segments with brown on the dorsal surface and then brown segments. The dorsal surface of the head and thorax has three bluish-white lines. The basal ½ of the forewing has four silver-white lines of which the posteromedial line is slightly more distad and longer than the anteromedial line. The transverse fascia is bordered basally by two shining silver spots of which the subcostal spot is slightly basad of the posterior spot. Distally, the fascia is bordered by a pair of shining silver spots of which the subcostal spot is much smaller than the posterior spot. The basal, costal spot is followed by a few black scales, and a few dark-brown scales precede the posterior, distal spot. The fascia extends between the basal and distal pairs of spots. A white line runs from the distal extension of the fascia to the apex of the wing along the posterior margin. The dorsal surface of the hindtarsus is bluish white from the apex of the first segment to the end of the tarsus. Wing length varies from 3.8 to 4.3 mm. The fascia varies from pale yellow to yellowish orange, and the silver spots following the fascia may be separate or rarely joined. The dorsal surface of the abdomen is yellowish brown medially, shining yellowish gray to almost white ventrally and sublaterally except for the seventh abdominal segment in the female which is brown. *Lespedezae* is very similar to *abdita* but can be separated by the discrete silver spots bordering the fascia basally and by the bluish-white dorsal surface of the hindtarsus.

Food plants of *lespedezae* are various species of legumes including *Lespedeza* and *Desmodium*. Braun (1930: 9) described the larval habits as follows: "The mine begins at the midrib, spreading outward and upward between two lateral veins; in that part of the early mine adjacent to the lower of the lateral veins between which it lies, the parenchyma is left in narrow transverse bars, giving the mine a very characteristic aspect; as the mine increases, it extends beyond the confines of the two veins, with irregular projections and parenchyma all consumed. Pupa formed within the mine beneath the transverse bars, which are here lined with silk forming a tubular pupal chamber."

*Lespedezae* may be confined to the eastern deciduous forest. It has been taken in South Carolina, Kentucky, Ohio, Arkansas and Texas. Specimens have been collected from early July through mid-September.

### *Cosmopterix opulenta* Braun

PL. 2, FIGS. 39, 40. TEXT FIGS. 5 *n*; 20 *e* (McD. 7681).

*Cosmopteryx* [sic] *opulenta* Braun, 1919, *Ent. News*, **30**: 260.
Type-locality: Rivera, Los Angeles Co., California. [ANSP]

*Opulenta* has the apical $\frac{1}{2}$ segment of the antenna gray brown preceded by $3\frac{1}{2}$ white segments, three brown segments, one white segment, three white segments with brown patches then mainly brown segments. The dorsal surface of the head, thorax and forewings is mainly brown with shining yellowish reflections. The dorsal surface of the head has three white lines, and the dorsal surface of the thorax has a medial-white line. The basal $\frac{1}{2}$ of the forewing has four white lines of which the anteromedial line extends from the base to $\frac{3}{4}$ the distance to the transverse fascia, and the posteromedial line starts farther out on the wing and extends nearer to the transverse fascia. The transverse fascia is yellow or yellowish orange and is bordered basally by two shining silver spots. The subcostal spot precedes the posterior spot and is followed by a small number of black scales. Distally, two shining silver spots, equidistant from the base, border the fascia. The fascia extends between the basal spots for a short distance and broadly beyond the distal spots. A bluish-white line is on the posterior margin of the forewing from beyond the yellow fascia to the apex of the wing. The dorsal surface of the abdomen is grayish brown with the posterior margin of individual segments grayish white. The ventral surface of the abdomen is pale yellowish gray with gray-brown and white blotches. The last segment is yellowish gray to gray white. Wing length varies from 3.2 to 4.6 mm. A short series of specimens from Oklahoma differs from California specimens as follows: apical segment of antenna white preceded by one brown segment or two gray-brown segments preceded by two white segments; the silver spots on the forewing are not as large or as intensely colored; and the yellow fascia extends from the costal margin to the posterior margin, whereas the yellow area is reduced beyond the distal silver spots. *Opulenta* is closely related to *chisosensis*, *diaphora* Walsingham and *quadrilineella*. It can be separated from *chisosensis* and *diaphora* by the antennal color pattern. The dorsal surface of the head and thorax of *diaphora* lacks white lines; these are present in *opulenta*, *chisosensis* and *quadrilineella*. *Quadrilineella* is distinct because the yellow fascia starts at $\frac{2}{5}$ the wing length and surrounds the basal silver spots, and the three anterior white lines at the base of the wing extend to the fascia.

Additional study is required to determine whether the population occurring in Oklahoma is distinct.

Braun (1919: 260) reared *opulenta* from western ragweed, *Ambrosia psilostachya* D.C., and said, concerning the mine, ". . . The mines extend principally along the midrib, with irregular projections branching out on either side. The larva spins a cocoon on the densely pubescent under side of the leaf, constructed of silk, and the whitish pubescence of the leaf. Mines collected in October; imagoes in April of the following year." Opler (from specimen data) has reared *opulenta* from *Artemisia douglasiana* Besser.

*Opulenta* occurs from central Oklahoma west to California. Specimens from Oklahoma have been collected in July and September. One from southern Arizona was collected at light in mid-October, and specimens from central California were collected as larvae in late June and emerged in late July.

### *Cosmopterix chisosensis* Hodges, NEW SPECIES

PL. 6, FIG. 2. TEXT FIG. 6 *a*.

*Cosmopterix chisosensis* Hodges.
Type-locality: Green Gulch, 5500′, Chisos Mts., Big Bend National Park, Brewster Co., Texas. [USNM]

Upper surface as illustrated. Head: tongue shining white; frons and lower part of vertex shining white with yellow and blue reflections; dorsal surface of vertex and occiput brown with shining yellow reflections, a white line above each eye and one medially on vertex and occiput; labial palpus mainly white, second and third segments with a medial and lateral brown line extending from base to apex; apical four segments of antenna white preceded by three brown segments, one white segment, one brown segment, one white segment, one brown segment, four mainly white with some brown scales then brown segments. Thorax: brown with shining yellow and blue reflections, a medial white line present, no lateral white lines. Forewing brown with shining yellow and blue reflections on basal $\frac{1}{2}$ and on apical $\frac{1}{4}$; basal $\frac{1}{2}$ with four or five white lines, costal line extending from base to $\frac{2}{5}$ wing length, anteromedial line extending from base to $\frac{2}{5}$ wing length, posteromedial line beginning at $\frac{1}{2}$ length of brown area and extending to transverse yellow patch, a row of white scales on posterior margin from base to $\frac{1}{3}$ wing length and a row of white scales on costal margin from $\frac{1}{4}$ to $\frac{1}{2}$ wing length; transverse yellow fascia from $\frac{1}{2}$ to $\frac{3}{4}$ wing length bordered basally by two silver spots, the subcostal one anterad of the posterior one and followed by a few black scales, posterior spot preceded by a few black scales, distally two silver spots are surrounded by the fascia; a white line on the posterior margin of the wing extends from the yellow fascia to the apex of the wing. Hindwing gray brown. Foreleg: coxa shining gray brown with yellowish-gray scales at base and apex; femur, tibia and tarsus mainly darker gray brown, a white line extending from base of tibia to apex of tarsus sometimes interrupted on third tarsal segment. Midleg: coxa and femur shining yellowish gray; tibia dark brown with three bluish-white streaks, one from base to $\frac{1}{2}$ length, one at middle and one at apex; tarsus shining yellowish gray ventrally, streaked with brown and white dorsally. Hindleg: coxa and femur shining yellowish gray; tibia brown with a bluish-white line extending from base to $\frac{1}{2}$ length on ventral margin and moving to dorsal margin at $\frac{3}{4}$ length, a band of bluish-white scales just before apex; tarsus mainly yellowish gray with some brown scales on dorsal surface of first, second and third segments. Abdomen not observed before dissecting. Wing length: 3.5–3.8 mm. Male genitalia: as illustrated.

The immature stages are unknown.

TYPES: Holotype: ♂. Texas: Brewster Co., [Big Bend National Park], Chisos Mts., Green Gulch, 5500′; 6 VI 1973; R. W. Hodges; USNM slide 3890. USNM. Paratype: 1 ♂. Texas: Brewster Co., Chisos Mts., Panther Pass, 6000′; 4 VI 1973; R. W. Hodges. Paratype to USNM.

*Chisosensis* is similar to *opulenta* but can be separated by the four apical white segments of the antenna; the apex of the antenna is gray brown in *opulenta*. The male genitalia are extremely similar to those of *opulenta* but differ by the elliptical, heavily sclerotized part of the right brachium.

### *Cosmopterix quadrilineella* Chambers

PL. 2, FIGS. 41, 42. TEXT FIGS. 5 *o*; 20 *c* (McD. 7680).

*Cosmopteryx* [sic] *4-lineella* Chambers, 1878, *Bull. U.S. Geol. Geog. Surv. Terr.*, **4**: 95.
Type-locality: Bosque Co., Texas. [MCZ]

The dorsal surface of the head, thorax and forewings is gray brown with yellowish reflections giving the moth a pale appearance. The apical two segments of the antenna are brown preceded by two white segments, five brown segments, one white segment, one brown segment, one white segment then brown segments. Three white lines are on the dorsal surface of the head and thorax. The base of the forewing has four white lines with the costal and anteromedial lines running from the base to the yellow fascia, the posteromedial line running from $\frac{1}{4}$ the wing length to the transverse fascia and the line on the posterior margin running from the base nearly to the transverse fascia. The transverse fascia extends from $\frac{1}{2}$ to $\frac{3}{4}$ the wing length, completely surrounds the two basal silver spots and passes as a broad band between the distal silver spots. The basal subcostal spot is followed by a few black scales and precedes the posterior basal spot. The distal spots are nearly opposite one another, and the posterior one is preceded by black scales. A white line extends along the posterior margin of the wing from the yellow fascia to the apex. The dorsal surface of the abdomen is yellowish gray medially and shining off white laterally. The ventral surface is shining off white. Wing length varies from 3.2 to 4.6 mm. The maculation is generally constant, but the distal part of the yellow fascia may be broad or narrow beyond the distal silver spots. *Quadrilineella* can be recognized by the yellow fascia on the forewing extending basad of the basal silver spots and by the anterior three lines at the base of the forewing extending to the fascia.

The immature stages are unknown.

*Quadrilineella* is known from the type-locality in eastern Texas, west Texas and the Santa Rita Mountains in southern Arizona. Arizona specimens have been collected from 10 July to 26 October. Most moths were collected in the early evening at the bottom of a light sheet.

*Cosmopterix minutella* Beutenmüller
PL. 2, FIG. 43. TEXT FIG. 6 *b* (McD. 7684).

*Cosmopteryx* [sic] *minutella* Beutenmüller, 1889, *Ent. Americana*, **5**: 10.
Type-locality: Central Florida. [USNM]

The dark areas of the head, thorax and forewings are brown with shining yellow and gold reflections. The apical four segments of the antenna are white preceded by three brown segments, one white segment, one brown segment, one white segment, one brown segment, two white segments with brown scales, alternating $\frac{1}{2}$ segments of brown and white then brown segments with a white line. Dorsal surface of head and thorax has three white lines. On the forewing the anteromedial line extends from the base to the transverse fascia, and the posteromedial line extends from $\frac{3}{5}$ the length of the brown area to the transverse fascia. The transverse fascia is yellow to yellow orange and is bordered basally by two silver spots of which the subcostal spot precedes the posterior one and is followed by a few black scales; black scales precede the posterior spot. Distally two silver spots are nearly surrounded by the fascia. A white line is on the posterior margin of the forewing from the yellow fascia to the apex. The dorsal surface of the abdomen is yellowish gray to pale yellowish brown medially; laterally the scales are shining silver white; ventrally the surface is pale yellowish white. Wing length varies from 3.5 to 4.1 mm, and development of the transverse fascia beyond the distal spots varies. *Minutella* differs from *opulenta* and *chisosensis* by the complete anteromedial line at the base of the forewing and by the shape of the distal $\frac{1}{2}$ of the right brachium.

The immature stages are unknown.

Specimens have been collected in central and southern Florida between March and early May.

*Cosmopterix abdita* Hodges
PL. 3, FIGS. 1, 2. TEXT FIGS. 6 *c*; 20 *g*.

*Cosmopterix abdita* Hodges, 1962, *Ent. Americana*, **42** (n.s.): 40, figs. 36. 151.
Type-locality: Archbold Biological Station, L. Placid, Highlands Co., Florida. [CU]

*Abdita* is a dark-brown moth with a transverse yellow or orange band on the forewing. Apical four segments of the antenna are white preceded by five brown segments, one white segment, one brown segment, one segment white with some brown scales then mainly brown segments. Dorsal surface of head and thorax has three shining bluish-white lines. Anteromedial and posteromedial lines at base of forewing are short with the posteromedial line longer than the anteromedial line. The transverse fascia is bordered basally by a band of shining silver scales extending from near the costal margin to the posterior margin of the forewing. Distally the fascia is bordered by a large posterior silver spot and a small subcostal spot, and it extends between them as a narrow band. A complete or interrupted shining bluish-white line extends from the extension of the fascia to the apex on the posterior margin of the wing. The dorsal surface of the abdomen is dark yellowish brown or grayish brown medially, shining silver white laterally. Ventrally, the abdomen is shining silver gray medially, dark gray laterally. Wing length varies from 3.5 to 4.5 mm. The transverse fascia varies from bright yellow and yellow orange to dark orange and orange brown, and sometimes the posterior $\frac{1}{2}$ of the fascia is darker than the anterior $\frac{1}{2}$. *Abdita* is similar to *lespedezae* and *inopis*. It is separated from *lespedezae* by the solid band of silver scales preceding the fascia and from *inopis* by the posteromedial line being longer than the anteromedial line. In the male genitalia the offset right brachium is distinctive.

The immature stages are unknown.

*Abdita* occurs from Florida, south to Key Largo, and Louisiana to northwest Arkansas. It probably is multiple brooded; specimens have been collected in Florida from January through August and in Louisiana from March through October.

*Cosmopterix inopis* Hodges
PL. 3, FIG. 3. TEXT FIGS. 6 *e*; 20 *f*.

*Cosmopterix inopis* Hodges, 1962, *Ent. Americana*, **42** (n.s.): 41, figs. 39, 103, 152.
Type-locality: Madera Canyon, 4880′, Santa Rita Mts., Santa Cruz Co., Arizona. [CU]

*Inopis* is dark brown with a yellow-orange transverse fascia on the forewing. Apical three segments of the antenna are white preceded by five brown segments, one white segment, one brown segment, two white segments, $4\frac{1}{2}$ brown segments, $\frac{1}{2}$ white segment, $\frac{1}{2}$ brown segment, $\frac{1}{2}$ white segment then brown segments with a white dorsal line. Dorsal surface of head and thorax has three white lines. Anteromedial line at base of forewing is slightly longer and precedes the posteromedial line, and both are short. The dark areas have shining gold reflections. Basally the transverse fascia is bordered by a silver band running from near the costal margin to the posterior margin of the forewing. Distally a small, silver costal spot and a slightly more basad, larger, silver posterior spot border the fascia. The fascia extends between the distal spots and is followed by a white line running to the apex on the posterior margin. Two specimens of *inopis* are known, and they are very similar. Wing lengths are 3.8 and 4.0 mm. *Inopis* is similar to *lespedezae* and *abdita* but can be separated as indicated under *abdita*. In the male genitalia the quadrate distal part of the valva separates *inopis* from *lespedezae* and *abdita* that have a rounded distal part of the valva.

The immature stages are unknown.

*Inopis* has been collected in the Santa Rita Mountains in southern Arizona on 2 and 23 September.

### *Cosmopterix chalybaeella* Walsingham

PL. 3, FIG. 4 (McD. 7682).

*Cosmopteryx* [sic] *chalybaeella* Walsingham, 1889, *Insect Life*, **1**: 289.
Type-locality: Southwest Texas. [BMNH]

Only the wings and pro- and mesothorax remain of the single specimen of *chalybaeella*, so the description is a composite of Walsingham's original description and what remains of the type. Dorsal surface of the head, thorax and forewings is dark brown with shining gold reflections. Antenna is white apically and has two additional white rings. The dorsal surface of the head and thorax has three white lines. The four white lines at the base of the forewing are short; the costal line is short and slightly sinuous; the posteromedial line is slightly longer than the anteromedial line. The yellow-orange transverse fascia is bordered basally by two offset shining silver spots of which the subcostal spot is basad of the posterior spot. Distally a small subcostal spot and a larger posterior spot, nearer the base, border the fascia, and the fascia does not extend beyond these spots. A white line is on the posterior margin of the forewing from $\frac{5}{6}$ the wing length to the apex. Wing length is 2.9 mm. Additional specimens of *chalybaeella* should be recognized by the combination of cited characters.

The immature stages are unknown.

The type specimen was collected in southwest Texas, probably in the Davis Mountains.

### *Cosmopterix gemmiferella* Clemens

PL. 3, FIG. 5. TEXT FIGS. 6 *d*; 21 *a* (McD. 7676).

*Cosmopteryx* [sic] *gemmiferella* Clemens, 1860, *Proc. Acad. Nat. Sci. Philadelphia*, [**12**]: 10.
Type-locality: none given [Pennsylvania?]. [ANSP]

*Gemmiferella* is dark brown with yellowish or yellow-gold reflections. Apical four segments of the antenna are white preceded by five brown segments, four white segments then brown segments. The dorsal surface of the head has three, very slender white lines, and the medial line is continued on the thorax ending in a white spot. The base of the forewing has four bluish-white lines each of which is short. The subcostal line is slightly curved; the anteromedial line slightly precedes the equal-lengthed posteromedial line. The orange transverse fascia is bordered basally by a shining silver band and distally by a small subcostal silver spot and a larger posterior silver spot. The costal spot begins approximately where the posterior spot ends, and the fascia stops at the distal spots. A patch of white scales is on the posterior margin of the forewing at the apex, and a few shining bluish-white scales are on the posterior margin half way between the fascia and apex. The dorsal surface of the abdomen usually is shining yellowish gray brown, and the ventral surface is shining yellow gray medially, pale gray on the posterior margin of the segments and dark gray elsewhere. Wing length varies from 4.5 to 5.3 mm. The apical three or four segments of the antenna may be white, and the preceding white band may have brown scales on the second segment from the end. The anteromedial and posteromedial lines on the forewing vary in length, and sometimes the posteromedial line is very short. The posterior $\frac{1}{2}$ of the transverse fascia may have black-tipped scales. *Gemmiferella* is very similar to *bacata* but can be separated by the banding of white and brown scales on the antenna. In *bacata* the apical seven or eight white segments are preceded by brown segments to the base. Also, the brown areas on the forewing of *bacata* are darker brown with less shining yellow-gold reflections.

Braun (1923: 115) gave the following on larval habits: "The larva mines the small basal leaves of *Panicum dichotomum* Linnaeus, in the spring, eating out almost the entire substance of the leaf. Just before pupation, it enters one of the lower stem leaves, in which it makes a small inconspicuous mine, scarcely larger than the larva, but broadening at its anterior end toward the tip of the leaf, slightly inflated, and showing as a convexity on the upper surface of the leaf. Within this cavity, which is silk-lined, pupation takes place. Beyond the pupation chamber, the mine extends a short distance forwards, but is scarcely visible except at its end, where the epidermis is almost eaten through, permitting the emergence of the imago."

*Gemmiferella* occurs from Maine, southern Quebec, Illinois and northwest Arkansas south to central Florida and southern Louisiana. Adults occur in June and July in the North, in early April and May farther south.

### *Cosmopterix bacata* Hodges

PL. 3, FIG. 6. TEXT FIG. 6 *h*.

*Cosmopterix bacata* Hodges, 1962, *Ent. Americana*, **42** (n.s.): 45, figs. 42, 154.
Type-locality: Leroy, Alabama. [CU]

*Bacata* is dark brown with an intense-orange transverse fascia. The apical seven segments of the antenna are white preceded by brown segments. The dorsal surface of the head has three white lines, and the dorsal surface of the thorax has a medial white line. The base of the forewing has four bluish-white lines of which the subcostal line is short and slightly curved, the anteromedial line is short and usually shorter than the posteromedial line. The transverse fascia is bordered basally by a band of shining silver scales and distally by two, offset silver spots. The subcostal spot commences about where the posterior spot ends, and each is preceded by a row of dark-brown scales. The fascia does not extend beyond the distal spots. A few bluish-white scales are on the posterior margin of the wing at $\frac{1}{2}$ the distance between the posterodistal silver spot and the apex, and a row of white scales is on the posterior margin at the apex. The dorsal surface of the abdomen is greasy yellow brown; the lateral surface has shining silver reflections; and the ventral surface is mainly pale yellowish gray. Wing length varies from 3.5 to 4.2 mm. The transverse fascia may be yellow to yellow orange, and the two distal silver spots may be separate or contiguous. *Bacata* is very similar to *gemmiferella* but can be separated by the antennal color pattern.

The immature stages are unknown.

*Bacata* occurs from northern Florida and southwest Alabama south to central Florida. Adults have been taken from late April through mid-July.

### *Cosmopterix damnosa* Hodges

PL. 3, FIG. 7. TEXT FIG. 6 *g*.

*Cosmopterix damnosa* Hodges, 1962, *Ent. Americana*, **42** (n.s.): 46, figs. 43, 155.
Type-locality: Archbold Biological Station, Highlands Co., Florida. [CU]

*Damnosa* is dark brown with shining gold reflections dorsally. The apical three segments of the antenna are brown preceded by two white segments, five brown segments, one white segment, one brown segment, one white segment then mainly brown segments. The dorsal surface of the head and thorax has three white lines. The costal line starts at the base of the forewing and moves slightly posteriorly to $\frac{1}{4}$ the wing length. The anteromedial and posteromedial lines are nearly equal in length with the anteromedial line slightly shorter and somewhat more basad than the posteromedial line. The yellow to yellow-orange transverse fascia is bordered basally by a silver band formed by the fusion of two discrete spots; the subcostal part is followed by a few black scales. Distally, the fascia is bordered by a pair of opposite silver spots, and the fascia passes narrowly between them. A silver-white line on the posterior margin of the forewing extends from the fascia to the apex. A small number of brown scales precedes each of the distal spots, and a silver-white line is on the costal margin of the wing from $\frac{2}{5}$ to $\frac{2}{3}$ the wing length. Wing length is 3.3 mm. *Damnosa* is similar to *clemensella* and *attenuatella*, but can be separated by the difference in antennal color pattern and by the basal silver band on the forewing being nearly divided into two spots in *damnosa*, solid in *clemensella* and *attenuatella*. The male genitalia of *damnosa* are not consistently separable from some specimens of *clemensella*.

The immature stages are unknown.

The single specimen of *damnosa* was collected in south-central Florida in late March.

### *Cosmopterix clemensella* Stainton

PL. 3, FIGS. 8, 9. TEXT FIGS. 6 *i*; 21 *c* (McD. 7675).

*Cosmopteryx* [sic] *clemensella* Stainton, 1860, *Ent. Week. Int.*, **9**: 31.
Type-locality: none given [Pennsylvania?]. [BMNH]

*Cosmopteryx* [sic] *hermodora* Meyrick, 1919, *Exotic Microlepidoptera*, **2**: 282.
Type-locality: Toronto, Canada. [BMNH]

*Clemensella* is dark brown with shining gold reflections and with a yellow to yellow-orange transverse fascia on the forewing. The apical four segments of the antenna are white preceded by five brown segments, one white segment, one brown segment, two white segments, one brown with white segment then brown segments. The dorsal surface of the head and thorax has three bluish-white to white lines. The base of the forewing has four bluish-white lines. The costal line starts at the base and extends to ¼ the wing length; the anteromedial line is normally shorter than the posteromedial line and slightly precedes the posteromedial line. The transverse fascia is bordered basally by a shining silver band. A few black scales follow this band at ⅓ the distance from the costal margin to the posterior margin. Distally, two shining silver spots, opposite one another, border the fascia, and each of these is preceded by a few brown scales. The fascia extends between the distal silver spots but usually not beyond them. A shining bluish-white line is on the posterior margin from just beyond the fascia to the apex. The dorsal surface of the abdomen is brown. Laterally, the abdominal scales have shining silver reflections. Ventrally, the posterior margins of the scales are shining yellowish gray. Wing length varies from 3.5 to 5.0 mm. The apical part of the antenna may have a color pattern of four white segments preceded by four brown segments, four white segments, one brown with white segment then brown segments, or the first brown area may have five segments. In most specimens the silver band basally bordering the transverse fascia on the forewing is solid, but in some it tends to separate into two spots. The distal silver spots usually are opposite, but the costal spot may be farther from the base than the posterior spot. The bluish-white line on the posterior margin of the wing near the apex usually is complete but may be separated into a dot at ½ the distance from the fascia to the apex and a short dash at the apex. The male genitalia vary in the shape of the apex of the right brachium and the shape of the distal ½ of the valva. In the female the lamella antevaginalis can be broad or narrow with an abrupt constriction at the basal ⅓, and the apophyses anteriores are one to two times the length of the sclerotized part of the eighth tergum.

*Clemensella* can be separated from *damnosa* by the color pattern at the apex of the antenna, the basal lines on the forewing, the solid silver band basally bordering the transverse fascia and the white line at the apex of the forewing not extending to the transverse fascia.

The larva is a leaf miner on *Carex* species.

*Clemensella* has been collected from Maine and south-central Manitoba south to the mountains of southern North Carolina, but most commonly in New York and Ohio. Adults have been collected from mid-May through early August.

### *Cosmopterix scirpicola* Hodges

PL. 3, FIGS. 10, 11. TEXT FIGS. 6*f*; 21 *b*.

*Cosmopterix scirpicola* Hodges, 1962, *Ent. Americana*, **42** (n.s.): 49, figs. 44, 105, 157.
Type-locality: Oklahoma City, Oklahoma. [USNM]

*Scirpicola* is grayish orange with shining yellow to gold reflections dorsally. The extreme apex of the antenna may be white preceded by 2½ brown segments, one white segment, five brown segments, one white segment, one brown segment, one white segment then gray-brown segments. The dorsal surface of the head and thorax has three bluish-white lines. The base of the forewing has four white lines of which the costal line extends from the base to ⅔ the length of the basal dark area, the anteromedial and posteromedial lines are about ¼ the length of the forewing, and the anteromedial line usually is longer and slightly basad of the posteromedial line. The posterior margin of the forewing usually is white from the base to ¾ the length of the basal brown area. The transverse fascia is pale yellow to yellowish orange and is bordered basally and distally by two pairs of shining silver spots. The basal spots are opposite one another, and the subcostal distal spot is distad of the posterior distal spot. The fascia extends between the distal spots for a short distance and is followed by a white line on the posterior margin of the wing that runs to the apex. The subcostal basal spot is followed by a series of black scales. The dorsal surface of the abdomen is shining yellowish gray, somewhat dull medially and brighter laterally; the ventral surface is shining pale yellowish gray and paler on the posterior margins of the segments. The hindtibia is yellowish brown and has three white streaks, one running from the base to ½ the length and from the ventral to dorsal margin, a shorter one at the base of the first pair of tibial spurs, and the third at the apex of the segment. Wing length varies from 5.0 to 6.5 mm. The antennal pattern of white and gray-brown scales varies.

*Scirpicola* is similar to *fernaldella* and *ebriola*. Differences are slight but appear to be constant. The distal silver spots bordering the transverse fascia are preceded by brown scales in *fernaldella*; brown scales are usually absent in *scirpicola* and *ebriola*. On the hindtibia *fernaldella* has a few white scales dorsally at the base of the segment and a diffuse medial band of white scales; *scirpicola* has the basal band running from the base to ½ the length of the segment bordered dorsally by brown scales and a well-defined medial band; *ebriola* has the basal band as for *scirpicola* but bordered only at the extreme base with brown scales and a well-defined medial band. The genital differences are given in the key. *Scirpicola* has the widest geographic range of the three species. It can be confused with *fernaldella* in the North and with *ebriola* in Florida.

Larvae have been reared as stem miners on *Scirpus* species.

*Scirpicola* occurs from Maryland and the Black Hills of eastern Wyoming to central Florida, southwest Louisiana and San Diego, California. In the North specimens have been collected in June and July, in Florida from late March through early June, in Louisiana in May and June and in California in late August.

### *Cosmopterix ebriola* Hodges

PL. 3, FIG. 12. TEXT FIGS. 7 *a*; 21 *d*.

*Cosmopterix ebriola* Hodges, 1962, *Ent. Americana* **42** (n.s.): 50, figs. 46, 100, 158.
Type-locality: Archbold Biological Station, Highlands Co., Florida. [CU]

The color pattern for *ebriola* is much as for *scirpicola* with the major difference in the hindtibia as noted under *scirpicola*. Wing length varies from 5.0 to 5.7 mm. Some specimens have the transverse fascia nearly concolorous with the basal ½ of the forewing and a large number of dark scales following the basal subcostal silver spot. *Ebriola* can be separated from *scirpicola* and *fernaldella* as indicated under *scirpicola*.

The immature stages are unknown, but the larva is likely to feed on *Scirpus* or *Carex*.

*Ebriola* occurs from northern Florida and southern Mississippi to central Florida. Moths have been collected from March through early June.

### *Cosmopterix fernaldella* Walsingham

PL. 3, FIG. 13. TEXT FIGS. 7 *b*; 21 *e*, *f* (McD. 7685, in part).

*Cosmopteryx* [sic] *fernaldella* Walsingham, 1882, *Trans. Amer. Ent. Soc.*, **10**: 97.
Type-locality: unknown. [USNM]

The color pattern of *fernaldella* is as given for *scirpicola*. Wing length varies from 4.5 to 5.5 mm. The transverse fascia on the forewing may be yellow orange to gray orange and concolorous with the basal part of the forewing. The silver spots bordering the transverse fascia usually are preceded by brown scales, and the white line on the posterior margin of the forewing usually starts at the extension of the fascia or may start beyond the extension of the fascia. Characters separating *fernaldella* from *scirpicola* and *ebriola* are given under *scirpicola* and in the key.

The larva is a leaf miner on *Carex* species.

*Fernaldella* has been collected from Maine, southern Ontario and the Lower Peninsula of Michigan south to New Jersey and Pennsylvania. Adults have been collected from early June through July with most records from mid- to late July. This is a northern species and, perhaps, may be most easily separated from *scirpicola* and *ebriola* on geographic distribution alone.

### *Cosmopterix floridanella* Beutenmüller

PL. 3, FIG. 14. TEXT FIGS. 7 *c*; 22 *a* (McD. 7685, in part).

*Cosmopteryx* [sic] *floridanella* Beutenmüller, 1889, *Ent. Americana*, **5**: 10.
Type-locality: Central Florida. [USNM]

*Cosmopteryx* [sic] *nigrapunctella* Busck, 1900, *Proc. U.S. Natl. Mus.*, **23**: 235.
Type-locality: Palm Beach, Florida. [USNM]

The dorsal surface of the head, thorax and forewings is dark yellowish gray to gray brown, and the transverse fascia on the forewing is yellow. The apical four segments of the antenna are white preceded by one brown segment, one white segment, three brown segments, one white segment, one brown segment, four or five alternating ½ segments of white and brown then brown segments with a dorsal white line basally. The dorsal surface of the head and thorax has three white lines. The basal ½ of the forewing has four white lines of which the costal line extends from the base to ¼ the wing length, the anteromedial line extends from near the base to near the transverse fascia, and the posteromedial line extends from ¼ the wing length to the transverse fascia. The transverse fascia is bordered basally by two offset silver spots; the subcostal spot is slightly basad of the posterior spot and is followed by a series of black scales, and the posterior spot is preceded by black scales. Distally, the transverse fascia is bordered by two opposite silver spots, and the fascia extends between them and is joined by a white line on the posterior margin that runs to the apex of the wing. A series of white scales on the costal margin extends from the distal silver spot to $\frac{3}{4}$ or $\frac{5}{6}$ the wing length. The abdomen is yellowish gray dorsally, shining off white to yellowish white laterally and shining pale yellowish gray ventrally. Wing length varies from 3.6 to 4.5 mm. The transverse fascia extends basad between the two silver spots and in some specimens nearly surrounds the posterior spot. The general color and pattern of *floridanella* is similar to that of *scirpicola*, *ebriola* and *fernaldella*, but it differs from them by the longer anteromedial line and the posteromedial line extending to the transverse fascia on the forewing.

The immature stages are unknown.

Nearly all specimens of *floridanella* have been collected in Florida, but the species has been taken in northwest Arkansas and southern Mississippi. Floridian specimens have been collected from February through early June, and the specimen from Arkansas was collected in early August.

### *Cosmopterix facunda* Hodges

PL. 3, FIG. 15. TEXT FIG. 22 *d*.

*Cosmopterix facunda* Hodges, 1962, *Ent. Americana*, **42** (n.s.): 55, figs. 106, 161.
Type-locality: Brownsville, Texas. [USNM]

The dorsal surface of the head, thorax and forewings is mainly pale grayish orange. The apical $4\frac{1}{2}$ segments of the antenna are white preceded by $4\frac{1}{2}$ brown segments, one white segment, one brown segment, five white segments then brown segments. The dorsal surface of the head and thorax has three white lines. The basal $\frac{1}{2}$ of the forewing has four white lines of which the costal line extends from the base to $\frac{2}{5}$ the wing length, the anteromedial and posteromedial lines are nearly equal in length, and neither extends to the transverse fascia, and the line on the posterior margin extends from the base nearly to the transverse fascia. The yellow transverse fascia is bordered basally by two offset shining silver spots; the subcostal spot is basad of the posterior spot and is followed by black scales. Distally, the transverse fascia is bordered by two opposite silver spots. Each of these is preceded by a few yellowish-brown scales, and the fascia extends distally between them and is followed by a white line on the posterior margin of the wing that runs to the apex. Wing length is 5.4 mm. *Facunda* might be confused with *ebriola*, *scirpicola*, *fernaldella* and *floridanella* but can be recognized by the antennal color pattern and the white lines at the base of the forewing.

The immature stages are unknown.

*Facunda* is known from a single specimen collected on 27 March at Brownsville, Texas.

GENUS
## *Pebops* Hodges, NEW GENUS

Type-species: *Cosmopterix ipomoeae* Busck, 1900.

*Pebops* is a monobasic genus whose single species occurs in southern Florida. It is closely related to *Cosmopterix* but is distinct on male genital characters. The male has slender valvae, the aedoeagus a single unit, the lobes of the juxta fused with the manica basally, the tegumen relatively long and the right brachium uniformly sclerotized. *Cosmopterix* males have broad valvae, a tripartite aedoeagus, the lobes of the juxta free to the base, the tegumen very short and the right brachium heavily sclerotized distally. *Pebops* is separable from *Tanygona* by hindwing veins $M_2$, $M_3$, $Cu_1$ and $Cu_2$ being developed; these veins are not developed in *Tanygona*.

Characters of the genus are: head smooth scaled; tongue well developed, scaled basally; labial palpus upturned, smooth scaled, third segment longer than second; basal segment of antenna with pecten a single seta, antenna about $\frac{3}{4}$ length of forewing; ocelli lacking. Hindtibia with long slender scales dorsally. Forewing narrowly lanceolate, anterior and posterior margins moving toward each other at $\frac{3}{4}$ wing length to very slender apex; 12 veins present; $R_4$, $R_5$ and $M_1$ stalked, $R_4$ arising before $M_1$; $M_2$ connate with $R_4+R_5+M_1$; cell weakly closed; $Cu_1$ arising opposite $R_1$; 1A absent; 2A and 3A forked basally; ventral surface with series of short setae on posterior margin near base a part of secondary wing coupling mechanism. Hindwing linear; eight veins present but weakly developed; cell open; $R_s$ and $M_1$ stalked; $M_2$, $M_3$, $Cu_1$ and $Cu_2$ apparently arising from a common stalk; one anal vein present; a tuft of stout scales on anterior margin at $\frac{1}{5}$ wing length serving as secondary wing coupling mechanism. Male genitalia as figured. Female genitalia as figured.

The larvae are leaf miners on morning glory, *Ipomoea* species.

### *Pebops ipomoeae* (Busck), NEW COMBINATION

PL. 3, FIG. 17. TEXT FIGS. 1 *h*; 7 *d*; 22 *b* (McD. 7678).

*Cosmopteryx* [sic] *ipomoeae* Busck, 1900, *Proc. U.S. Natl. Mus.*, **23**: 235.
Type-locality: Palm Beach, Florida. [USNM]

*Ipomoeae* is similar to many *Cosmopterix* species in forewing color pattern but can be separated by the blue-white line at the base of the wing being a single row of scales, and the anteromedial and posteromedial lines being interrupted by dark scales. The transverse median fascia extends broadly between the distal silver spots and is broader beyond them, running from the anterior to posterior margins. At $\frac{3}{4}$ the wing length the fascia becomes narrow, and a concolorous line on the posterior margin of the wing extends from the fascia to the apex. This line is bluish white and contrasts with the color of the transverse fascia in *Cosmopterix*. The apical four segments of the antenna are white preceded by five brown segments, four white segments then brown segments. The second white zone has a few brown scales dorsally on the second and fourth segments. The dorsal surface of the thorax has a median bluish-white line. The base of the forewing has four bluish-white lines, two nearly on the costal margin, an anteromedial line extending from near the base to $\frac{1}{3}$ the wing length and a shorter posteromedial line starting at $\frac{1}{4}$ the wing length and running nearly to the transverse fascia. A pale yellowish line is on the posterior margin from near the base to $\frac{1}{5}$ the wing length. The pale yellow transverse fascia is bordered basally by two silver spots, of which the subcostal spot is basad of the posterior spot and is preceded and followed by brown scales, and the posterior spot is preceded by brown scales. The two distal silver spots are surrounded by the transverse fascia. The lateral surface of the hindtibia is brown with a bluish-white line running from the ventral margin near the base to the dorsal margin beyond $\frac{1}{2}$ the length, and a band of white scales is at the apex. The base of the first tarsal segment is pale gray, and the apex of each tarsal segment is yellowish gray.

The larva is a leaf miner on morning glory.

*Ipomoeae* has been collected from Palm Beach to Key Largo, Florida during October through February and in May.

GENUS
## *Tanygona* Braun

*Tanygona* Braun, 1923, *Trans. Amer. Ent. Soc.*, **49**: 115.
Type-species: *Tanygona lignicolorella* Braun, 1923. Original designation.

*Tanygona* is a monobasic genus whose single species occurs in eastern North America. It is closely related to *Cosmopterix* and *Pebops* but can be separated by venation. In the forewing vein $Cu_2$ is absent, and in the hindwing veins $M_2$, $M_3$, $Cu_1$ and $Cu_2$ are not distinguishable, whereas in *Cosmopterix* and *Pebops* $Cu_2$ is present in the forewing, and $M_2$, $M_3$, $Cu_1$ and $Cu_2$ are present in the hindwing.

Characters of *Tanygona* are: labial palpus smooth scaled, third segment slightly longer than second; basal segment of antenna with pecten a slender seta, antenna $\frac{4}{5}$ forewing length; ocelli absent. Forewing narrowly lanceolate, anterior and posterior margins moving toward each other at $\frac{3}{4}$ length; $R_4$, $R_5$, $M_1$ and $M_2$ stalked, $M_1$ arising at $\frac{1}{2}$ distance between end of cell and end of wing, $M_2$ short stalked with other veins; $Cu_1$ arising between $R_2$ and $R_3$; $Cu_2$ absent; 2A and 3A forked basally; ventral surface with strong setae on posterior margin near base acting as part of secondary wing coupling mechanism. Hindwing linear; cell open; $R_s$ and $M_1$ stalked; $M_2$, $M_3$, $Cu_1$ and $Cu_2$ not distinguishable; one anal weakly present; a tuft of stout setae on costal margin near base of wing acting as part of secondary wing coupling mechanism. Male genitalia as figured. Valva slender, bent at right angle at $\frac{1}{2}$ length; lobes of juxta fused with manica basally, free distally; aedoeagus ankylosed with a distinct basal and distal part; tegumen moderately long with a heavily sclerotized anterior rim; right brachium heavily sclerotized. Female as figured. Lamella antevaginalis a rectangular plate; ductus bursae twisted at least once, shorter than corpus bursae; ductus seminalis arising from base of corpus bursae; corpus bursae without signa; apophyses posteriores longer than apophyses anteriores, approximately two to three times length of corpus bursae. Hindtibia with long setae on dorsal surface.

The immature stages are unknown.

### *Tanygona lignicolorella* Braun

PL. 3, FIG. 16. TEXT FIGS. 1 *b*; 7 *e*; 22 *e* (McD. 7691).

*Tanygona lignicolorella* Braun, 1923, *Trans. Amer. Ent. Soc.*, **49**: 116.
Type-locality: Cincinnati, Ohio. [ANSP]

The combination of color pattern and size should serve for recognition of *lignicolorella*. The dorsal surface of the head and thorax has three bluish-white lines. Wing length varies from 3.6 to 3.9 mm.

The immature stages are unknown.

*Lignicolorella* occurs from Washington, D.C., central Illinois and northwest Arkansas south to west-central Florida and the southern Appalachians. Moths have been collected from late June through early August except for the Floridian specimens collected in early May.

GENUS

## *Eralea* Hodges

*Eralea* Hodges, 1962, *Ent. Americana*, **42** (n.s.): 59, figs. 3, 8.
Type-species: *Eriphia albalineella* Chambers, 1878. Original designation.

The two species of *Eralea* occur in the southern United States from Florida to California. *Eralea* and *Melanocinclis* are closely allied. These two genera can be separated from *Cosmopterix*, *Pebops* and *Tanygona* by maculation and by the antennal pecten consisting of six to eight setae. *Eralea* differs from *Melanocinclis* by having complete venation in the forewing and having veins $M_1$ through $Cu_2$ weakly developed in the hindwing. *Melanocinclis* lacks veins $M_2$ and $Cu_1$ in the forewing, and veins $M_1$ through $Cu_2$ in the hindwing are indistinguishable.

Characters of *Eralea* are: labial palpus upcurved, third segment slightly longer than second; antennal scape short with pecten of six to eight slender setae, antenna $\frac{3}{4}$ to $\frac{4}{5}$ length of forewing; ocelli lacking. Dorsal surface of hindtibia with numerous, long slender scales. Forewing narrowly lanceolate, apex acute; $R_1$ arising at $\frac{1}{2}$ length of cell; $R_4$, $R_5$, $M_1$ and $M_2$ stalked, $M_1$ arising approximately $\frac{1}{2}$ distance between end of cell and end of wing; $Cu_2$ arising slightly before point of origin of $R_2$; 2A and 3A forked basally; ventral surface with series of strong setae near posterior margin and base acting as part of secondary wing coupling mechanism. Hindwing linear or nearly so, apex acute, cell open; veins other than Sc, $R_s$ and base of Cu very weak to scarcely defined; a tuft of stout scales on costal margin near base a part of secondary wing coupling mechanism. Male with eighth abdominal sternum a pair of lobes; valva lobate; lobes of juxta bulbous, free for distal $\frac{3}{5}$; aedoeagus somewhat angulate at $\frac{1}{2}$ or $\frac{2}{5}$ length, ankylosed; tegumen moderately long; brachia evenly sclerotized, right brachium longer than left brachium. Female with ostium bursae and lamella antevaginalis in membrane between seventh and eighth abdominal sterna; ductus bursae extremely slender basally, broadening slightly to corpus bursae; ductus seminalis arising from lobe at base of corpus bursae; paired signa thumb-tack shaped; apophyses anteriores shorter than apophyses posteriores; eighth abdominal segment with an incomplete, lightly sclerotized band, membranous ventrally.

The immature stages are unknown.

### KEY TO SPECIES OF *ERALEA*

1. Forewing shining dark gray with white band anteriorly bordering fold from base of wing to end of fold....... *albalineella* (Chambers) this page
— Forewing shining dark gray, without white band anterad of fold.......... *abludius* Hodges this page

### *Eralea albalineella* (Chambers)

PL. 3, FIGS. 18, 19. TEXT FIGS. 1 *d*; 7 *f*; 22 *f* (McD. 7692).

*Eriphia* (?) *albalineella* Chambers, 1878, *Bull. U.S. Geol. Geog. Surv. Terr.*, **4**: 95.
Type-locality: Bosque Co., Texas. [MCZ]

*Eralea striata* Hodges, 1962, *Ent. Americana*, **42** (n.s.): 62, figs. 50, 165. NEW SYNONYMY.
Type-locality: Siesta Key, Sarasota Co., Florida. [CU]

*Albalineella* can be recognized by the combination of dark-gray dorsal surface with a white band anteriorly bordering the fold from the base of the forewing to the end of the fold and with a series of three white lines extending from the costal margin toward the posterior margin. Antenna with dorsal surface of scape dark gray to brown, ventral surface of scape and entire shaft cream white. Wing length varies from 3.5 to 3.8 mm. The forewing varies in shade of dark gray depending on degree of wear, and the white band varies in width.

The immature stages are unknown.

*Albalineella* occurs from western Florida to San Diego County, California, but very few specimens have been collected. Adults have been taken in mid-May and early November in Florida, early to mid-July in the mountains of southern Arizona and mid-June at La Jolla, California.

### *Eralea abludius* Hodges, NEW SPECIES

PL. 3, FIG. 20. TEXT FIG. 7 *h*.

*Eralea abludius* Hodges.
Type-locality: Archbold Biological Station, L. Placid, Florida. [USNM]

Upper surface as figured. Head: mainly yellowish white, a few dark-brown scales at base of tongue and on outer surfaces of first and base of second segments of labial palpus; a row of dark-brown scales around eye; antennal scape dark brown except for anterior surface, shaft with a few scattered dark-brown scales on posterior surface. Thorax dark brown. Foreleg: shining dark gray brown, tarsus mainly yellowish white with dark gray-brown scales on first three segments. Mid- and hindlegs: much as for foreleg but somewhat paler, apex of midtibia yellowish white, hindtibia with pale yellowish-gray scales at $\frac{1}{2}$ length and at apex. Abdomen: very pale yellowish gray, apices of segments yellowish white, ventral surface darker than dorsal surface. Wing length: 2.8–3.2 mm. Male genitalia: as in text figure 7 *h*. Valva with saccular margin angulate at $\frac{1}{2}$ length, two or three long setae (longer than maximum width of valva) on distal $\frac{1}{3}$ of costal margin; distal $\frac{1}{3}$ to $\frac{1}{4}$ of each brachium slender, apices rounded. Female genitalia: no specimens available.

The immature stages are unknown.

TYPES: Holotype: ♂. Archbold Biological Station, L. Placid, Florida; 23–31 May 1964; R. W. Hodges; RWH slide 3495. Type no. 70,566, USNM. Paratypes: 2 ♂♂. Same locality as for holotype; 1–22 May 1964; R. W. Hodges. Paratypes to USNM.

*Abludius* can be separated from *albalineella* by the forewing lacking a white line along the fold.

GENUS
## *Melanocinclis* Hodges

*Melanocinclis* Hodges, 1962, *Ent. Americana*, **42** (n.s.): 62, figs. 8, 15.
Type-species: *Melanocinclis lineigera* Hodges, 1962. Original designation.

The five species of *Melanocinclis* occur from New Jersey, south to Florida and west to southern Arizona. Three species occur in Florida. *Melanocinclis* is nearest *Eralea*, but the species can be separated by the reduced forewing venation in which veins $M_2$ and $Cu_1$ are absent and by the third segment of the labial palpus being shorter than the second segment. In *Eralea* veins $M_2$ and $Cu_1$ are present in the forewing, and the second segment of the labial palpus is shorter than the third segment.

Characters of the genus are: second segment of labial palpus longer than third segment; antennal scape short, pecten of six to eight setae located on middle $\frac{1}{2}$ of anterior margin; ocelli absent. Dorsal surface of hindtibia with numerous long slender scales. Forewing narrowly lanceolate, apex slender and acute; $R_1$ arising just before $\frac{1}{2}$ length of cell; $R_4$, $R_5$ and $M_1$ stalked with $M_1$ separating from $R_5$ beyond $\frac{1}{2}$ distance between end of cell and apex of wing; $M_2$ and $Cu_1$ absent; $Cu_2$ arising from cell between $R_2$ and $R_3$; ventral surface with series of setae on posterior margin near base as part of secondary wing coupling mechanism. Hindwing linear; Sc, $R_s$ and base of cubitus well developed, other veins scarcely distinguishable; 2A weak; a tuft of stout scales on anterior margin near base. Male: eighth abdominal sternum bilobate, valva straight, usually expanded apically; lobes of juxta short to long, normally free from base to apex; aedoeagus short and angulate to relatively long, formed of distinct basal and distal parts; tegumen with anterior margin heavily sclerotized; brachia evenly sclerotized, right brachium longer than left brachium. Female: ostium bursae in membrane between seventh and eighth abdominal sterna; lamellae antevaginalis and postvaginalis usually forming a cylinder; ductus bursae very slender; corpus bursae membranous, without signa; ductus seminalis arising from lobe at base of corpus bursae; apophyses anteriores shorter than apophyses posteriores; eighth abdominal segment lightly sclerotized, membranous ventrally.

Larvae have been reared from pine cones, *Pinus* species, and the flowers of saw palmetto, *Serenoa repens* (Bart.) Small.

### KEY TO SPECIES OF *MELANOCINCLIS*

1. Forewing with a nearly continuous, broad line of brown scales on costal margin from base to apex . . . . . . . . *lineigera* p. 42
— Forewing without a continuous line from base to apex . . . . . . . . 2

2. Forewing with pair of dark-brown marks posterad of fold . . . . . . . . *vibex* p. 43
— Forewing without distinct dark marks posterad of fold, scales usually tipped with brown giving area suffused appearance . . . . . . . . 3

3. Forewing with longitudinal bars, or an oblique fascia on basal $\frac{2}{5}$ . . . . . . . . 4
— Forewing lacking distinct marks on basal $\frac{2}{5}$ . . . . . . . . *gnoma* p. 43

4. Hindtibia with outer spurs dark gray . . *nigrilineella* this page
— Hindtibia with outer spurs yellowish white . . . . . . . . . . . . . . . . . . . . . . . . . . . . . . . . . . *sparsa* this page

### *Melanocinclis lineigera* Hodges

PL. 3, FIG. 21. TEXT FIGS. 1 *e*; 7 *i*; 22 *c*.

*Melanocinclis lineigera* Hodges, 1962, *Ent. Americana*, **42** (n.s.): 64, figs. 57, 116, 168.
Type-locality: Archbold Biological Station, Highlands Co., Florida. [CU]

*Lineigera* is nearest *nigrilineella* but can be separated by the numerous red-brown scales on the forewings; *nigrilineella* lacks red-brown scales on the forewings. In the male genitalia the left brachium is bilobed in *lineigera*, single lobed in *nigrilineella*. Additional characters are: second segment of labial palpus is dark gray brown from the base to $\frac{3}{5}$ the length followed by a white ring, a dark gray-brown ring and then a white ring at the apex; the third segment is white with dark gray-brown rings at $\frac{1}{5}$ and $\frac{3}{5}$ the length; the lower part of the frons is shining gray; the vertex and occiput are shining white, occasionally with dark gray-brown scales above the eye; the base of the antenna is shining cream white with dark-gray scales on the dorsal surface of the scape. Outer surface of hindtibia is mainly white with a transverse, dark-brown band from ventral margin at first pair of spurs to dorsal margin at $\frac{5}{6}$ length. Hindtarsus is dark brown at base and on dorsal surface of third segment. Wing length varies from 2.5 to 3.2 mm. Color pattern varies in the amount of dark-brown scales on antenna, thorax and forewings. The red-brown scales may be lost, so worn specimens have few of them.

Larvae have been reared from cones of loblolly pine, *Pinus taeda* L., and slash pine, *P. elliottii* Engelm.

*Lineigera* occurs from the New Jersey pine barrens and northwest Arkansas south to southern Florida and southern Louisiana. Adults have been collected from the end of June through early August in New Jersey, early August in North Carolina, November and January through May in Florida, early July in Arkansas and mid-April through early June in Louisiana.

### *Melanocinclis nigrilineella* (Chambers)

PL. 3, FIG. 22. TEXT FIG. 7 *g* (McD. 7713).

*Eriphia* (?) *nigrilineella* Chambers, 1878, *Bull. U.S. Geol. Geog. Surv. Terr.*, **4**: 96.
Type-locality: Bosque Co., Texas. [MCZ]

*Nigrilineella* is nearest *lineigera* but can be separated as indicated under *lineigera*. In addition the second segment of the labial palpus is white on the distal $\frac{1}{3}$; the frons is shining gray to the base of the antennae; and the forewing is brown and white. Wing length of the single available specimen is 3.3 mm.

The immature stages are unknown.

*Nigrilineella* is known from Bosque County, Texas and the Santa Rita Mountains of southern Arizona. The latter specimen was collected on 30 June.

### *Melanocinclis sparsa* Hodges, NEW SPECIES

PL. 3, FIGS. 23, 24. TEXT FIGS. 8 *a*; 23 *a*.

*Melanocinclis sparsa* Hodges.
Type-locality: Devil's Den State Park, Washington Co., Arkansas. [USNM]

Upper surface as figured. Head: yellowish white; first segment of labial palpus brown, basal $\frac{1}{3}$ of second segment and a preapical, incomplete ring brown, third segment with a brown ring at $\frac{1}{3}$ and $\frac{2}{3}$ length; lower $\frac{1}{2}$ of frons shining gray; a few scales above eye tipped with brown; antenna with dorsal surface of scape with several brown-tipped scales, shaft alternating brown and yellowish gray on basal $\frac{2}{3}$, apical $2\frac{1}{2}$ segments yellowish white preceded by $1\frac{1}{2}$ brown segments, $2\frac{1}{2}$ yellowish-white segments, $1\frac{1}{2}$ brown segments, $2\frac{1}{2}$ yellowish-white segments then one brown segment. Thorax: mottled brown and pale yellowish gray, tegula brown at base and apex, yellowish white medially. Foreleg: coxa yellowish white with some dark yellowish-gray scales at base; femur, tibia, and tarsus mainly brown, yellowish white at apex of first tarsal segment and on all of fifth tarsal segment. Midleg: mainly yellowish white; femur with some yellowish-gray scales; tibia with an oblique dark-brown fascia at $\frac{1}{5}$ and $\frac{2}{3}$ length; tarsus with a brown saddle at apex of first segment, one at apex of third segment and on entire fourth segment. Hindleg: yellowish white; tibia with an oblique brown fascia from near base to $\frac{2}{5}$ length and another from $\frac{1}{2}$ length nearly to apex; tarsus with a broad, brown saddle on middle; apex of second segment and base of third segment brown. Abdomen: dorsal surface yellowish gray medially, yellowish white laterally; ventral surface yellowish white. Wing length: male, 6.0–6.5 mm; female, 7.5–8.0 mm. Male genitalia as figured. Distal $\frac{1}{2}$ of aedoeagus relatively stout; anellar lobes stout, gradually tapering to blunt apices; right brachium expanded slightly before apex, left brachium slender, gradually tapering to acute apex. Female as figured. Apex of seventh sternum incurved; lamellae vaginales forming an inwardly projecting, V-shaped plate; ductus bursae with spicules on distal $\frac{1}{3}$.

The immature stages are unknown.

TYPES: Holotype: ♂. Devil's Den State Park, Washington Co., Arkansas; 6 Aug. 1966; R. W. Hodges; RWH slide 4065. Type no. 70,567, USNM. Paratypes: 4 ♂♂, 2 ♀♀. Same locality as for holotype; 22 May–23 July 1966; R. W. Hodges. Paratypes to University of Arkansas; USNM.

*Sparsa* can be separated from other *Melanocinclis* species by the incomplete, pale fascia running from the costa to the fold at $\frac{1}{3}$ the length of the forewing.

The species may be multiple brooded with an emergence occurring from mid-May to early June and another in July. It is dimorphic in size with males being smaller than females. As adults age, they lose many of the brown-tipped scales, particularly on the distal part of the forewing, and become very pale. Fresh specimens are relatively dark.

### *Melanocinclis gnoma* Hodges, NEW SPECIES

PL. 3, FIG. 25. TEXT FIGS. 8 *c*; 23 *b*.

*Melanocinclis gnoma* Hodges.
Type-locality: Archbold Biological Station, L. Placid, Florida. [USNM]

Upper surface as figured. Head: yellowish white; lateral surface of labial palpus olive gray to brown on first segment and basal $\frac{1}{3}$ of second segment, a brown anellus just before apex of second segment and one at base and at $\frac{2}{3}$ length of third segment; antenna with scales on dorsal surface of scape tipped with brown, apex of scape yellowish white, apical $\frac{1}{2}$ segment of shaft without scales, preceded by alternating series of $1\frac{1}{2}$ brown segments and $\frac{1}{2}$ off-white segments for a total of five white elements, remaining segments mainly brown except at base; a few brown and brown-tipped scales above eye. Thorax: mottled yellowish gray and brown, scale bases pale. Proleg: coxa pale yellow; base of femur yellowish gray at base becoming brown distally; tibia brown, a few off-white scales at apex; tarsal segments one, two and three mainly brown, a few white scales on first and second segments, fourth and fifth segments mainly yellowish gray. Midleg: coxa and femur mainly pale yellow to yellowish gray; tibia white with an oblique brown fascia at $\frac{1}{4}$ and $\frac{2}{3}$ length; tarsal segments one and two white, an oblique brown fascia at middle of first segment, distal $\frac{1}{2}$ of second and all of third segment brown, fourth and fifth segments pale yellow. Hindleg: much as for midleg with brown areas more restricted. Abdomen: varying shades of pale yellowish gray. Wing length: 2.3–2.5 mm. Male genitalia as figured. Distal part of aedoeagus approximately $\frac{1}{2}$ width of basal part, right brachium decidedly sinuous. Female as figured. Lamella antevaginalis roughly I-shaped with basal arms broader and longer then distal arms.

The immature stages are unknown.

TYPES: Holotype: ♂. Archbold Biological Station, L. Placid, Florida; 8–15 May 1964; R. W. Hodges; RWH slide 3767. Type no. 70,568, USNM. Paratypes: 1 ♂, 5 ♀♀. Same locality as for holotype; 1 May–1 June 1964; R. W. Hodges. Paratypes to USNM.

The small size, generally diffuse pattern on the forewings and antennal color pattern will separate *gnoma* from the other species of *Melanocinclis*.

### *Melanocinclis vibex* Hodges, NEW SPECIES

PL. 3, FIGS. 26, 27. TEXT FIGS. 8 *b*; 23 *c*.

*Melanocinclis vibex* Hodges.
Type-locality: Archbold Biological Station, L. Placid, Florida. [USNM]

Upper surface as figured. Head: tongue pale yellow; labial palpus yellowish white, outer surface of first segment and second segment at base with brown scales, an incomplete brown fascia at $\frac{2}{3}$ length on outer surface of second segment, third segment with a brown anellus just beyond base and another just before apex, distal one darker than basal one; antenna light orange and brown, ventral surface light orange, scape with brown on posterior $\frac{1}{2}$ except at apex, dorsal surface of shaft with alternating bands of light orange and brown; frons, vertex and occiput shining pale yellow. Thorax: mottled shining orange and brown. Wings: upper surfaces as illustrated: ventral surface of forewing dark gray brown, costal margin from $\frac{2}{3}$ length nearly to apex pale yellow, apex dark gray brown, posterior cilia pale yellow to gray. Foreleg: coxa yellowish white, a lateral brown band running from base nearly to apex; femur and tibia brown, each with scattered yellowish-white scales laterally; first three tarsal segments mainly brown, first and second segments with yellowish white at base, fourth and fifth segments yellowish white. Midleg: mainly yellowish white; femur with a few gray-brown scales; tibia with two oblique brown fasciae, one just beyond base, second at $\frac{2}{3}$ length; tarsus with a brown saddle at middle of first segment and one extending from distal $\frac{1}{3}$ of second segment to base of fourth segment. Hindleg: maculation much as for midleg but generally paler, tibia and tarsus off white, tibial spurs with some gray-brown scales, inner midtibial spur nearly twice length of outer one. Abdomen: dorsal surface yellowish gray, tergites yellowish white laterally and apically; ventral surface yellowish white, a medial brown stripe on segments three through seven. Wing length: male, 6.0–7.5 mm; female, 8.0–9.0 mm. Male genitalia as figured. A long-handled, dipper-shaped structure arising in saccal area; anellar lobes long, slender, slightly expanded before apices, aedoeagus long, slender, apex drawn to fine point; brachia subequal in length, apices blunt, right brachium with short ventral lobe at apex. Female as figured. Tube formed from lamellae vaginales drawn into slender lobe on left side; ductus bursae lightly sclerotized on basal $\frac{1}{2}$, heavily sclerotized on distal $\frac{1}{2}$; wall of corpus bursae spiculose, more heavily anteriorly than posteriorly; ductus seminalis arising from base of corpus bursae.

One larva has been found in the flowers of saw palmetto, *Serenoa repens* (Bart.) Small.

TYPES: Holotype: ♂. Archbold Biological Station, L. Placid, Florida; 16–22 May 1964; R. W. Hodges; RWH slide 3493. Type no. 70,569, USNM. Paratypes: 9 ♂♂, 2 ♀♀. Same locality as for holotype; 16–31 May 1964; R. W. Hodges (7 ♂♂, 1 ♀). Same locality as for holotype; collected 1–8 June 1964 on *Serenoa repens* flowers, issued 10 July 1964; R. W. Hodges (1 ♀). Parker Islands, Highlands Co., Florida; 26 May–7 June 1964; R. W. Hodges (2 ♂♂). Paratypes to USNM.

*Vibex* can be recognized by the pair of brown spots on the posterior margin of the forewing. Males are smaller than females.

GENUS
## *Stagmatophora* Herrich-Schäffer

*Stagmatophora* Herrich-Schäffer, [1853], *Systematische Bearbeitung der Schmetterlinge von Europa*, **5**: 49.
Type-species: *Oecophora heydeniella* Fischer von Röslerstamm, 1841. Monotypy.

*Eteobalea* Hodges, 1962, *Ent. Americana*, **42** (n.s.): 66, figs. 2, 20. NEW SYNONYMY.
Type-species: *Gelechia sexnotella* Chambers, 1878. Original designation.

*Parastagmatophora* Riedl, 1965, *Polskie Pismo Ent.*, **35**: 433. NEW SYNONYMY.
Type-species: *Oecophora serratella* Treitschke, 1833. Original designation.

*Stagmatophora* is a holarctic genus with approximately 50 species. None are known to occur in the Neotropical Region, and four species occur in North America north of Mexico. These occur from New Hampshire and Washington to Florida and southern California. Our species can be recognized by forewing pattern which consists of a nearly unicolorous ground color with three triangular white marks on the costal margin and a series of metallic silver spots near the posterior margin. The antennal pecten is composed of six to eight slender setae. The presence of metallic scales on the forewing separates *Stagmatophora* from *Limnaecia*, *Pyroderces*, *Melanocinclis*, *Eralea* and *Tanygona*. The antennal pecten of six to eight setae will separate *Stagmatophora* from *Cosmopterix* and *Pebops* in which the antennal pecten is a single seta.

Characters of the genus are as follows: tongue moderately short; second segment of labial palpus smooth or rough scaled, shorter than, equal to or longer than third segment; antennal scape short, pecten of six to eight setae, antenna $\frac{3}{4}$ to $\frac{4}{5}$ forewing length; ocelli present or absent. Forewing lanceolate with acute apex, 12 veins present; $R_1$ arising from before $\frac{1}{2}$ length of cell; $R_4$, $R_5$ and $M_1$ usually stalked, $M_1$ arising before or after point of separation of $R_4$; $M_2$ connate or separate with $R_4+R_5+M_1$; $Cu_2$ arising before origin of $R_2$; fold strong; 2A and 3A forked basally. Hindwing narrowly lanceolate, apex acute, cell open; $R_s$ and $M_1$ stalked; $M_2$, $M_3$, $Cu_1$ and $Cu_2$ separate and arising from stalk of Cu; 2A weakly developed. Dorsal surface of hindtibia with numerous slender scales. Male genitalia with valva broadly lobate, lightly sclerotized; vinculum very short and weakly connected medially; aedoeagus with a bulbous distal lobe and a flangelike basal lobe; lobes of juxta closely associated with aedoeagus, right lobe longer than left lobe; tegumen well developed; apex of right brachium expanded, truncated; eighth sternum weakly bilobed. Female with ostium bursae in membrane between seventh and eighth abdominal sterna; lamella antevaginalis an irregular plate; ductus bursae extremely slender; corpus bursae membranous with two signa; ductus seminalis arising from lobe at base of corpus bursae; eighth abdominal segment an incomplete, weakly sclerotized ring; apophyses anteriores shorter than apophyses posteriores.

Larvae have been reared from roots of *Plantago* (Plantaginaceae), from leaf mines on *Linaria* and *Antirrhinum* (Scrophulariaceae) and from stem galls on *Teucrium* and *Trichostema* (Labiatae).

I have synonymized *Eteobalea* with *Stagmatophora* because the characters I used to separate them do not appear to be constant. The forewing pattern of *Stagmatophora heydeniella* is different from that of other species in the genus, but this alone is not adequate for recognition of a separate genus.

## KEY TO SPECIES OF *STAGMATOPHORA*

1. Forewing dark gray or black with white spots .................................... 2
— Forewing yellow, yellow brown or bronze...... 3

2. Hindwing light gray, base of tongue white, base of forewing usually black, occiput white .................................. *sexnotella* this page
— Hindwing dark gray, base of tongue black, base of forewing white, occiput with posterior margin dark brown or black. . *wyattella* this page

3. Labial palpus with base of second segment bronze black, ocelli absent............ *enchrysa* p. 46
— Labial palpus with base of second segment white, ocelli present.................... *iridella* this page

### *Stagmatophora sexnotella* (Chambers)

PL. 3, FIG. 28. TEXT FIGS. 1 *g*; 8 *d*; 23 *d* (McD. 7716).

*Gelechia sexnotella* Chambers, 1878, *Bull. U.S. Geol. Geog. Surv. Terr.*, **4**: 88.
Type-locality: Bosque Co., Texas. [MCZ]

The dorsal surface of the head, thorax and forewings of *sexnotella* is dark brown to black with shining white and silver spots. The moths are similar to those of *wyattella* but can be separated by having the base of the tongue white, the occiput uniformly shining white and the base of the forewing dark brown. In contrast *wyattella* has the base of the tongue dark brown, the posterior margin of the occiput dark brown and the base of the forewing white. Scales on the dorsal surface of the forewing lie flat. Wing length varies from 3.8 to 6.3 mm.

The larva is a stem gall maker on bluecurls, *Trichostema dichotomum* L. and *T. suffrutescens* Kearney.

*Sexnotella* occurs from New Hampshire and southern Ontario south to central Florida and the mountains of West Texas. Adults have been collected from late May through August in the northern part of the range, January through June in Florida and late May in northwest Arkansas and West Texas.

### *Stagmatophora wyattella* Barnes and Busck

PL. 4, FIG. 1. TEXT FIGS. 8 *e*, 23 *e* (McD 7717).

*Stagmatophora wyattella* Barnes and Busck, 1920, *Contrib. Nat. Hist. Lep. N. Am.*, **4**: 222.
Type-locality: Palos Park, Illinois. [USNM]

*Wyattella* is similar to *sexnotella* but can be separated as indicated under *sexnotella*. In addition many scales on the upper surface of the forewings are curved, giving a roughened appearance to the wings; and the hindwing is dark gray, contrasting with the pale-gray hindwing of *sexnotella*. Wing length varies from 4.5 to 8.0 mm. The major variation is the development of white areas on the forewing. The illustrated specimen has fairly extensive white marks. In others the second white mark on the costal margin is nearly absent as are those on the posterior margin.

The immature stages are unknown.

*Wyattella* occurs from Long Island, New York west to southeastern South Dakota, but it has been collected in these two states and Illinois only. All specimens have been collected at light in July and August.

### *Stagmatophora iridella* (Busck)

PL. 4, FIGS. 2, 3. TEXT FIGS. 8 *i*; 23 *f* (McD. 7715, 7718).

*Mompha iridella* Busck, 1907, *Proc. Ent. Soc. Washington*, **8**: 96.
Type-locality: Claremont, California. [USNM]

*Stagmatophora niphochrysa* Meyrick, 1930, *Exotic Microlepidoptera*, **3**: 546.
Type-locality: Fort Davis, Texas. [BMNH]

*Iridella* is shining pale white or shining dark yellow brown and white. It is similar to *enchrysa* but can be recognized by having ocelli and by having the base of the second segment of the labial palpus white. Wing length varies from 4.0 to 6.5 mm. *Enchrysa* is a larger species with wing length 6.0 to 7.5 mm. It is shining dark yellow brown, lacks ocelli and has the base of the second segment of the labial palpus dark brown. Pale specimens of *iridella* have the labial palpus, antenna and hindwing light in addition to the forewing. The second segment of the labial palpus may have a few yellow-brown scales on the distal $\frac{1}{2}$, and the third segment may have one or two poorly developed yellow-brown rings.

The immature stages are unknown.

*Iridella* occurs from the Davis Mountains in West Texas to Pullman, Washington and San Diego, California. Most specimens have been collected in California. Adults have been collected at light from February through August in California, mid-July in Washington and late May in Texas.

*Stagmatophora enchrysa* (Hodges), NEW COMBINATION

PL. 4, FIG. 4. TEXT FIGS. 8*f*; 23 *g*.

*Eteobalea enchrysa* Hodges, 1962, *Ent. Americana*, **42** (n.s.): 72, figs. 54, 115, 199.
Type-locality: Cajon Valley, San Bernardino Co., California. [USNM]

*Enchrysa* is similar to *iridella* but can be separated as indicated under *iridella*. *Enchrysa* is a larger species and usually is shining dark yellow brown with shining yellow-gold marks on the forewing. Ocelli are lacking. Wing length varies from 6.0 to 7.5 mm. The dark mark at the base of the second segment of the labial palpus varies and may be extensive.

The larva is a stem gall maker on bluecurls, *Trichostema lanatum* Ben.

*Enchrysa* has been taken in Cajon Valley, San Bernardino County and near Los Prietos, Santa Barbara County, California. In each instance specimens were collected as immatures. Galls collected on 25 June near Los Prietos produced adults from 6 July through 4 August.

GENUS

## *Pyroderces* Herrich-Schäffer

*Pyroderces* Herrich-Schäffer, [1853], *Systematische Bearbeitung der Schmetterlinge von Europa*, **5**: 47, pl. XIII, figs. 29, 30.
Type-species: *Pyroderces goldeggiella* Herrich-Schäffer, 1853, now considered to be a synonym of *Cosmopteryx* [sic] *argyrogrammos* Zeller, 1847. Monotypy.

NOTE—Herrich-Schäffer ([1854]: 212) attributed *goldeggiella* to Fischer von Röslerstamm and synonymized it with *argyrogrammos*. *P. goldeggiella* probably was a manuscript name which Herrich-Schäffer inadvertently validated.

*Anatrachyntis* Meyrick, 1915, *Exotic Microlepidoptera*, **1**: 325.
Type-species: *Gracilaria* [sic] ?*falcatella* Stainton, 1859. Original designation.

*Sathrobrota* Hodges, 1962, *Ent. Americana*, **42** (n.s.): 73, figs. 5, 13. NEW SYNONYMY.
Type-species: *Batrachedra rileyi* Walsingham, 1882. Original designation.

*Pyroderces* has more than 80 species occurring in the tropical, subtropical and warm temperate parts of the world. Most species occur in Australia and the Old World tropics. Three species occur in America north of Mexico from Maryland and northwest Arkansas south to Florida, Texas and Yuma, Arizona. Structurally *Pyroderces* is closest to *Stagmatophora*, but the species can be separated by the lack of metallic scales.

Characters of the genus are as follows: tongue moderately long; maxillary palpus folded over base of tongue; second segment of labial palpus with slight rough scaling on anteroventral surface or smooth, third segment smooth scaled, longer than second segment; antenna $\frac{3}{5}$ to $\frac{3}{4}$ length of forewing, basal segment with pecten of six to eight slender scales; ocelli absent. Forewing lanceolate with acute apex; 12 veins present; $R_4$, $R_5$, $M_1$ and $M_2$ stalked; $Cu_1$ and $Cu_2$ usually bent and slightly downcurved from end of cell, often parallel; $Cu_2$ arising slightly before or opposite $R_2$; 2A and 3A forked basally with 3A often weak. Hindwing narrowly lanceolate to linear; a row of stout scales on anterior margin beyond base acting as secondary wing-coupling mechanism; $R_s$ and $M_1$ stalked beyond what would be end of cell; cell open; $M_2$, $M_3$, $Cu_1$ and $Cu_2$ arising equidistant; anal veins absent or weakly developed. Male: eighth tergum a short, sclerotized band; eighth sternum a pair of lobes or a basal band with a pair of distal lobes; genitalia asymmetrical; valvae usually symmetrical, closely approximated basally, rarely with a saccular lobe at base; vinculum poorly developed; aedoeagus ankylosed, very short to as long as valva; lobes of juxta arising from near aedoeagus, right lobe very short, left lobe short to very long; tegumen deeply emarginate on anterior margin; right brachium much longer than left brachium, apex acute to blunt. Female: ostium bursae in membrane just beyond posterior margin of seventh sternum, often in an invaginated, slightly sclerotized pouch; base of ductus bursae usually heavily sclerotized, rest very slender and lightly sclerotized to membranous; corpus bursae nearly membranous, walls with fine lines; signum present or absent, a series of small, inwardly projecting points; ductus seminalis from corpus bursae near connection with ductus seminalis; apophyses anteriores shorter than apophyses posteriores; eighth segment slightly sclerotized ventrolaterally and somewhat dorsally, a row of setae on posterior margin.

*Pyroderces* larvae are scavengers on an array of plants and usually feed on developing or dried flowers or fruit.

Examination of several species from Japan, Oceania, the Palearctic Region and Central America has led me to conclude that the three genera, *Pyroderces*, *Sathrobrota* and *Anatrachyntis*, are synonymous. Three somewhat different color-pattern types occur. These are exemplified by *argyrogrammos*, *rileyi* and *albistrigella*. Characters of the labial palpus, forewing maculation, venation and male and female genitalia vary from one character state to another without disjunctions.

## KEY TO SPECIES OF *PYRODERCES*

1. Forewing with transverse marks, without a pale yellowish-white line on fold . . . . . . . . . . . . 2
— Forewing without transverse marks, with a pale yellowish-white line on fold . . . . . *albistrigella* p. 48

2. Hindtibia with median white streak having dorsolateral row of brown scales; male with left lobe of juxta blunt apically; female with posterior margin of seventh sternum emarginate . . . . . . . . . . . . . . . . . . . . . . . . . . *rileyi* this page
— Hindtibia with median white streak unicolorous; male with left lobe of juxta acute apically; female with posterior margin of seventh sternum convex . . . . . . . . . . . . . . . . . *badia* this page

### *Pyroderces rileyi* (Walsingham) (Pink Scavenger Caterpillar*)

PL. 3, FIGS. 29, 30. TEXT FIGS. 2 *d*; 8 *h*; 23 *h* (McD. 7723).

*Batrachedra rileyi* Walsingham, 1882, *Trans. Amer. Ent. Soc.*, **10**: 198.
Type-locality: Savannah, Georgia. [BMNH]

*Batrachedra stigmatophora* Walsingham, 1897, *Proc. Zool. Soc. London*, 1897: 104.
Type-locality: St. Thomas, West Indies. [BMNH]

*Rileyi* is very similar to *badia* but can be separated by the color pattern of the hindtibia as indicated in the key. Wing length varies from 4.2 to 6.8 mm. Moths vary in the amount of dark-gray overlay on the forewings and the intensity of dark-gray scales on the undersurface of the fore- and hindwings. The undersurface of the hindwing may be pale gray.

Larvae are scavengers on various plants including varieties of *Sorgum vulgare* Pers.; flowers of castorbean, *Ricinus* species; flowers of bushmint, *Hyptis* species; dasheen, *Colocasia esculenta* (L.) Schott.; corn stems, husks and tassels; cotton seed and rotten cotton bolls and pineapple. Chittenden (1916) reported the larvae as pests on stored corn. Busck (1917) and Heinrich (1921) have recorded *rileyi* as a secondary pest on cotton bolls, and each illustrated the larva and pupa.

*Rileyi* occurs from Washington, D.C. and Arkansas south to Florida, Texas and southwestern Arizona. Farther south it occurs in the Antilles and South America to Argentina. In the South adults have been collected in nearly every month.

### *Pyroderces badia* (Hodges), NEW COMBINATION

PL. 3, FIGS. 31–33. TEXT FIGS. 8 *g*; 23 *i*.

*Sathrobrota badia* Hodges, 1962, *Ent. Americana*, **42** (n.s.): 75, figs. 60, 118, 170.
Type-locality: South Florida. [USNM]

*Badia* can be separated from *rileyi* as indicated in the key. Wing length varies from 3.5 to 5.4 mm. General forewing pattern is as figured with variation in shade. Some specimens have shining, pale-yellow to gold-brown scales on the forewing; others have the forewing heavily overlaid with dark-gray scales. The undersurface of the wings usually has a heavy overlay of dark-gray scales.

Larvae have been reared from seed pods of coffee senna, *Cassia occidentalis* L.; peach and loquat mummies; fruits of lime, grapefruit and banana; cabbage; coconut blossoms; elm leaves; and cones of *Pinus* species.

*Badia* occurs from Washington, D.C., south to Florida and west to Louisiana. An apparently disjunct population occurs in coastal Orange County, California. Moths have been collected in June in the District of Columbia, February through September in Florida and in April and September in California. *Badia* has been introduced into Molokai and feeds on the staminate cones of *Pinus pinaster* Aitken.

*Pyroderces albistrigella* (Möschler), NEW COMBINATION

PL. 3, FIG. 34. TEXT FIGS. 9 *b*; 24 *a*.

*Batrachedra albistrigella* Möschler, 1890, *Abhandl. Senckenbergischen Naturforschenden Ges.*, **15**: 345. Type-locality: Puerto Rico. [Institut für Spezielle Zoologie und Zoologisches Museum der Humboldt-Universität zu Berlin]

*Albistrigella* can be recognized by the combination of forewing coloration, undersurface of forewing heavily overlaid with dark-gray scales except for pale gray on the fold, undersurface of hindwing dark gray on anterior $\frac{1}{3}$, and frons, vertex and lateral margins of occiput white. No other North American cosmopterigid looks like *albistrigella.* Its nearest allies are a species from Brazil and one from Japan. Wing length varies from 5.4 to 6.2 mm. The few specimens collected are nearly homogeneous in pattern.

The immature stages are unknown.

*Albistrigella* has been collected at Pensacola and the Archbold Biological Station in Florida and Baton Rouge, Louisiana. Moths were collected in late May and the last half of September.

GENUS

## *Limnaecia* Stainton

*Limnaecia* Stainton, 1851, *Supplementary Catalogue of British Tineidae and Pterophoridae*, 4. Type-species: *Limnaecia phragmitella* Stainton, 1851. Monotypy.

*Limnaecia* is a large genus with approximately 100 described species that occur mainly in Australasia and Oceania. One species occurs in North America. None are known from South America. Whether all species ascribed to *Limnaecia* are congeneric is problematical. None of the species described by Diakonoff (1954: 74–84) has massive apophyses in the female as does the type-species. *Limnaecia* can be separated from other North American cosmopterigids by the forewing with $R_4$ and $R_5$ stalked and $M_1$ arising separately and the hindwing with a closed cell and $R_s$, $M_1$ and $M_2$ arising approximately. *Triclonella*, *Teladoma* and *Anoncia* which have similar forewing venation have the hindwing with $R_s$, $M_1$ and $M_2$ generally parallel and arising distant from one another.

Characters of the genus are as follows: head extremely smooth scaled; labial palpus with second and third segments very slender, third segment longer than second segment; antenna $\frac{3}{5}$ length of forewing, basal segment with pecten of four slender, hairlike scales. Male with hair pencil arising from base of metathoracic pleuron. Forewing broadly lanceolate, 12 veins present; $R_1$ arising at $\frac{2}{3}$ length of cell; $R_4$ and $R_5$ stalked from end of cell; other veins separate. Hindwing lanceolate, eight veins present, Sc and $R_s$ separate, $R_1$ joining Sc near base of wing, cell closed, $M_1$ arising approximate to $R_s$ and diverging from $R_s$, a series of strong scales on costal margin near base acting as secondary wing coupling mechanism. Abdominal terga four through seven with broad patch of very short setae. Male with eighth abdominal sternum a pair of lobes; valvae asymmetrical, broadly joined at base to much reduced vinculum; juxta with left lobe large, right lobe reduced; aedoeagus heavily sclerotized; right brachium much larger than left brachium. Female with ostium bursae at apex of tubular structure formed by genital plates; ductus bursae heavily sclerotized on basal $\frac{1}{2}$, cylindrical, distal $\frac{1}{2}$ slender, membranous; corpus bursae with spiculose wall; signa paired, thumbtacklike; ductus seminalis arising from lobe at base or corpus bursae; apophyses massive, heavily sclerotized; eighth abdominal tergum heavily sclerotized on posterior margin; papillae anales finely spiculose.

Larvae of *phragmitella* feed on flowers and seeds of *Typha* species.

*Limnaecia phragmitella* Stainton (Shy Cosmet, Br.)

PL. 1, FIG. 17; PL. 4, FIGS. 5, 6. TEXT FIGS. 2 *a*; 9 *c*; 24 *c* (McD. 7744).

*Limnaecia phragmitella* Stainton, 1851, *Supplementary Catalogue of British Tineidae and Pterophoridae*, 4. Type-locality: [England]. [BMNH]

The combination of size, wing shape and color pattern will separate *phragmitella* from other North American cosmopterigids. The moths are remarkably uniform in color pattern with the major variation being development of a faint gray line extending from the dark spot at $\frac{3}{5}$ the length of the cell to the apex of the forewing. Wing length varies from 5.9 to 10.2 mm, and males are usually smaller than females.

Larvae feed on the flowers and developing seeds of cattail, *Typha* species. Claassen (1921: 487) gave the following information: "The larvae restrict their work to the head of the plant, except occasionally when they bore into the stem to transform. The young larvae feed on the tender styles of the pistillate flowers, but as these grow larger and become dry, the larvae move farther inward and eat the seeds of the plant. As cold weather approaches, they migrate still farther inward, and finally locate near the rachis of the flower spike, where they often eat away the basal part of the little stalks which bear the seeds. The larvae spin an abundance of silk with which they tie the down, or pappus, together, thus keeping it from being torn off or blown away.

"The cat-tail heads which are infested by these larvae present a striking appearance. The silk spun by the larvae holds the downy material together and does not allow the seeds to escape, but the heads fluff out. . . .

"The larvae overwinter in the half-grown stage in the head of the plant, the fluffy material of the fruiting spike being their protection.

"In the latter part of May or early June the larvae attain their full growth. Then, in the midst of the downy material, the larvae spin their thin, tough, white cocoons and transform to the pupal stage. Many of the larvae, leaving the heads, go down and bore into the stem of the cat-tail plant, forming burrows which they line with silk; and there they pupate."

*Phragmitella* occurs from Nova Scotia and eastern Wyoming south to northern Virginia and central Oklahoma. Moths are present in July and August. Elsewhere, *phragmitella* occurs in Europe, North Africa, Australia and New Zealand. The moths are extremely local and appear not to fly far from the food plant.

GENUS
## *Teladoma* Busck

> *Teladoma* Busck, 1932, *Proc. Ent. Soc. Washington*, **34**: 17.
> Type-species: *Teladoma helianthi* Busck, 1932. Original designation.

Eight species of *Teladoma* occur from the piedmont of North Carolina and southern South Dakota south and west to southern Arizona and Riverside, California. Additional species undoubtedly occur in Mexico, but none has been described. *Teladoma* is closely allied to *Triclonella* and *Anoncia*, but adults can be separated by forewing venation in which ten veins usually are present; $Cu_1$ and $Cu_2$ are absent. If $Cu_1$ and $Cu_2$ are present, then they are directed almost at a right angle with the cell. In *Anoncia* and *Triclonella* $Cu_1$ and $Cu_2$ are present and do not arise at a right angle from the cell. *Teladoma* has the hindwing with veins $M_3$ and $Cu_1$, when present, separate. In *Triclonella* and *Anoncia* $M_3$ and $Cu_1$ are connate or short stalked. *Teladoma* has the gnathos fused ventrad of the anal tube, and the aedoeagus has a dorsal arm or extension; whereas in *Triclonella* and *Anoncia* the brachia are not fused ventrad of the anal tube, and the aedoeagus lacks a dorsal extension.

Characters of the genus are as follows: tongue moderately short; labial palpus short, upcurved, second segment longer than third segment; ocelli present or absent; basal segment of antenna with pecten of numerous slender scales, antenna $\frac{3}{5}$ to $\frac{3}{4}$ length of forewing. Forewing broadly lanceolate; 10 to 12 veins present; $R_4$ and $R_5$ stalked; $M_2$ and $M_3$ usually approximate and distant from $M_1$; $Cu_1$ and $Cu_2$ usually absent, when present at right angle with cell. Hindwing narrowly lanceolate; seven or eight veins present; cell usually closed; $R_s$, $M_1$, $M_2$, $M_3$, $Cu_1$ and $Cu_2$ arising separately and often equidistant; $M_3$ sometimes absent. Male with seventh and eighth abdominal segments modified, seventh segment with posterior margin heavily sclerotized and produced medially, eighth tergum a pair of heavily sclerotized lateral lobes, eighth sternum a pair of heavily sclerotized basal bands; valvae usually symmetrical, lobate, broadly joined to reduced vinculum; aedoeagus straight or angulate with dorsal projection arising between $\frac{1}{2}$ and $\frac{3}{4}$ length; tegumen often asymmetrical with ventrocaudal margins having lobes on right and/or left sides; gnathos with arms fused ventrad of anal tube. Female with lamellae antevaginalis and postvaginalis modified as cylinder in membrane between seventh and eighth abdominal sterna; ductus bursae very slender, membranous; corpus bursae membranous, lacking signa; ductus seminalis arising from lobe at base of corpus bursae.

Larvae of *Teladoma helianthi* are blister miners on leaves of sunflowers, *Helianthus* species, and cocklebur, *Xanthium* species.

Females are known of one species, *helianthi*. All the species are remarkably uniform in appearance and essentially lack a pattern. Moths are shades of gray, yellowish gray and gray brown.

## KEY TO SPECIES OF *TELADOMA* BASED ON MALE GENITALIA

1. Gnathos with single terminal lobe (text figure 9 *a*) . . . . . 2

— Gnathos with pair of terminal lobes (text figure 10 *d*) . . . . . *exigua* p. 52

2. Left valva with lobe at base (text figure 9 *d*) . . . . 3

— Left valva without lobe at base (text figure 9 *e*) . . . . 4

3. Valva with margins nearly continuously rounded, lobe longer than basal width (text figure 9 *d*) . . . . . *nebula* p. 52

— Valva with distal margin nearly straight, lobe shorter than basal width (text figure 10 *c*) . . *habra* p. 51

4. Tegumen with lobe on ventrocaudal margin on left side (text figure 9 *e*) . . . . . *astigmatica* this page

— Tegumen without lobe on ventrocaudal margin on left side . . . . . 5

5. Tegumen with two lobes on ventrocaudal margin on right side (text figure 10 *k*) . . . . *murina* p. 51

— Tegumen with one lobe on ventrocaudal margin on right side . . . . . 6

6. Gnathos sinuous on distal $\frac{2}{3}$, apical region downcurved (text figure 10 *i*) . . . . . *incana* p. 51

— Gnathos straight or slightly curved on distal $\frac{2}{3}$, apical region straight or but slightly downcurved . . . . . 7

7. Aedoeagus nearly as long as length of valva, dorsal projection of aedoeagus about $\frac{1}{2}$ length of aedoeagus, small in relation to aedoeagus (text figure 9 *f*) . . . . . *tonia* this page

— Aedoeagus about $\frac{1}{2}$ length of valva, dorsal projection about $\frac{3}{4}$ length of aedoeagus, large in relation to aedoeagus (text figure 9 *a*) . . *helianthi* this page

### *Teladoma helianthi* Busck

PL. 3, FIGS. 35, 36. TEXT FIGS. 2 *c*; 9 *a*; 24 *b* (McD. 7694).

*Teladoma helianthi* Busck, 1932, *Proc. Ent. Soc. Washington*, **34**: 17, figs. 1–6.
Type-locality: East St. Louis, Illinois. [USNM]

*Helianthi* is nearest *astigmatica* in appearance and size but can be separated by the genital characters indicated in the key. Wing length varies from 3.6 to 5.2 mm, the ground color varies from yellowish gray to dark gray, and a dark spot may be present at $\frac{1}{2}$ the length of the cell and another at the end of the cell. In the male the caudal margin of the eighth sternum is deeply but narrowly indented; western specimens have the indentation less pronounced.

Larvae make blister mines on the leaves of sunflowers, *Helianthus* species, and cockleburs, *Xanthium* species. A full description of the immature stages was to have been given by Satterthwait, but it never appeared.

*Helianthi* occurs from the piedmont in North Carolina and extreme southeastern South Dakota southwest to the White Mountains in Arizona and San Diego, California. Most specimens have been collected in Illinois and Arkansas. Moths have been collected from late June through early August in the East and in early March in California.

### *Teladoma astigmatica* (Meyrick)

PL. 3, FIG. 37. TEXT FIG. 9 *e* (McD. 8368).

*Paratheta astigmatica* Meyrick, 1928, *Exotic Microlepidoptera*, **3**: 467.
Type-locality: Bent, New Mexico. [BMNH]

*Astigmatica* is close to *helianthi* in venation and color pattern but can be separated on genital characters. Also, *astigmatica* usually is larger, wing length 5.8 to 6.0 mm. In the forewing the forked part of veins 2A and 3A is nearly $\frac{1}{2}$ the length of the anal veins in *astigmatica*, $\frac{1}{3}$ the length of the anal veins in *helianthi*. Few specimens of *astigmatica* are known, and these vary little from one to another.

The immature stages are unknown.

*Astigmatica* occurs in the mountains of West Texas and southern New Mexico. Moths were collected in October.

### *Teladoma tonia* Hodges, NEW SPECIES

PL. 6, FIG. 3. TEXT FIGS. 9 *f*, *g*; 10 *a*.

*Teladoma tonia* Hodges.
Type-locality: Madera Canyon, 5600′, Santa Rita Mts., Arizona. [CU]

Upper surface as figured. Head: tongue with pale-gray scales basally; labial palpus dark gray, individual scales with pale-gray apices; frons shining pale gray, scales with darker gray just before apices; vertex and occiput mainly darker gray with same pattern on scales; ocelli absent; antenna dark gray, first segment relatively short, a row of pale-gray tipped scales at apex. Forewing mainly dark gray with a faint transverse fascia just beyond ½ the wing length. Hindwing uniformly dark gray. Undersurface of wings paler than upper surface, scales unicolorous, hindwing medium to pale gray on posterior ½. Foreleg: mainly dark gray, scales with pale-gray apices, particularly at apex of tibia and at apices of tarsal segments. Midleg: much as for foreleg with pale areas more intense. Hindleg: coxa, trochanter and femur mainly dark gray with numerous pale-gray tipped scales; tibia dark gray with a grayish-white band at base of each pair of spurs; apex of each tarsal segment pale. Wing length: 3.3 mm. Male: genitalia and seventh and eighth abdominal segments as in text figures 9*f*, *g*; 10 *a*. Valva moderately slender basally, somewhat expanded distally and angulate at about ½ length; aedoeagus curved, concave dorsally, approximately equal to length of valva; tegumen heavily sclerotized, deeply emarginate on anterior margin medially, ventrocaudal margin with expanded, blunt lobe on right side; gnathos nearly straight beyond anal tube, apical region slightly downcurved. Female: no specimens available.

The immature stages are unknown.

TYPE: Holotype: ♂. Madera Canyon, 5600′, Santa Rita Mts., Arizona; 1 Aug. 1959; R. W. Hodges; USNM genital slide 3969. CU.

*Tonia* is nearest *astigmatica* in genital characters but can be separated on color and by having a lobe from only the right side of the tegumen.

### *Teladoma incana* Hodges

PL. 3, FIGS. 41, 42. TEXT FIG. 10 *i*.

*Teladoma incana* Hodges, 1962, *Ent. Americana*, **42** (n.s.): 85, figs. 64, 175.
Type-locality: Putnam Co., Illinois. [USNM]

*Incana* is nearly uniformly dark gray. Many scales on the forewing have pale-gray apices, and a poorly developed dark spot usually is present at the end of the cell. The hindwing is paler than the forewing, and the scales are unicolorous. The moth lacks ocelli; the forewing has $M_2$ and $M_3$ connate, and $Cu_1$ and $Cu_2$ are present; and in the hindwing the cell is open, and all veins are present. Wing length varies from 3.8 to 4.5 mm.

The immature stages are unknown.

*Incana* is known from Putnam County, Illinois and Washington County, Arkansas. Moths have been collected from late May through early August with most collected in June and July.

### *Teladoma murina* Hodges

PL. 3, FIG. 38. TEXT FIG. 10 *k*.

*Teladoma murina* Hodges, 1962, *Ent. Americana*, **42** (n.s.): 84, figs. 66, 174.
Type-locality: Peña Blanca Canyon, Santa Cruz Co., Arizona. [CU]

*Murina* can be separated from other species in the genus as indicated in the key. Additionally, ocelli are lacking; forewing has $M_2$ and $M_3$ connate, $Cu_1$ present and $Cu_2$ absent; hindwing has the cell weakly closed, and all veins are present. Wing length varies from 3.6 to 4.2 mm. Specimens in the type series are dark gray; one specimen from Texas is yellowish brown with numerous pale yellowish-gray tipped scales.

The immature stages are unknown.

*Murina* is known from San Benito, Texas and Peña Blanca Canyon in southern Arizona. Moths were collected in late August and early September in Arizona.

### *Teladoma habra* Hodges, NEW SPECIES

PL. 3, FIG. 43. TEXT FIGS. 10 *b*, *c*.

*Teladoma habra* Hodges.
Type-locality: Devil's Den State Park, Washington Co., Arkansas. [USNM]

Upper surface as figured. *Habra* is nearly uniformly dark gray with many scales tipped with light gray, particularly the apex of the second and third segments of the labial palpus and apices of tarsal segments. Ocelli are present. Forewing with ten veins, $Cu_1$ and $Cu_2$ absent, $M_3$ slightly downcurved from end of cell. Hindwing with cell open, veins difficult to determine, $M_3$, $Cu_1$ or $Cu_2$ absent. Wing length: 3.2–3.5 mm. Male as in text figures 10 *b*, *c*. Valva short, lobate, almost as broad as long, setae arising from inner surface long, broadly curved apically, heavily sclerotized lobe from base of left valva; aedoeagus stout; ventrocaudal margins of tegumen lacking lobes; terminal part of gnathos straight. Female: no specimens available.

The immature stages are unknown.

TYPES: Holotype: ♂. Devil's Den State Park, Washington Co., Arkansas; 22 May 1966; R. W. Hodges; USNM genital slide 3980. Type no. 70,570, USNM. Paratype: 1 ♂. Same locality as for holotype; 20 May 1966; R. W. Hodges. Paratype to USNM.

*Habra* occurs sympatrically with *incana* and *helianthi*. It can be distinguished from them by venation, dark shade of the forewings and genitalia.

### *Teladoma nebula* Hodges, NEW SPECIES

PL. 3, FIG. 39. TEXT FIG. 9 *d*.

*Teladoma nebula* Hodges.
Type-locality: Peña Blanca Canyon, Santa Cruz Co., Arizona. [CU]

Upper surface as figured. Head: base of tongue with dark-gray scales; labial palpus mainly dark gray on outer surface, base and apex of third segment with numerous white-tipped scales, inner surface of second segment pale grayish white from base to ⅔ length and third segment pale grayish white at base and apex; frons, vertex and occiput mainly pale gray, scales pale tipped and darker before apices; ocelli absent. Foreleg: coxa, trochanter, femur, tibia and first tarsal segment dark gray; apices of tibia and first tarsal segment and rest of tarsus pale gray. Midleg: mainly medium to pale gray, apices of tarsal segments very pale gray. Hindleg: coxa, trochanter and femur mainly pale gray; tibia with pale-gray fascia at base of first pair of spurs and pale gray at apex; tarsus with dark-gray saddle on basal ½ of first segment and extreme base of second through fifth segments. Forewing with $Cu_1$ and $Cu_2$ absent. Hindwing with cell open, $Cu_1$ and $Cu_2$ stalked. Wing length: 3.7 mm. Male: as in text figure 9 *d*. Valva broadly lobate, relatively short, base of left valva with broad lobe extending from costal margin at base; tegumen broadly emarginate at base, ventrocaudal margins lacking lobes; distal part of gnathos stout, straight; aedoeagus stout, nearly straight, approximately equal in length to valva. Female: no specimens available.

The immature stages are unknown.

TYPE: Holotype: ♂. Peña Blanca Canyon, Santa Cruz Co., Arizona; 26 Aug. 1959; R. W. Hodges; RWH slide 4725; RWH wing slide 5014. CU.

In addition to the genital characters given in the key, *nebula* can be distinguished by its lack of ocelli, forewing having $M_2$ and $M_3$ connate and $Cu_1$ and $Cu_2$ absent, and hindwing with cell open, $M_1$ and $M_2$ stalked and $M_3$ absent. *Nebula* is nearest *habra* in venation and represents a stage before *habra* in reduction of veins.

### *Teladoma exigua* Hodges, NEW SPECIES

PL. 3, FIG. 40. TEXT FIGS. 10 *d*, *f*, *j*.

*Teladoma exigua* Hodges.
Type-locality: Madera Canyon, 4880′, Santa Rita Mts., Arizona. [CU]

Upper surface as figured. Head: tongue pale gray basally; labial palpus medium gray, scales tipped with pale gray; frons, vertex and occiput mainly pale gray; basal segment of antenna pale gray to ⅔ length, darker distally, shaft unicolorous gray; ocelli absent. Foreleg mainly dark gray, many scales tipped with pale gray, apices of tibia and tarsal segments with pale-tipped scales. Mid- and hindlegs too poor to determine color pattern. Forewing with $M_2$ and $M_3$ connate, $Cu_1$ and $Cu_2$ absent. Hindwing with cell closed, $M_3$ or $Cu_1$ absent. Wing length: 3.9 mm. Male: as in text figures 10 *d*, *f*, *j*. Valva short, very broad; aedoeagus much longer than valva, strongly sinuous; tegumen broadly emarginate on anterior margin, ventrocaudal margins lacking lobes; gnathos fused ventrad of anal tube, then separated as two, heavily sclerotized projections. Female: no specimens available.

The immature stages are unknown.

TYPES: Holotype: ♂. Madera Canyon, 4880′, Santa Rita Mts., Arizona; 11 July 1959; R. W. Hodges; RWH slide 4759. CU. Paratype: 1 ♂. Madera Canyon, 5600′, Santa Rita Mts., Arizona; 1 Aug. 1959; R. W. Hodges. Paratype to USNM.

*Exigua* is superficially close to *nebula* but can be separated by genital characters as indicated in the key and by the closed cell of the hindwing.

GENUS

## *Triclonella* Busck

*Triclonella* Busck, 1900 [1901], *Jour. New York Ent. Soc.*, **8**: 236.
Type-species: *Triclonella pergandeella* Busck, 1901. Original designation.

*Anorcota* Meyrick, 1920, *Exotic Microlepidoptera*, **2**: 365. NEW SYNONYMY.
Type-species: *Promalactis platyxantha* Meyrick, 1909. Original designation.

*Pharmacoptis* Meyrick, 1932, *Exotic Microlepidoptera*, **4**: 277. NEW SYNONYMY.
Type-species: *Pharmacoptis breviramus* Meyrick, 1932, considered to be a synonym of *Triclonella rhabdophora* Forbes, 1930. Monotypy.

*Triclonella* is a New World genus with approximately 20 species that occur from Washington, D.C. and northwest Arkansas south through Central America and the Antilles to northern Argentina. Five species occur in America north of Mexico. *Triclonella* is closely allied with *Teladoma* and *Anoncia*. Characters separating it from *Teladoma* are indicated under *Teladoma*. Structurally, *Triclonella* and *Anoncia* are extremely close; each is similar in venation and genital characters, and constant differences have not been found. But, in *Triclonella* species the ocelli are present, the antennal pecten consists of one to four slender scales near the base of the first antennal segment, the antenna is dark with a white line or white bars, the labial palpus is long, slender and sickle shaped with the third segment usually longer than the second, the foreleg has the tibia and tarsus dark brown with one or two rows of white scales, and the forewing usually has a distinct pattern. In contrast *Anoncia* species lack ocelli, the antennal pecten consists of numerous, slightly broader scales that extend from the base nearly to the apex of the first antennal segment, the labial palpus is relatively short with the third segment usually shorter than the slightly rough-scaled second segment, the foreleg has the tibia and tarsus gray or brown with individual scales pale tipped, and the forewing lacks a distinct pattern and is somberly colored. *Triclonella* is also very similar to the Mexican genus *Ecballogonia* Walsingham. *Ecballogonia* differs in lacking the antennal pecten, the foreleg with the tibia and tarsus dark brown with transverse, white bands, the hindwing with a transverse, white band at $\frac{1}{2}$ its length on dorsal and ventral surfaces, the upper surface of both wings with streaks of metallic-blue scales, and the costal margin of the hindwing decidedly sinuous.

Characters of the genus are as follows: tongue well developed; labial palpus long, sickle shaped, slender, third segment usually longer than second; ocelli present, small; first segment of antenna with pecten of one to four slender scales at base, shaft with broad sensory areas on ventral surface, sensory setae very short, antenna approximately $\frac{3}{4}$ forewing length. Hindtibia with numerous long scales on dorsal surface, male of one species with dorsal hair pencil from base. Forewing broadly lanceolate; 12 veins present; $R_4$ and $R_5$ stalked; $M_1$, $M_2$, $M_3$, $Cu_1$ and $Cu_2$ arising separately; $Cu_2$ arising well before end of cell; base of 3A weak, often not developed as a vein. Hindwing broadly lanceolate, costal margin sinuous to slightly convex; cell usually closed; seven or eight veins present; $M_1$ arising closer to $R_s$ than to $M_2$; $M_2$ separate, connate or stalked with $M_3+Cu_1$, usually arched toward costal margin; $M_3$ and $Cu_1$ connate, stalked or fused; $Cu_2$ separate; anal veins weakly developed; a tuft of scales on dorsal surface of 2A. Male with seventh and eighth abdominal segments modified as asymmetrical bands and plates; genitalia asymmetrical, valvae basically symmetrical with lobe or lobes from costal margin; aedoeagus heavily sclerotized, often with one or more possibly deciduous cornuti; tegumen heavily sclerotized, anterior margin emarginate medially; brachia heavily sclerotized, right brachium longer than left. Female with ostium bursae in membrane between seventh and eighth abdominal sterna surrounded by heavily sclerotized plate; ductus bursae membranous and linear or heavily sclerotized and curved with large outgrowth; corpus bursae membranous with spiculose walls; signa absent or a pair of series of inwardly projecting points; apophyses anteriores shorter than apophyses posteriores.

Larvae of *Triclonella pergandeella* have been reared from several genera of legumes. They tie leaflets together and skeletonize the dorsal surfaces.

## KEY TO SPECIES OF *TRICLONELLA*

1\. Forewing with bright-yellow or yellow-orange patches contrasting with medium- to dark-brown or dark-gray to black patches . . . . . 2
— Forewing mainly light brown with some darker gray or gray-brown scales on outer margin . . . . . . . . . . . . . . . . . . . . . . . . *bicoloripennis* p. 55

2\. Forewing yellow to yellow orange basally . . . . . . 3
— Forewing dark brown basally . . . . . . *determinatella* p. 54

3\. Forewing dark gray to nearly black with some white scales on distal $\frac{1}{3}$ . . . . . . . . . . . . . . . . . 4
— Forewing with a pair of brown patches at $\frac{2}{3}$ length followed by pale yellowish-orange to pale-yellow scales on costal margin and with pale fringe . . . . . . . . . . . . . . . . . . . . . . . . *antidectis* p. 54

4\. Forewing with some orange scales in dark distal part . . . . . . . . . . . . . . . . . . . . . . . . . . *xuthocelis* p. 54
— Forewing without orange scales in dark distal part . . . . . . . . . . . . . . . . . . . . . . . *pergandeella* p. 54

### *Triclonella pergandeella* Busck

PL. 4, FIG. 7. TEXT FIGS. 2 *b*; 10 *g*; 24 *e* (McD. 8354).

*Triclonella pergandeella* Busck, 1900 [1901], *Jour. New York Ent. Soc.*, **8**: 237, pl. 9, fig. 2.
Type-locality: District of Columbia. [USNM]

NOTE—The lectotype, ♂, present designation, bears the following labels: 1. "AB 12 iss. Aug 21.98"; 2. "Type No. 5354 U.S.N.M."; 3. "*Triclonella pergandeella* type! Busck"; 4. "Male genitalia on slide and figured. AB Jun 1921"; 5. "Genitalia Slide By AB ♂ USNM 3994"; 6. "LECTOTYPE *pergandeella* Busck By R. W. Hodges '74".

*Pergandeella* is nearest *xuthocelis* in color pattern but can be separated from it as indicated in the key. In addition *pergandeella* is smaller (wing length 5.2 to 6.2 mm), has different male and female genitalia and occurs in the East. *Xuthocelis* has a wing length of 6.3 to 7.7 mm and occurs in the Southwest. Variation is expressed in wing length, intensity of orange scales on the forewings, development of a white patch on the tornal area of the forewing, the relative amounts of dark-gray and black scales on the distal $\frac{1}{3}$ of the forewing, and in the hindwing vein $M_2$ is separate to short stalked with $M_3+Cu_1$. Male and female genitalia are as figured.

Busck (1901: 237, 238) gave information on the life history. The larva has been found on *Lespedeza* and *Clitoria* in the Leguminosae. It ties together a pair of leaflets and skeletonizes the dorsal surface of the leaflets. Pupation occurs in a cocoon in the larval cell. Two or three generations occur as far north as Washington, D.C., and overwintering occurs in the adult stage.

*Pergandeella* occurs from Washington, D.C. and northwest Arkansas south to central Florida, southern Louisiana and eastern Texas. Moths collected from March to mid-May tend to be worn. Adults are attracted to light.

### *Triclonella xuthocelis* Hodges

PL. 1, FIG. 18; PL. 4, FIGS. 8, 9. TEXT FIGS. 10 *e*, *h*; 24 *d*.

*Triclonella xuthocelis* Hodges, 1962, *Ent. Americana*, **42** (n.s.): 89, figs. 68, 68a, 125, 177.
Type-locality: Madera Canyon, 4880', Santa Rita Mts., Arizona. [CU]

*Xuthocelis* is most closely related to *pergandeella* and two Central American species. It can be separated from *pergandeella* as indicated under *pergandeella*. Variation is expressed in wing length (6.3–7.7 mm), development of white scales in the tornal area of the forewing and intensity of the orange scales on the forewing. Superficially, most of the moths are very similar. Male and female genitalia are as figured.

The immature stages are unknown.

*Xuthocelis* has been collected in the Chiricahua, Huachuca and Santa Rita mountain ranges in southern Arizona. Adults have been collected at light from 30 June through 23 September.

### *Triclonella antidectis* (Meyrick)

PL. 4, FIG. 10. TEXT FIGS. 10 *l*; 24 *h* (McD. 8363).

*Epicallima antidectis* Meyrick, 1914, *Exotic Microlepidoptera*, **1**: 218.
Type-locality: Patagonia Mts., Arizona [probably Patagonia on western side of Huachuca Mts.]. [BMNH]

*Antidectis* can be recognized by the forewing color pattern which is unique in the genus. Wing length varies from 5.3 to 6.1 mm. Forewing color pattern varies little, but as the moths become worn, the orange scales are lost. Male and female genitalia are as figured.

The immature stages are unknown.

*Antidectis* occurs from the mountains of southern Arizona south to Hidalgo, Mexico. Moths have been collected in July and August. Specimens in the type series have cryptic information, but may have been collected in April and June.

### *Triclonella determinatella* (Zeller)

PL. 4, FIG. 11. TEXT FIGS. 11 *a*; 24 *f* (McD. 8355).

*Oecophora determinatella* Zeller, 1873, *Verh. K.-K. Zool.-Bot. Ges. Wien*, **23**: 289, fig. 31.
Type-locality: Texas. [MCZ]

*Oecophora australisella* Chambers, 1875, *Cincinnati Quart. Jour. Sci.*, **2**: 114.
Type-locality: Kentucky. [MCZ]

*Determinatella* can be recognized by the forewing pattern; the round, orange spot at $\frac{2}{3}$ the length of the forewing is distinctive. Wing length varies from 3.8 to 4.7 mm. Most northern specimens have the circular spot on the forewing incomplete, opening to the series of white scales on the costal margin. Southern specimens tend to have the circular spot smaller, farther from the costal margin and separate from the white costal patch. No additional differences are associated with this difference in maculation. Male and female genitalia are as figured. Wing venation is as described for the genus with $M_3$ and $Cu_1$ connate in the hindwing.

The immature stages are unknown.

*Determinatella* occurs from northwest Illinois and central Oklahoma south to central Louisiana and extreme southern Texas. The species is multiple brooded occurring in early June and mid-September in Illinois, May, June and July in northwest Arkansas and March, April and June in Texas.

### *Triclonella bicoloripennis* Hodges

PL. 4, FIGS. 12, 13. TEXT FIGS. 11 *b*; 24 *g*.

> *Triclonella bicoloripennis* Hodges, 1962, *Ent. Americana*, **42** (n.s.): 93, figs. 71, 180.
> Type-locality: Corpus Christi, Texas. [CU]

*Bicoloripennis* has a color pattern different from other species in the genus, but the pattern on the head and forewing is characteristic of *Triclonella*. It is most nearly allied to *rhabdophora* Forbes from Puerto Rico and the Virgin Islands. The hindwing is distinct with the costal margin evenly convex and veins $M_3$ and $Cu_1$ fused. The somewhat Y-shaped lamella antevaginalis is also distinctive. Wing length varies from 4.2 to 5.3 mm. The hindwing is sexually dimorphic: males are dark gray, nearly black dorsally; females are paler gray brown. The male and female genitalia are as figured.

The immature stages are unknown.

*Bicoloripennis* has a limited distribution from central and southern Louisiana west around the Gulf Coast to Brownsville, Texas. Adults have been collected from April through September.

GENUS

## *Anoncia* Clarke

> *Anoncia* Clarke, 1941, *Proc. U.S. Natl. Mus.*, **90**: 268, pl. 3, fig. 20; pl. 7, fig. 52; pl. 13, figs. 85, 85a; pl. 19, fig. 114.
> Type-species: *Hypatopa conia* Walsingham, 1907. Original designation.

*Anoncia* is a New World genus with 31 species that are mainly confined to the southwest United States, Mexico and Honduras. Until relatively recently these species have been poorly collected, and adequate information concerning distribution is lacking. However, of these species 13 are known only from California, six from Arizona and one from Texas. Three species, *leucoritis*, *longa* and *diveni*, occur widely. Recognition of 13 new species at this time is an indication of the generally poor knowledge of the genus. [*Triclonella chordostoma* (Meyrick), new combination, was transferred from *Cryptolechia* Zeller to *Anoncia* by Clarke (1965: 478), but genital, labial palpal and antennal characters associate it with *Triclonella*.] *Anoncia* is closely related to *Teladoma* and *Triclonella*. Characters differentiating *Anoncia* from them are given in the respective generic discussions.

Characters of the genus are as follows: tongue well developed; labial palpus recurved, second segment longer than third segment and often with slightly roughened scales on anteroventral surface; ocelli absent; first segment of antenna with pecten of numerous slender scales from base nearly to apex; shaft with broad sensory areas on ventral surface in male, sensory cilia short to nearly equal depth of antennal segments, sensory areas reduced and sensory cilia very short in female, antenna $\frac{1}{2}$ to $\frac{3}{4}$ length of forewing. Hindleg with long scales on dorsal surface of tibia. Forewing broadly lanceolate, apex acute to rounded; 12 veins present; $R_1$ from before $\frac{1}{2}$ length of cell; $R_4$ and $R_5$ stalked; remaining veins arising separately from cell; base of 3A well developed. Hindwing broadly lanceolate, apex acute to rounded, anterior margin slightly convex; eight veins present but anal veins very weakly developed; base of $R_s$ weakly developed; $M_2$ arising separately and distant from $M_3$; $M_3$ usually connate or short stalked with $Cu_1$, rarely separate; $Cu_2$ arising from $\frac{3}{4}$ to $\frac{4}{5}$ length of cell. Male with eighth abdominal segment and often with seventh abdominal segment modified as sclerotized bands, rods or plates; valvae usually asymmetrical, lobes (often heavily sclerotized) arising from base of either or both valvae; aedoeagus heavily sclerotized; cornuti absent; vinculum usually reduced but sometimes broad in saccal region; tegumen heavily sclerotized, deeply emarginate anteromedially; right brachium often with a short lobe from near base. Female with ostium bursae asymmetrically situated on ventral, lateral or dorsal surface; ductus bursae membranous or heavily sclerotized basally, rarely with an accessory pouch from base; corpus bursae membranous with spiculose walls, usually with an accessory lobe at base; ductus seminalis usually arising from accessory lobe; signa usually absent, but when present, paired, relatively diffuse sclerotized patches with an inwardly directed, heavily sclerotized fold; apophyses anteriores arising from lateral margins of eighth abdominal tergum to arising medially and fused basally, often slightly asymmetrical; apophyses posteriores straight and longer than apophyses anteriores; seventh abdominal segment with varying degrees of sclerotization and with heavily sclerotized projections.

Larvae of four species have been reared from plants in the Labiatae, Loasaceae and Verbenaceae. They tie leaves together and mine individual leaves or feed in the immature ovaries of the developing fruit.

Some species can be associated in groups on the basis of male and female genital characters, but many are somewhat indeterminate. For this reason I have not recognized species groups in the genus. Recognition of some species is easy on coloration and maculation, but most species are very similar, and characters of the genitalia and male abdomen must be studied for determination. Most diagnostic characters vary in differing degrees as observed in the study of more than 500 dissections. Specimens usually can be placed to species by the characters indicated in the key and species discussions and the illustrations.

## KEY TO SPECIES OF *ANONCIA*

1. Males . . . . . . . . . . . . . . . . . . . . . . . . . . . . . . . . . . 2
— Females . . . . . . . . . . . . . . . . . . . . . . . . . . . . . . . 28

2. Left valva with sclerotized lobe or lobes from basal region (text figure 16 *d*) . . . . . . . . . . 6
— Left valva without sclerotized lobes from basal region (text figure 12 *a*) . . . . . . . . . . . . . . 3

3. Right valva with a pair of unequal, slender lobes . . . . . . . . . . . . . . . . . . . . . . . . . . . . . . . . . 4
— Right valva with a single, massive lobe (text figure 16 *a*) . . . . . . . . . . . . . . . . . . . . . . *longa* p. 66

4. Right brachium a single lobe; right valva with longer basal lobe slender to apex (text figure 12 *a*) . . . . . . . . . . . . . . . . . . . . . . . . . . . . . 5
— Right brachium with a second lobe from dorsomedial margin; right valva with longer lobe expanded on distal $\frac{1}{4}$ to blunt apex (text figure 11 *h*) . . . . . . . . . . . . . *brunneipes* p. 59

5. Right brachium with a right-angled cross piece apically (text figure 12 *a*); posterior margin of seventh sternum with heavily sclerotized, triangular projection (text figure 12 *e*) . . . . . . . . . . . . . . . . . . . . . . . . *sphacelina* p. 60
— Right brachium extremely broad for basal $\frac{2}{3}$, then constricted and curved to acute apex (text figure 11 *c*); posterior margin of seventh sternum with heavily sclerotized, broadly lobate projection (text figure 11 *e*) . . . *conia* p. 59

6. Right valva with single, massive, basal lobe; left valva with single reduced lobe (text figure 13 *k*) . . . . . . . . . . . . . . . . . . . . . . . . . 7
— Right valva with two basal lobes; left valva with massive basal lobe or with two small lobes (text figure 13 *e*) . . . . . . . . . . . . . . . . . . . . 13

7. Basal lobe on right valva with acute apex; seventh abdominal sternum with symmetrical posterior margin . . . . . . . . . . . . . . . . . . . . . . 8
— Basal lobe on right valva with blunt apex (text figure 13 *k*); seventh abdominal sternum with asymmetrical posterior margin (text figure 14 *e*) . . . . . . . . . . . . . . . . . *aciculata* p. 63

8. Saccal region of vinculum well developed, long (text figure 14 *b*) . . . . . . . . . . . . . . . . . . . . 10
— Saccal region of vinculum not well developed, a slender, sclerotized band (text figure 16 *d*) . . . . . . . . . . . . . . . . . . . . . . . . . . . . 9

9. Aedoeagus angulate just before apex; forewing blotched brown and yellow brown, a well-developed brown spot in cell . . . . . *furvicosta* p. 66
— Aedoeagus uniformly curved; forewing yellowish white, without brown spot in cell . . . . . . . . . . . . . . . . . . . . . . . . . . . . . . . . . *leucoritis* p. 66

10. Forewing off white with pale-gray streaks parallel to veins, a single pale-gray spot at end of cell; hindwing pale gray . . . . . . . . . . *callida* p. 64
— Forewing pale yellow or yellow brown and brown, hindwing medium to dark yellowish gray . . . . . . . . . . . . . . . . . . . . . . . . . . . . . . . . 11

11. Forewing immaculate pale yellow, valva with concave costal margin (text figure 14 *i*) . . . . . . . . . . . . . . . . . . . . . . . . . . . . . . . . . . *naclia* p. 64
— Forewing blotched yellowish gray and brown, valva with convex costal margin (text figure 15 *e*) . . . . . . . . . . . . . . . . . . . . . . . . . 12

12. Posterior margin of seventh tergum symmetrical, broadly rounded, stout setae numerous, distributed over ½ width of segment (text figure 15 *d*); apex of lobe at base of right valva blunt (text figure 15 *c*).... *nocticola* p. 65

— Posterior margin of seventh tergum slightly asymmetrical, narrowly rounded, stout setae few on apical lobe (text figure 15 *g*); apex of lobe at base of right valva acute (text figure 15 *e*).................... *nebritis* p. 65

13. Seventh abdominal sternum with asymmetrical posterior margin.................. 14

— Seventh abdominal sternum with symmetrical posterior margin.................. 18

14. Aedoeagus tapering to acute apex........... 15

— Aedoeagus with blunt, rounded apex........ 17

15. Right valva with a pair of slender basal lobes (text figure 13 *g*)..................... 16

— Right valva with a long, broad lobe and a short, slender lobe (text figure 13 *e*).... *porriginosa* p. 62

16. Seventh abdominal sternum with long, projecting, lateral extension, sclerotized plate on posterior margin ½ width of segment (text figure 13 *m*); forewing with brown transverse fascia at ⅓ wing length .................................... *flegax* p. 62

— Seventh abdominal sternum with short, projecting, lateral extension, sclerotized plate on posterior margin very narrow (text figure 13 *d*); forewing without transverse fascia.......................... *bitoqua* p. 61

17. Left valva with apex of basal lobe straight, right valva with basal lobes approximately equal in length and diameter (text figure 12 *j*).............................. *fasciata* p. 60

— Left valva with apex of basal lobe turned up, right valva with basal lobes disparate in length and width (text figure 12 *h*)..... *alboligula* p. 60

18. Aedoeagus with three toothlike projections on left side (text figure 12 *o*)........... *glacialis* p. 61

— Aedoeagus without such projections......... 19

19. Seventh abdominal sternum with symmetrical caudolateral extensions, each with long, apical setae (text figure 14 *d*)........ *mones* p. 63

— Seventh abdominal sternum without such projections............................. 20

20. Left valva with heavily sclerotized, basal lobe partially encased by a second, partial lobe (text figure 17 *c*)...................... 21

— Left valva with a single, basal lobe........... 22

21. Seventh abdominal sternum with narrow medial projection on caudal margin (text figure 17 *o*), (known from California)..... *noscres* p. 68

— Seventh abdominal sternum with broad medial projection on caudal margin (text figures 17 *h*, *l*), (known from eastern Arizona).............................. *loexya* p. 68

22. Left valva with very small basal lobe (text figure 17 *a*)........................... *orites* p. 67

— Left valva with large basal lobe (text figure 17 *b*).................................. 23

23. Valvae short, broadly joined basally, left valva with basal lobe broad apically (text figure 17 *b*).......................... *diveni* p. 66

— Valvae longer, less broadly joined, left valva with basal lobe tapering to acute apex (text figure 18 *d*)............................ 24

24. Left valva with basal lobe tapering gradually from base to apex (text figure 18 *d*)....... 25

— Left valva with basal lobe broad, nearly cylindrical basally, abruptly decreasing in diameter, then gradually tapering to apex (text figure 18 *f*)......................... 26

25. Right valva with dorsal (costal) basal lobe longer than ventral basal lobe (text figure 18 *d*); forewing with very poorly developed transverse fascia at ⅓ length (plate 6, figure 12) .............................. *smogops* p. 72

— Right valva with dorsal (costal) basal lobe shorter than ventral basal lobe (text figure 18 *m*); forewing with well-defined, brown transverse fascia at ⅓ length (plate 4, figure 22)................................. *mosa* p. 71

26. Right valva with basal lobes separate nearly to surface of valva, right brachium with short medial projection (text figure 18 *f*); forewing without transverse fascia at $\frac{1}{4}$ length . . . . . . . . . . *piperata* p. 71

— Right valva with basal lobes separate for short distance at apex of conical projection, right brachium without a medial projection (text figure 18 *a*); forewing with transverse fascia at $\frac{1}{4}$ length . . . . . . . . . . 27

27. Abdominal sterna heavily sclerotized, western California . . . . . . . . . . *episcia* p. 69

— Abdominal sterna not heavily sclerotized; Inyo County, California to New Mexico . . . *venis* p. 70

28. Apophyses anteriores widely separated basally, arising from lateral margin of eighth abdominal tergum (text figure 27 *e*) . . . . 29

— Apophyses anteriores approximate or fused basally, arising from middle of eighth abdominal tergum (text figures 26 *b*, *d*) . . . . . . . 40

29. Eighth abdominal tergum with pair of submedial lobes (text figures 27 *e*; 28 *d*) . . . . . . . . . . 30

— Eighth abdominal tergum without pair of submedial lobes (text figure 25 *a*) . . . . . . . . . . 33

30. Posterior margin of eighth abdominal sternum rounded, indented medially (text figure 27 *e*) . . . . . . . . . . *loexya* p. 68

— Posterior margin of eighth abdominal sternum straight (text figure 28 *d*) . . . . . . . . . . 31

31. Genital plate with medial margin S-shaped (text figure 28 *e*) . . . . . . . . . . 32

— Genital plate with medial margin rounded (text figure 28 *d*) . . . . . . . . . . *psentia* p. 73

32. Abdominal sterna sclerotized . . . . . . . . . . *episcia* p. 69

— Abdominal sterna not sclerotized . . . . . . . . . . *venis* p. 70

33. Ductus bursae heavily sclerotized basally . . . . . 34

— Ductus bursae membranous throughout . . . . . . 37

34. Ostium bursae on eighth abdominal sternum (text figure 25 *d*) . . . . . . . . . . 48

— Ostium bursae on eighth abdominal tergum . . . 35

35. Eighth abdominal tergum with pair of heavily sclerotized curved bands medially (text figure 25 *a*) . . . . . . . . . . *conia* p. 59

— Eighth abdominal tergum without such bands (text figure 27 *a*) . . . . . . . . . . 36

36. Forewing with diffuse pattern of pale-gray and dark-gray scales, hindwing gray . . . . . . *orites* p. 67

— Forewing and hindwing nearly immaculate pale yellow . . . . . . . . . . *fregeis* p. 67

37. Seventh abdominal sternum with broadly rounded (semicircular) posterior margin (text figure 28 *b*) . . . . . . . . . . *mosa* p. 71

— Seventh abdominal sternum with concave to slightly convex posterior margin . . . . . . . . . . 38

38. Ductus bursae with sclerotized, blind pouch at base (text figure 28 *c*) . . . . . . . . . . *piperata* p. 71

— Ductus bursae without such a pouch . . . . . . . . . 39

39. Eighth abdominal tergum concavely excavated, genital plate at least $\frac{1}{2}$ width of eighth abdominal tergum (text figure 28 *a*) . . . . . . . . . . *smogops* p. 72

— Eighth abdominal tergum concavely excavated, genital plate less than $\frac{1}{3}$ width of eighth abdominal tergum (text figure 26 *f*) . . . . . . . . . . *diveni* p. 66

40. Apophyses anteriores separate basally (text figure 26 *d*) . . . . . . . . . . *psepsa* p. 65

— Apophyses anteriores fused basally . . . . . . . . . . 41

41. Signa present . . . . . . . . . . 42

— Signa absent . . . . . . . . . . 46

42. Signum a rounded or elliptical plate with an inwardly projecting, medial flange . . . . . . . . 43

— Signum ellipsoidal with numerous inwardly projecting teeth or points . . . . . . . . . . 44

43. Abdominal sterna three, four and five membranous . . . . . . . . . . *nebritis* p. 65

— Abdominal sterna three, four and five sclerotized . . . . . . . . . . *nocticola* p. 65

44. Eighth abdominal tergum subdivided into two plates (text figure 26 *c*); forewing off white with a gray mark at end of cell and another in tornal area................*leucoritis* p. 66

— Eighth abdominal tergum subdivided into three plates (text figure 26 *e*); maculation otherwise..............................45

45. Head yellowish white, forewing usually grayish yellow to yellowish white (plate 4, figure 36)............................*longa* p. 66

— Head brown, scales tipped with yellowish white, forewing yellowish brown and brown (plate 4, figure 39)..................*furvicosta* p. 66

46. Eighth abdominal tergum with posterior margin produced submedially, excavated medially (text figure 25 *c*).............*fasciata* p. 60

— Eighth abdominal tergum otherwise.........47

47. Ostium bursae on left margin, apophyses anteriores fused for very short distance basally (text figure 25 *e*)..............*alboligula* p. 60

— Ostium bursae slightly left of midline on dorsal surface, apophyses anteriores fused for ¼ length basally (text figure 25 *f*).......*slales* p. 61

48. Ostium bursae lateral, heavily sclerotized part of ductus bursae directed medially, membranous part running anteriorly to corpus bursae; posterior margin of eighth abdominal tergum with shallow invagination on left side (text figure 25 *b*)......*brunneipes* this page

— Ostium bursae mediolateral; ductus bursae expanded basally and directed anteriorly throughout; posterior margin of eighth abdominal tergum without an invagination (text figure 25 *d*)....................*sphacelina* p. 60

### *Anoncia conia* (Walsingham)

PL. 4, FIGS. 14–16. TEXT FIGS. 2 *e*; 11 *c–e*; 25 *a* (McD. 8372).

*Hypatopa conia* Walsingham, 1907, *Proc. U.S. Natl. Mus.*, **33**: 212.
Type-locality: Blue L., Lake Co., California. [BMNH]

NOTE—The lectotype, ♂, present designation, bears the following labels: 1. "Lectotype"; 2. "Type"; 3. "Blue Lake, Lake Co., California, 15–16.VI.1871, Wlsm. 90405"; 4. "Walsingham Collection, 1910–427."; 5. "*Hypatopa conia*, Wlsm., Type ♂ descr."; 6. "Lectotype ♂, *Hypatopa conia* Wlsm., teste K. Sattler, 1974".

"*Borkhausenia*" *marinensis* Keifer, 1935, *Mo. Bull. Dept. Agric. California*, **24**: 215, pl. II, figs. 5A–E.
Type-locality: Phoenix L., Marin Co., California. [CAS]

*Conia* is closely allied to *brunneipes* and *sphacelina*. Genital characters should be studied for specific recognition, and the characters are as given in the key. Wing length varies from 6 to 8 mm. The forewing color pattern, when well developed, consists of a dark-brown oblique streak at ⅔ the length of the cell, a dark spot at the end of the cell and a series of alternating pale and dark marginal streaks on the outer ⅓ of the wing. Nearly immaculate specimens have a slight pale area in the cell and a dark spot at the end of the cell of the forewing. Color varies from blue gray to brownish gray. In males the heavily sclerotized structures on the posterior margin of the seventh abdominal tergum vary in shape with the large lobe asymmetrical to nearly quadrate, and the eighth tergum varies in length of lobes and degree of curvature. The base of the left valva rarely has a very short, pointed lobe, and the shape of the right brachium varies. In females the position of the ostium bursae varies, and the pair of curved, heavily sclerotized bands on the middle of the seventh tergum are variously developed.

Larval food plants are *Lepechinia calycina* (Benth.) Epl. and *Salvia* species. Keifer (1935: 218) said that the larva mines individual leaves and may mine several leaves. He also presented habitus and setal drawings of the larva and pupa.

*Conia* occurs in California from Siskiyou County south to San Diego. Moths have been collected from late May through early August with most specimens collected in June and July.

### *Anoncia brunneipes* Hodges

PL. 4, FIG. 18. TEXT FIGS. 11 *f–h*; 25 *b*.

*Anoncia brunneipes* Hodges, 1962, *Ent. Americana*, **42** (n.s.): 99, figs. 75, 183.
Type-locality: San Diego, California. [USNM]

*Brunneipes* is similar to *conia* and *sphacelina* but can be separated as indicated in the key. For males the combination of shape of the posterior margin of the seventh abdominal tergum, reduced lobes on the left valva, long, stout aedoeagus with paired, pointed flanges near the apex and the bilobed right brachium are diagnostic. Female abdominal characters are very similar to those of *sphacelina*, but the ostium bursae is more lateral, the base of the ductus bursae lacks a broad pouch, and the posterior margin of the eighth tergum has an invagination on the right side. Wing length varies from 6.8 to 7.5 mm. Moths are very consistent in color pattern (or lack of pattern), having a pale zone from $\frac{1}{3}$ the length of the cell to the end of the cell on the forewing.

The immature stages are unknown.

*Brunneipes* has been collected in San Francisquito Canyon, Los Angeles County and San Diego, California from mid-April through mid-June.

### *Anoncia sphacelina* (Keifer)

PL. 4, FIG. 17. TEXT FIGS. 12 *a*, *b*, *d*, *e*; 25 *d*.

"*Borkhausenia*" *sphacelina* Keifer, 1935, *Mo. Bull. Dept. Agric. California*, **24**: 214, pl. V.
Type-locality: 3 mi W Shingle Springs, El Dorado Co., California. [CAS]

Characters for separating *sphacelina* from *conia* and *brunneipes* are given in the key and under *brunneipes*. Wing length varies from 7.5 to 9.5 mm. The forewing color pattern is indistinct. Some specimens have a better defined pattern than the one illustrated. The number of stout setae on the strong projection of the seventh abdominal tergum varies, and the length of the projections on the eighth tergum varies.

Keifer (1935: 214) reared larvae on leaves of *Lepechinia calycina*. They made globular chambers among bunched leaves and mined the individual leaves. Pupation occurred within the larval chamber.

*Sphacelina* occurs in California from El Dorado County south to San Diego. Larvae and pupae have been collected from 17 April through 19 May, adults from late April through June.

### *Anoncia fasciata* (Walsingham)

PL. 4, FIG. 21. TEXT FIGS. 12 *c*, *f*, *g*, *i*, *j*, *l*; 25 *c* (McD. 8374).

*Hypatopa fasciata* Walsingham, 1907, *Proc. U.S. Natl. Mus.*, **33**: 213.
Type-locality: Head of Noyo R., Mendocino Co., California. [BMNH]

Adults of *fasciata* can be confused with those of *porriginosa*, *noscres*, *episcia*, *venis* and *mosa* in maculation, but characters of the male and female genitalia ally them most closely with *alboligula*. Males of *fasciata* and *alboligula* are alike in the seventh and eighth abdominal segments and the entire genital capsule. In *fasciata* the posterior lobes on the seventh tergum are broadly rounded with the most posterior point somewhat medial, and the right brachium is J-shaped. In *alboligula* the most posterior point on the seventh sternum is lateral, and the right brachium has a short, basal, free extension. Females of the two species are similar in the fused basal part of the apophyses anteriores, the dorsal position of the ostium bursae and the broadly rounded anterior margin of the eighth abdominal sternum. In *fasciata* the ostium bursae has a trilobed structure associated with it (text figure 25 *c*) which is unique in the genus. The ostium bursae of *alboligula* lacks such an associated structure. Wing length varies from 6.0 to 8.5 mm. Maculation in well-marked individuals is as figured, but individual parts of the pattern vary in intensity. In males the shape of the seventh abdominal tergum varies somewhat but generally is as figured; the lobes on the eighth tergum may be shorter or longer than figured; and the two lobes at the base of the right valva may be unequal in length. In females the major variation is in the shape of the trilobed structure associated with the ostium bursae: the lateral lobes are rounded to subquadrate apically.

The immature stages are unknown.

*Fasciata* occurs in California from Glenn County south to San Bernardino County in June and July.

### *Anoncia alboligula* Hodges

PL. 4, FIG. 25. TEXT FIGS. 12 *h*, *k*, *m*, *p*; 25 *e*.

*Anoncia alboligula* Hodges, 1962, *Ent. Americana*, **42** (n.s.): 102, figs. 76, 187.
Type-locality: Olancha, Inyo Co., California. [USNM]

*Alboligula* can be confused with *loexya*, *noscres*, *episcia* and *smogops* in maculation, but it is nearest *fasciata* in genital and abdominal characters. It can be separated from *fasciata* as indicated in the key and under *fasciata*. Adults are highly variable in maculation, genitalia and abdomen, particularly the seventh and eighth segments in the male. To the unaided eye the moths appear gray or gray brown, but when examined under the microscope, they are distinctly white and dark brown or gray brown. An incomplete, transverse white fascia usually is present just beyond the base of the forewing, but even this mark tends to disappear in some specimens. The forewing may be streaked marginally on the distal ⅓. Wing length varies from 5.3 to 6.7 mm. In males the seventh abdominal tergum varies in the shape and development of the two lateral lobes and the shape of the marginal sclerite associated with the larger lobe; the lobes on the eighth tergum vary in length; the lobes at the base of the right valva tend to be extremely disparate in size. In females the position of the ostium bursae relative to the lateral margin varies as does the shape of the seventh abdominal sternum.

The immature stages are unknown.

*Alboligula* occurs from Redington, Arizona west to Riverside and Kern Counties, California. Moths have been collected at light from early April through the end of June with an apparent peak from May to mid-June.

### *Anoncia slales* Hodges, NEW SPECIES

PL. 6, FIG. 4. TEXT FIG. 25*f*.

*Anoncia slales* Hodges
Type-locality: Mt. Hamilton, Santa Clara Co., California. [UCB]

Upper surface as figured. Head: most scales pale gray to nearly white basally with dark brown before pale-gray apices; second segment of labial palpus mainly dark brown, many scales pale tipped particularly on dorso-medial surface, apex white, base and apex of third segment white, middle part mainly dark brown with individual scales having white apices; scales on frons, vertex and occiput dark brown just before apices; antennal scape dark brown on dorsal surface, many pale-gray to white scales on ventral surface, scales of pecten mainly pale with dark-gray to brown band before apices. Foreleg mottled pale gray and dark brown; coxa and femur with more pale scales than dark scales; tibia mainly dark with more pale scales at middle and apex; tarsal segments dark brown with pale apices. Midleg: much as for foreleg but with more scales having pale apices. Hindleg: mainly pale gray to nearly white; second, third, fourth and fifth tarsal segments with darker scales basally. Wing length: 5.8–6.0 mm. Male: no specimens available. Female: as in text figure 25 *f*; ostium bursae on posterior part of eighth tergum, nearly medial; base of ductus bursae heavily sclerotized, rest membranous; corpus bursae membranous, with a basal lobe or accessory pouch; signa absent; eighth abdominal sternum broadly rounded posteriorly, a semicircular, heavily sclerotized ring extending from near right margin around the left margin to near the midline of the eighth tergum; posterior margin of eighth tergum asymmetrical; apophyses anteriores fused basally for ¼ to ⅓ length then diverging.

The immature stages are unknown.

TYPES: Holotype: ♀. Mt. Hamilton, Santa Clara Co., California; 11 July 1967; W. C. Gagné; SBA genitalia slide 35. UCB. Paratype: 1 ♀. Same data as for holotype; SBA genitalia slide 2. Paratype to USNM.

*Slales* is similar to some specimens of *alboligula*, *loexya*, *noscres*, *episcia* and *mosa*, but the female genital characters given in the key are diagnostic. The semicircular, rounded structure on the anterior margin of the eighth abdominal segment is unique within the genus.

### *Anoncia glacialis* Hodges

PL. 4, FIG. 19. TEXT FIGS. 12 *n*, *o*, *q*.

*Anoncia glacialis* Hodges, 1962, *Ent. Americana*, **42** (n.s.): 100, figs. 74, 184.
Type-locality: Riverside, California. [CU]

*Glacialis* is nearest *bitoqua*. It can be separated from *bitoqua* by having the labial palpus short, barely reaching the vertex with the third segment ⅔ the length of the second segment and having a well-developed, dark-brown band medially and by having the scales of the antennal pecten dark gray before the apices. In *bitoqua* the labial palpus is much longer, well exceeding the vertex, the third segment is nearly as long as the second segment and the third segment is mainly white with very few dark-brown tipped scales on the middle ⅓, and the scales of the antennal pecten are uniformly pale gray to white. Wing length is 7.9 mm.

The immature stages are unknown.

The single specimen of *glacialis* was collected at Riverside, California on 8 July.

### *Anoncia bitoqua* Hodges, NEW SPECIES

PL. 4, FIG. 20. TEXT FIGS. 13 *a–d*.

*Anoncia bitoqua* Hodges.
Type-locality: Surprise Canyon, Inyo Co., Panamint Mts., California. [UCB]

Upper surface as figured. Head, thorax, legs and upper surface of forewings white with dark-brown tipped scales. Head: base of tongue white with very few dark-brown tipped scales; maxillary palpus with dark-brown tipped scales at apex; labial palpus with lateral surface of first and second segments having many dark-tipped scales, base and apex of second segment white, inner surface of first and second segments mainly white with scattered dark-tipped scales, third segment mainly white with few dark-tipped scales on anterior surface; lower part of frons with numerous dark-tipped scales, rest mainly white; vertex and occiput mainly white with few dark-tipped scales; antenna with scape white ventrally and with a few dark-tipped scales dorsally, scales of pecten white, shaft white with brown-tipped scales on first four segments then becoming brown with yellowish-gray scales to apex. Foreleg: coxa and femur white heavily overlaid with dark-tipped scales; tibia mainly dark with numerous white scales at ⅜ length and at apex, epiphysis with pale scales toward apex; base and apex of first tarsal segment and apices of remaining segments white, ventral surface of second through fifth segments mainly dark brown. Midleg: much as for foreleg but with reduction of dark-tipped scales. Hindleg: mainly white, all segments with a few dark-tipped scales, dorsal scales on tibia mainly white, tarsus with brown scales on ventral surface of third, fourth and fifth segments. Abdomen shining white to yellowish gray dorsally, ventral surface slightly darker. Wing length: 6.2–8.0 mm. Male: as figured; seventh abdominal tergum with slight posterolateral extensions, a very short sclerite on posterior margin near longer extension; eighth abdominal tergum with a pair of slender, posteriorly directed lobes; genitalia with slightly asymmetrical valvae; right valva with pair of slender lobes from base, left valva with a very stout, broad-based lobe at base; aedoeagus gradually tapering to acute apex; right brachium with a long slender lobe and with a short basal projection. Female: no specimens available.

The immature stages are unknown.

TYPES: Holotype: ♂. Surprise Canyon, Inyo Co., Panamint Mts., California; 24 April 1957; J. Powell. UCB. Paratypes: 16 ♂♂. All with same data as for holotype. Paratypes to UCB; USNM.

*Bitoqua* appears similar to *glacialis* but can be separated from that species as indicated under its description. The maculation consists of a series of indistinct streaks running generally parallel to the veins. Specimens vary in the amount of dark-tipped scales.

### *Anoncia porriginosa* Hodges

PL. 4, FIG. 23. TEXT FIGS. 13 *e, h, i.*

*Anoncia porriginosa* Hodges, 1962, *Ent. Americana* **42** (n.s.): 102, figs. 85, 186.
Type-locality: Claremont, California. [USNM]

*Porriginosa* appears as a faded specimen of *fasciata* with the transverse fascia at ⅓ the wing length and a pair of dark spots, one in the cell and one at ⅔ the length of the fold. Genital characters ally *porriginosa* with *bitoqua* and *flegax*. The three species can be separated as indicated in the key. Also, *bitoqua* is white and dark brown, lacks transverse marks and normally has the forewing marks elongated. In *porriginosa* the third segment of the labial palpus is much shorter than the second segment, the palpus is dark with several pale-tipped scales, and the apex of the second segment and base and apex of the third segment are white. Individual scales of the antennal pecten are medium gray basally becoming pale gray then very dark gray for ⅝ the length and very pale gray apically. Wing length varies from 6.5 to 8.0 mm. Moths vary in the amount of pale or dark areas. Fresh specimens have a fairly well-developed dark transverse fascia at ⅓ the wing length. In males the long posterolateral extension on the seventh abdominal tergum varies in length and the number of stout setae; the dorsal lobe at the base of the right valva usually is completely reduced.

The immature stages are unknown.

*Porriginosa* has been collected in San Louis Obispo County and at San Bernardino, Claremont and San Diego, California. Moths have been collected at light from mid- to late April.

### *Anoncia flegax* Hodges, NEW SPECIES

PL. 6, FIG. 5. TEXT FIGS. 13*f, g, l, m.*

*Anoncia flegax* Hodges.
Type-locality: Mt. Shasta City, Siskiyou Co., California. [UCB]

Upper surface as figured. Head: base of tongue and maxillary palpus gray brown; second segment of labial palpus mainly dark brown with some white scales on inner surface, apex white, third segment white at base and apex and with numerous dark scales medially, third segment slender and slightly shorter than second segment; frons gray brown on lower part; vertex and occiput white, some individual scales with gray brown before apices; antenna mainly dark brown with some pale-tipped scales toward base, individual scales of antennal pecten pale gray on basal ½ to ⅗ and at apex, dark gray between. Foreleg: mainly dark gray to brown with several pale-gray tipped scales, apices of tarsal segments pale gray. Midleg: much as for foreleg but somewhat paler, particularly on coxa. Hindleg: much paler than foreleg; dorsal surface of femur pale yellowish gray to white; tibia pale gray, spurs darker gray and contrasting with tibia; first tarsal segment nearly concolorous with tibia, apex paler, remaining tarsal segments dark gray, each slightly darker than the preceding, apices of second, third and fourth segments pale. Abdomen: lustrous yellowish gray, somewhat darker ventrally, posterior margins of individual segments pale. Wing length: 8.0–10.2 mm. Male: as in text figures; posterior margin of seventh abdominal tergum produced as a long, fingerlike lobe with several stout setae apically, a sclerite on posterior margin from base of long lobe to middle of segment; eighth abdominal tergum somewhat H-shaped; genitalia with valvae symmetrical, pair of lobes at base of right valva subequal in length and development, medial lobe often slightly longer than dorsal lobe; large lobe at base of left valva broad basally then constricted to more slender, distal part; aedoeagus tapering from before ½ length to very slender apex; right brachium long, slightly downcurved at apex, a short, stout projection at base. Female: no specimens available.

The immature stages are unknown.

TYPES: Holotype: ♂. Mt. Shasta City, Siskiyou Co., California; 3 July 1958; light trap; J. Powell. UCB. Paratypes: 9 ♂♂. Same locality as for holotype; 20 June–18 July 1958; J. Powell (4 ♂♂). Kern Co., 7 mi NW Frazier Park, California; 1 July 1965; J. Powell (1 ♂). Chuchupate Rgr. Sta., base Frazier Mtn., Ventura Co., California; 5–9 May 1959; J. Powell (3 ♂♂). San Diego Co., Mt. Palomar, California; 18 July 1963; J. Powell (1 ♂). Paratypes to UCB; USNM.

*Flegax* is very closely allied with *porriginosa* but can be separated on maculation by the streaking parallel to the veins on the distal ⅖ of the forewing, by the more contrasting pattern of the forewing and by the lengthening of the two spots to become dashes. The third segment of the labial palpus is moderately long and slender in *flegax* as contrasted with the somewhat shorter and stouter third segment in *porriginosa*. Variation is expressed in the relative amounts of white and dark-brown scales on the head, thorax, legs and forewings, the degree of lengthening of the spot in the cell from circular to a long dash and in the development of the fingerlike lobe on the seventh abdominal tergum, in degree of sclerotization and number of stout setae.

### *Anoncia aciculata* (Meyrick)

PL. 4, FIG. 42. TEXT FIGS. 13 *j*, *k*; 14 *e* (McD. 8379).

*Borkhausenia aciculata* Meyrick, 1928, *Exotic Microlepidoptera*, **3**: 468.
Type-locality: Fort Davis, 5000′, Texas. [BMNH]

*Aciculata* is the most distinct species in the genus and can be distinguished by maculation and male genitalia. It is associated with *fasciata*, *alboligula*, *bitoqua*, *glacialis*, *porriginosa* and *flegax* on the characters of the seventh and eighth abdominal terga in which the seventh segment has asymmetrical, posterior lobes and a separate sclerite on the posterior margin. Wing length is 7.8 mm.

The immature stages are unknown.

*Aciculata* has been collected at Fort Davis, Texas in October and Clark County, Nevada in mid-April.

### *Anoncia mones* Hodges, NEW SPECIES

PL. 6, FIG. 6. TEXT FIGS. 14 *a*, *c*, *d*.

*Anoncia mones* Hodges.
Type-locality: Cherry Canyon, 5096′, Guadalupe Mts., Culberson Co., Texas. [USNM]

Upper surface as figured. Head: scales on base of tongue and maxillary palpus pale gray tipped with white; labial palpus exceeding frons, third segment moderately slender, about ¾ length of second segment, lateral surface of second segment mainly dark brown, several scales tipped with white, inner surface mottled dark brown, medium gray and white, base and apex of third segment white, middle part dark brown laterally and brown and white on inner surface; frons mottled dark brown and pale gray to white, mainly dark; vertex and occiput mainly pale, scales dark brown to gray before apices; antenna nearly denuded of scales, individual scales of pecten pale throughout but slightly paler at apex. Foreleg: mainly dark brown, individual scales tipped with pale gray; apices of coxa, femur, tibia and tarsal segments pale gray to white. Midleg: much as for foreleg with more pale scales on coxa; tibia with a white streak from base to ¼ and a patch of pale scales on dorsal surface at

$\frac{3}{8}$ and at apex. Hindleg: absent. Wing length: 6.5 mm. Forewing: streaked with white, medium-gray and dark-brown scales dorsally. Hindwing: yellowish gray. Male: as in text figures 14 *a*, *c*, *d*; seventh abdominal tergum with a pair of symmetrical lateral extensions bearing fused setae at apex; eighth abdominal tergum with broad base and pair of mediolateral extensions; eighth abdominal sternum with a long medial projection bearing stout setae inside a fold; valvae nearly symmetrical, paired lobes from base of right valva with costal lobe longer than medial lobe; lobe from base of left valva heavily sclerotized, gradually tapering from base to apex; aedoeagus curved with distal $\frac{1}{3}$ excavated and running to slender apex; right brachium very heavily sclerotized and J-shaped; vinculum with saccus developed as somewhat triangular projection.

The immature stages are unknown.

TYPE: Holotype: ♂. Texas, Culberson Co., Guadalupe Mts., Cherry Canyon, 5096′; 22 May 1973; R. W. Hodges; USNM slide 4624. USNM.

*Mones* is nearest *aciculata* in maculation and to *naclia* and *callida* on genital characters. It is separated from all three species as indicated in the key. The symmetrical, paired, posterolateral projections on the seventh abdominal tergum are unique in the genus as are the heavily sclerotized lobe at the base of the left valva and the extremely heavily sclerotized right brachium.

## *Anoncia naclia* Hodges, NEW SPECIES

PL. 6, FIG. 7. TEXT FIGS. 14*f*, *g*, *i*.

*Anoncia naclia* Hodges.
Type-locality: Baboquivari Mts., Pima Co., Arizona. [USNM]

Upper surface as figured. Head: tongue and maxillary palpus yellowish white; labial palpus mainly white to pale yellowish gray, outer surface of second segment heavily dusted with medium gray, apex white, outer surface of third segment heavily dusted with yellowish gray to yellowish brown, particularly toward apex, inner surface of first and second segments mainly white; frons, vertex and occiput pale yellowish white to grayish white, scales above eye heavily marked with yellowish gray to yellowish brown; antenna grayish white on basal $\frac{1}{8}$ dorsally, ventral surface with scape darker, scales of pecten pale grayish white. Foreleg: coxa, trochanter and femur grayish white dusted with yellowish brown, apex of coxa, entire trochanter and base of femur uniformly pale; tibia and tarsus very heavily dusted with yellowish brown to brown, scale bases pale, particularly on tarsus. Midleg: much as for foreleg but with all surfaces paler. Hindleg: coxa, trochanter, femur and tibia pale grayish white to yellowish white, outer surface of tibial spurs medium brown, contrasting with segment; tarsal segments dusted with medium gray to grayish brown, particularly on distal part of individual segments. Wing length: 7.8 mm. Forewing: mainly pale yellowish gray to yellowish white lightly dusted with darker yellowish gray. Hindwing: medium gray, contrasting with paler forewing. Male: as in text figures 14*f*, *g*, *i*; seventh abdominal tergum with row of stout setae on posterior margin; eighth abdominal tergum with pair of heavily sclerotized, somewhat outwardly curved, posterior projections; eighth abdominal sternum with posterior margin deeply excavated, margin of excavation heavily sclerotized; valvae symmetrical, right valva with a single, heavily sclerotized, large basal lobe, left valva with a smaller, somewhat conical basal lobe; aedoeagus curved (apex broken); right brachium relatively straight, slightly downcurved at apex; vinculum broadly expanded in saccal region. Female: no specimens available.

The immature stages are unknown.

TYPE: Holotype: ♂. Baboquivari Mts., Pima Co., Arizona; 1–15 July 1924; O. C. Poling; USNM slide 4112. USNM.

*Naclia* is allied with *mones* and *callida* and can be separated as indicated in the key. Superficially, *naclia* is similar to *leucoritis* from which it can be separated by the lack of distinct marks on the forewing; *leucoritis* tends to have a dark mark at the end of the cell that may extend to the tornal area.

## *Anoncia callida* Hodges

PL. 4, FIG. 33. TEXT FIGS. 14 *b*, *h*; 15 *a*, *b*.

*Anoncia callida* Hodges, 1962, *Ent. Americana*, **42** (n.s.): 106, figs. 84, 191.
Type-locality: Madera Canyon, 4880′, Santa Rita Mts., Santa Cruz Co., Arizona. [CU]

*Callida* is most closely related to *naclia*, *nocticola*, *nebritis* and perhaps *psepsa*. Only males of *callida* are known, so direct comparison with *psepsa* is impossible. On maculation it is closest to *naclia* from which it can be distinguished by the distinctly gray, somewhat lined forewing, shorter labial palpi and paler hindwings. The characters given in the key should be studied for positive identification. Wing length varies from 6.8 to 9.6 mm. Specimens from West Texas tend to have less yellowish gray on the forewings and legs than do those from Arizona. The shape of the seventh and eighth abdominal terga and the eighth abdominal sternum vary somewhat among specimens from the same locality.

The immature stages are unknown.

*Callida* occurs from the Chisos and Sierra Diablo mountains in West Texas west to the west fork of Oak Creek and the mountains of southern Arizona. Texas specimens were collected from 26 May to 6 June and Arizona specimens were collected from 13 July to 13 August.

### *Anoncia nocticola* Hodges

PL. 4, FIG. 41. TEXT FIGS. 15 *c, d, i*; 26 *a*.

*Anoncia nocticola* Hodges, 1962, *Ent. Americana*, **42** (n.s.): 109, figs. 128, 196.
Type-locality: Madera Canyon, 4880′, Santa Rita Mts., Santa Cruz Co., Arizona. [CU]

*Nocticola* is very close to *nebritis* in maculation and genital characters. It can be separated from *nebritis* by the generally paler forewing lacking a transverse fascia at ½ the wing length. Male and female characters are as given in the key. Wing length of the two known specimens is 5.0 mm. These two specimens differ slightly in the amount of gray-brown dusting on the forewings and legs.

The immature stages are unknown.

*Nocticola* has been collected in the Santa Catalina Mountains and Santa Rita Mountains of southern Arizona in the first week of August.

### *Anoncia nebritis* Hodges

PL. 4, FIG. 26. TEXT FIGS. 15 *e–h, l*; 26 *b*.

*Anoncia nebritis* Hodges, 1962, *Ent. Americana*, **42** (n.s.): 103, figs. 81, 132, 188.
Type-locality: Madera Canyon, 4880′, Santa Rita Mts., Santa Cruz Co., Arizona. [CU]

*Nebritis* is closely related to *nocticola* and can be separated from *nocticola* as indicated under that species. Three specimens of *nebritis* are known, and these vary in wing length from 4.8 to 5.4 mm and in coloration in the dusting of gray-brown scales on forewing, head, thorax and legs.

The immature stages are unknown.

*Nebritis* has been collected in Madera Canyon, Santa Rita Mountains, Arizona from 11 to 25 July.

### *Anoncia psepsa* Hodges, NEW SPECIES

PL. 6, FIG. 8. TEXT FIG. 26 *d*.

*Anoncia psepsa* Hodges.
Type-locality: White Mts., 7000′, Arizona. [USNM]

Upper surface as figured. Head: tongue pale yellowish gray mottled with medium gray; maxillary palpus mainly pale yellowish gray; frons, vertex and occiput yellowish white, a few yellowish-gray scales on anterior margin of eye; labial palpus pale yellowish gray to white, outer surface of second segment medium gray with apex pale, inner surface mainly white to pale yellowish gray, dusted with some dark-tipped scales, third segment mainly pale, some darker scales on anterior surface; antenna pale yellowish gray to white basally becoming darker yellowish gray distally, scales of pecten unicolorous off white. Foreleg: mottled pale yellowish gray and darker yellowish gray, apex of coxa unicolorous pale yellowish gray; middle and apex of tibia with numerous pale-tipped scales; apices of first four tarsal segments pale. Midleg: much as for foreleg but with more numerous pale scales. Hindleg: very pale; second through fifth tarsal segments medium to pale gray basally, pale at apices. Wing length: 6.0 mm. Male: no specimens available. Female: as in text figure 26 *d*; ostium bursae mediolateral on eighth abdominal tergum; ductus bursae membranous; corpus bursae membranous; paired signa present; ductus seminalis arising from near posterior end of corpus bursae; no accessory pouch present; apophyses anteriores arising from near midline of eighth abdominal tergum, nearly parallel for short distance, then directed toward lateral margins of body; eighth abdominal tergum asymmetrical, posterior margin with series of rounded lobes.

The immature stages are unknown.

TYPE: Holotype: ♀. White Mts., Arizona, el. 7000′; 5–15 July 1925; O. C. Poling; USNM slide 4117. USNM.

*Psepsa* is closely related to *nocticola* and *nebritis* but can be separated from them as indicated in the key. Additionally, *psepsa* is a larger species, and the forewings have a pale-yellow aspect as contrasted with the more nearly gray forewings of *nebritis* and *nocticola*.

### *Anoncia leucoritis* (Meyrick)

PL. 4, FIG. 35. TEXT FIGS. 15*j*, *k*, *m*; 26 *c* (McD. 8378).

*Borkhausenia leucoritis* Meyrick, 1927, *Exotic Microlepidoptera*, **3**: 381.
Type-locality: Alpine, 7500′, Texas. [BMNH]

*Anoncia mentzeliae* Clarke, 1942, *Can. Ent.*, **74**: 18, figs. 1–3.
Type-locality: Snake R., Whitman Co., Washington, opposite Clarkston. [USNM]

*Leucoritis* is allied with *longa* and *furvicosta* on genital characters and the male seventh and eighth abdominal segments. Fresh specimens are nearly immaculate white on the head, thorax and forewings. The forewings have a dusting of gray or brown scales on the outer margin and base of the fringe on the distal ¼ of the wing, and a spot at the end of the cell and sometimes another in the tornal area. A light dusting of pale yellow is on the head, thorax and forewings. Wing length varies from 7.0 to 8.8 mm.

Larvae have been reared from *Mentzelia laevicaulis* (Douglas) T. & G. in Washington and Butte Counties, California. According to Clarke (1942: 19) the larvae feed on the immature ovaries of the food plant, and pupation occurs in a cocoon in the ovary.

*Leucoritis* occurs from southeastern Washington south into northern California and east to the Big Bend country of Texas. Reared adults emerged from 7 to 24 August in Washington, and moths were collected from 20 to 22 May in West Texas. One adult from Dewey, Arizona was collected from 16 to 23 June.

### *Anoncia longa* (Meyrick)

PL. 4, FIGS. 36–38. TEXT FIGS. 16 *a–c*, *e*, *g*; 26 *e* (McD. 8377).

*Borkhausenia longa* Meyrick, 1927, *Exotic Microlepidoptera*, **3**: 381.
Type-locality: Alpine, 7500′, Texas. [BMNH]

Adults can usually be recognized by the combination of size and forewing color pattern, but some specimens are similar to adults of *leucoritis* and require study of genital characters for determination. These characters are given in the key. *Longa* is closely related to *leucoritis* and *furvicosta* on genital characters and male abdominal characters. Wing length varies from 8.8 to 13.0 mm. Moths vary in the ground color and amount of gray-brown to brown overlay on the head, thorax and forewings. The moths illustrated indicate the extremes. Most specimens have a dark spot in the cell, one at the end of the cell and one at ⅔ the length of the fold on the forewing. Males vary considerably in the shape of the posterior margin of the seventh abdominal tergum and the entire eighth abdominal tergum.

The immature stages are unknown.

*Longa* occurs from southeastern Colorado and north-central Utah south to West Texas and southeastern Arizona. Adults have been collected from 1 June to 7 October.

### *Anoncia furvicosta* Hodges

PL. 1, FIG. 20; PL. 4, FIGS. 39, 40. TEXT FIGS. 16 *d*, *f*, *h*; 26 *g*.

*Anoncia furvicosta* Hodges, 1962, *Ent. Americana*, **42** (n.s.): 108, figs. 80, 131, 195.
Type-locality: Madera Canyon, 4880′, Santa Rita Mts., Santa Cruz Co., Arizona. [CU]

*Furvicosta* is closely related to *leucoritis* and *longa* in genital characters and male abdominal characters. It can be separated from them as indicated in the key. *Furvicosta* is similar to *nocticola* in maculation, but *furvicosta* is a much larger species. Wing length varies from 6.4 to 10.0 mm. The elements of the color pattern are relatively consistent and vary basically in the amount of dusting of brown scales on the forewing, head and thorax. Worn specimens are pale. In males the shape of the posterior margin of the seventh abdominal tergum and the entire eighth tergum vary slightly.

The immature stages are unknown. Many specimens in the type series have become greasy, an indication that the larva is likely to be an internal feeder.

*Furvicosta* has been collected in the Chiricahua, Huachuca, Santa Rita and Baboquivari mountains in southern Arizona from 11 to 30 July.

### *Anoncia diveni* (Heinrich)

PL. 4, FIG. 27. TEXT FIGS. 17 *b*, *e*, *f*; 26 *f* (McD. 8373).

*Borkhausenia diveni* Heinrich, 1921, *Jour. Agric. Research*, **20**: 814, pl. 96, figs. C–F.
Type-locality: Brownsville, Texas. [USNM]

*Diveni* and *orites* are closely related on genital characters. In males the seventh abdominal segment is unmodified; the eighth abdominal segment is modified much as for *leucoritis*, *longa* and *furvicosta*; the valvae are very short and broad; the aedoeagus is relatively short, ending in a blunt apex; and the right brachium has a long basal extension that nearly joins the left brachium. In females the ostium bursae is dorsolateral on the eighth abdominal tergum which is a broad, irregular rhomboid with a semicircular excavation around part of the ostium bursae; the ductus bursae is membranous; the corpus bursae has an accessory pouch arising from the base; signa are lacking; and the apophyses anteriores arise from the lateral margins of the eighth abdominal tergum. *Diveni* can be separated from *orites* by characters given in the key. Superficially, *diveni* is similar to some specimens of *noscres* and *mosa*, but *diveni* has the forewings brown to yellow brown as contrasted with the gray and bluish-gray forewings of *noscres* and *mosa*. The patch of yellowish-brown scales in the middle of the cell on the forewing and the brown, transverse fascia at $\frac{3}{5}$ the forewing length in combination are distinct within the genus. Wing length varies from 5.0 to 7.5 mm. Some specimens are pale with the distal $\frac{1}{2}$ of the forewing very lightly dusted with brown scales; others are quite dark with the distal $\frac{1}{2}$ of the forewing heavily dusted with brown scales. The hindwing varies from medium to dark gray. In males the long process from the base of the left valva varies in diameter.

Larvae have been reared from two species of *Lantana*. They make a narrow blotch mine at the edge of the leaf and curl the leaf. Pupation occurs within the mine.

*Diveni* has been reared at Brownsville, Texas, Cuernavaca, Mexico, and Zamorana, Honduras. Specimens in the type series were collected as larvae on 22 April and emerged from 27 April to 5 May.

### *Anoncia orites* (Walsingham)

PL. 4, FIG. 32. TEXT FIGS. 17 *a*, *d*, *i*; 27 *a* (McD. 8375).

*Hypatopa orites* Walsingham, 1907, *Proc. U.S. Natl. Mus.*, **33**: 213.
Type-locality: Mt. Shasta, Siskiyou Co., California. [BMNH]

*Orites* can be recognized by the genital characters given in the key. It is nearest *diveni* in structural characters, but superficially it is distinct. The forewing color pattern, although diffuse, is often characteristic of the species. In particular the oblique fascia running from the end of the cell to the tornal area is distinctive. Wing length varies from 6.0 to 7.8 mm. Maculation varies in the number of gray-brown scales on the forewings, head and thorax, but the scales on the base of the tongue, maxillary palpus and lower part of the frons are dark gray brown.

The immature stages are unknown.

*Orites* has been collected in Mohave County, Arizona and Placer, Santa Clara, Siskiyou and Modoc Counties, California. The Arizona specimen was collected from 8 to 15 May, and those from California were collected from 11 to 26 July. Although adequate data are lacking, *orites* appears to occur at higher elevations in mountain ranges.

### *Anoncia fregeis* Hodges, NEW SPECIES

PL. 6, FIG. 9. TEXT FIG. 27 *b*.

*Anoncia fregeis* Hodges.
Type-locality: Carrizo Plains, Kern Co., California. [USNM]

Upper surface as figured. All body surfaces nearly uniformly pale yellowish white; labial palpus short, scarcely exceeding frons, third segment much shorter than second. Forewing: apparently overlaid with darker yellow, but specimens in type series worn. Wing length: 8.8–9.3 mm. Male: no specimens available. Female: as in text figure 27 *b*; ostium bursae on left side of eighth abdominal tergum; base of ductus bursae heavily sclerotized, somewhat convoluted linearly, membranous part of ductus bursae shorter than heavily sclerotized part; corpus bursae with an accessory pouch from base, ductus seminalis arising from this pouch; paired signa present, each a somewhat poorly defined, moderately heavily sclerotized plate with several inwardly directed, pointed projections; posterior margin of eighth abdominal sternum a heavily sclerotized band; eighth abdominal tergum with semicircular emargination around ostium bursae, immediately posterad of ostium bursae is an inwardly directed, heavily sclerotized pouch; apophyses anteriores arising from lateral margins of eighth abdominal sternum and directed laterad from point of origin.

The immature stages are unknown.

TYPES: Holotype: ♀. Carrizo Plains, Kern Co., California; 20 Aug. 1935; G. D. Hanna; USNM slide 6129. USNM. Paratypes: 2 ♀♀. Same data as for holotype. Paratypes to USNM.

The female genitalia ally *fregeis* with *orites*, *loexya*, *noscres*, *episcia*, *venis*, *piperata*, *mosa*, *smogops* and *psentia*. *Fregeis* is nearest *naclia* in appearance, but the hindwings of *fregeis* are shining, pale yellowish white; those of *naclia* are gray brown.

### *Anoncia loexya* Hodges, NEW SPECIES

PL. 4, FIG. 29. TEXT FIGS. 17 *c*, *g*, *h*, *k–n*; 27 *e*.

*Anoncia loexya* Hodges.
Type-locality: Pine, Gila Co., Arizona. [UCB]

Upper surface as figured. Head: base of tongue pale gray, nearly white; labial palpus white, heavily mottled with dark brown, lateral surface of second segment mainly dark brown with some pale-tipped scales and with apex white, inner surface with some brown-tipped scales on anteroventral surface, apex white, third segment white at base and apex, many brown-tipped scales in between; frons, vertex and occiput mainly pale gray with numerous scales brown just before apices; antenna mottled pale gray and dark brown, scales of pecten pale at base and apex, darker at $\frac{3}{4}$ length. Foreleg: coxa mottled pale gray to white and dark brown; trochanter mainly pale; femur mainly dark brown, mottled with pale gray; tibia mainly dark brown with several pale-tipped scales, particularly at $\frac{3}{5}$ length and at apex; tarsal segments mainly dark brown basally with white apices, fifth segment unicolorous brown. Midleg: much as for foreleg but with more numerous pale scales. Hindleg: pale scales predominant; tibia with dorsal tuft pale yellowish gray, nearly unicolorous pale gray at base and apex; tarsus with first segment pale, apex paler than rest, second through fifth segments progressively darker brown, fifth segment unicolorous brown. Wing length: 6.0–6.8 mm. Male: as in text figures 17 *c*, *g*, *h*, *k–n*; valvae nearly symmetrical, right valva with pair of long lobes from base, costal lobe more slender than medial lobe, left valva with basal lobe very broad basally, then reduced in diameter at $\frac{2}{5}$ length to slender apex; aedoeagus with slender apex; right brachium J-shaped. Female genitalia: as in text figure 27 *e*; ostium bursae nearly at middle of eighth abdominal tergum; base of ductus bursae lightly sclerotized; corpus bursae lacking signa but with paired regions having numerous, inwardly directed bulges, a secondary pouch from near base of corpus bursae, ductus seminalis arising from apex of pouch; eighth abdominal tergum with a medial excavation and a medio-anterior projection; apophyses anteriores arising from near middle of anterior margin of eighth abdominal tergum then directed toward lateral margins, a pair of short lobes between bases.

The immature stages are unknown.

TYPES: Holotype: ♂. Arizona, Pine, Gila Co.; 4 June 1968; at black and white lights; P. Opler & J. Powell; USNM Gel. slide 1193. UCB. Paratypes: 8 ♂♂, 1 ♀. Same data as for holotype (3 ♂♂, 1 ♀). White Mts., Arizona, Apache Ind. Res., el. 7000′; 1–15 June 1925; O. C. Poling (5 ♂♂). Paratypes to UCB; USNM.

*Loexya* is very closely related to *noscres* but can be separated as indicated in the key based on genital characters; by being generally less contrasting pale gray and dark brown, *noscres* usually is pale gray and very dark brown; and by distribution, *loexya* is known from northern Arizona, *noscres* is known from San Bernardino and Kern Counties, California.

### *Anoncia noscres* Hodges, NEW SPECIES

PL. 6, FIG. 10. TEXT FIGS. 17 *j*, *o*, *p*; 27 *c*.

*Anoncia noscres* Hodges.
Type-locality: Piute Mts., 4350′, Rancheria Creek, T29S, R33E, S23, Kern Co., California. [LACM]

Upper surface as figured. Head: base of tongue white; maxillary palpus mottled white and dark brown; labial palpus white and dark brown, second segment with lateral surface mainly dark brown except white at apex, inner surface with dark-brown scales more numerous toward anteromedial margin, third segment white at base and apex, mainly dark brown at middle; frons, vertex and occiput white and dark brown, many scales with dark brown just before apices; antenna with scape white on ventral surface and anterior margin and dark brown on most of dorsal surface, shaft mottled white and dark brown, becoming dark by $\frac{1}{5}$ length, scales of pecten pale at base and apex, dark brown from $\frac{3}{5}$ to $\frac{4}{5}$ length. Foreleg: coxa and femur dark brown mottled with white, apex of coxa and entire trochanter white, apex of femur mainly white; tibia mainly dark brown with several white scale bases visible, an incomplete, white streak at $\frac{2}{3}$ length, apex white; tarsal segments mainly dark, apices of first, second and third segments white, fourth and fifth segments nearly unicolorous dark. Midleg: coxa and trochanter mainly white with a few gray scales; femur white at base and apex and along dorsal margin; tibia mainly dark with a transverse, white irregular band at $\frac{2}{3}$ length, apex white; tarsus with white on apices of first four segments, individual segments becoming progressively darker, ventral surface dark brown to black. Hindleg: coxa and trochanter mainly white with some pale-gray tipped scales; femur white on dorsal margin and at apex, mottled gray and white on outer surface laterally; tibia pale, a white streak at $\frac{2}{3}$ length and apex white, darker scales tipped with medium to dark gray, shorter tibial spurs dark gray, contrasting with tibia, dorsal tuft nearly unicolorous pale gray; first tarsal segment white at base and apex, mottled gray and white medially, remaining tarsal segments progressively darker, fifth segment dark, ventral surface dark brown. Wing length: 6.3–7.0 mm. Male: as in text figures 17 *j*, *o*, *p*; seventh abdominal tergum sclerotized, posterior margin with medial, triangular projection; eighth abdominal tergum a U-shaped band, posterior margin of this band very heavily sclerotized; seventh abdominal sternum with posterior margin concave, sometimes with slight posterolateral projections and these with stout scales; eighth abdominal sternum yoke shaped with posterior projections varying in length and shape; valvae nearly symmetrical, base of right valva with pair of slender lobes, costal lobe shorter than medial lobe, base of left valva with broad-based projection that diminishes in diameter at $\frac{1}{3}$ length, rest a dorsally open, heavily sclerotized fold containing a heavily sclerotized spinelike structure; aedoeagus somewhat curved diminishing in cross sectional area before slender apex; right brachium broadly J-shaped. Female: as in text figure 27 *c*; ostium bursae nearly at middle of eighth tergum, margined on left side by heavily sclerotized fold; ductus bursae evenly membranous throughout, a small lobe at $\frac{1}{5}$ length; corpus bursae with broad-based accessory pouch from base; signa absent, but paired series of concentric circles of inwardly directed bulges present; ductus seminalis arising from accessory pouch; apophyses anteriores arising from near midline on anterior margin of eighth abdominal tergum, a pair of irregular lobes associated with left apophysis; eighth abdominal sternum with posterior margin straight to somewhat indented; an apparently free, sclerotized band extending anteriorly from anterior margin of eighth abdominal sternum.

The immature stages are unknown.

TYPES: Holotype: ♀. Kern Co., California, Piute Mts., 4350′, T29S, R33E, Sec. 23; 1–3 June 1973; J. P. & K. E. Donahue; SBA slide 142. LACM. Paratypes: 14 ♂♂, 5 ♀♀. Same locality and collectors as for holotype; 1–13 June (10 ♂♂, 5 ♀♀). Indian Wells, Kern Co., California; 18 April 1962; C. A. Toschi (1 ♂). 15 mi SW Idria, S. Benito Co., California; 28 May 1960; D. C. Rentz (1 ♂). Morongo Wash, San Bernardino Co., California; 12 May 1937; Lloyd M. Martin (1 ♂). Morongo Valley, California; 11 May 1937; Grace H. & John L. Sperry (1 ♂). Paratypes to LACM; USNM.

*Noscres* is very similar to *loexya* but can be separated from it as indicated under *loexya*. The differences used for recognition of these two species are consistent in the specimens examined, but examination of additional material may show the character states represent variation in one species. Males of *noscres* vary in the development of the seventh tergum and the posteriorly directed lobes on the eighth sternum.

## *Anoncia episcia* (Walsingham)

PL. 4, FIGS. 24, 28, 30, 31. TEXT FIGS. 18 *a–c*; 27 *d* (McD. 8371).

*Hypatopa episcia* Walsingham, 1907, *Proc. U.S. Natl. Mus.*, **33**: 211.
Type-locality: Russian R., Sonoma Co., California. [BMNH]

NOTE—The lectotype, ♀, present designation, bears the following labels: 1. "Lectotype"; 2. "Type"; 3. "Russian R., Sonoma Co., California, 19.V.1871, Wlsm. 90421"; 4. "Walsingham Collection, 1910—427."; 5. "*Hypatopa episcia*, Wlsm., Type ♀"; 6. "Lectotype ♀, *Hypatopa episcia* Wlsm., teste K. Sattler, 1974"; 7. "B.M. ♀ Genitalia Slide No. 19857".

Genital characters ally *episcia* most closely with *venis* and less so with *piperata*. *Episcia* and *venis* can be separated by the abdominal sterna being heavily sclerotized in *episcia*, membranous in *venis*. In males the abdomens and genitalia vary over the same range. In females the anterior margin of the eighth abdominal tergum of *episcia* has a pair of medial rods, whereas in *venis* the margin has a pair of anteriorly directed, broad lobes. *Episcia* lacks the conspicuous dashes, one in the cell and one on the fold, on the forewing that are present in *venis*. In maculation *episcia* can be confused with *loexya*, *noscres*, some specimens of *mosa* and *smogops*. Wing length varies from 5.8 to 8.0 mm. The forewing varies considerably in degree of development of pattern as illustrated. A series of specimens from

Pozo, San Louis Obispo County are consistently dark and have a well-developed pattern as contrasted with a series from Piute Mountains in Kern County in which the forewings are very pale and the pattern washed out. In males the eighth abdominal tergum varies in length and width of the pair of posteriorly directed lobes, and the pair of lobes at the base of the right valva varies in size and relative size; however, the more medial lobe consistently has more setae before the apex, and the costal lobe bears setae only at the apex.

The immature stages are unknown.

*Episcia* occurs in California from Mendocino and Sonoma Counties south to Kern and Riverside Counties. Moths have been collected from 18 April to 9 June.

*Anoncia venis* Hodges, NEW SPECIES
PL. 6, FIG. 11. TEXT FIGS. 18 *e*, *g*, *l*, *o*; 28 *e*.

*Anoncia venis* Hodges.
Type-locality: Quartzite, Yuma Co., Arizona. [USNM]

Upper surface as figured. Head: base of tongue pale gray to off white; maxillary palpus mottled very pale gray and medium gray; frons, vertex and occiput mainly pale gray to white, several scales tipped with darker gray; labial palpus with first segment mainly pale, lateral surface of second segment mainly dark with some pale scales at base and apex pale, inner surface pale posterodorsally from base to $\frac{2}{5}$ length, rest mainly dark except for pale apex, third segment pale at base and apex, dark medially on lateral surface, mainly pale on inner surface; antenna mottled pale grayish white on basal $\frac{1}{5}$ then becoming dark brown, ventral surface of scape mainly pale gray, scales of pecten pale gray at base and apex, medium gray at $\frac{2}{3}$ length. Foreleg: coxa mottled grayish white and brown; trochanter pale grayish white; femur pale gray and dark brown; tibia mainly dark brown, some scale bases pale gray, a row of pale-gray scales at base of epiphysis and another at apex of segment; tarsus mainly dark brown with some pale scales, apices of first four segments pale gray, ventral surface dark brown. Midleg: much as for foreleg but with more pale scales. Hindleg: coxa and trochanter mainly pale gray to yellowish white; femur pale yellowish gray to white on dorsal surface and at apex, many dark-gray scales on ventral $\frac{1}{2}$; tibia mainly pale with many dark scales, a transverse, pale streak at base of first pair of spurs and at apex, dorsal, long scales pale yellowish gray; tarsus mottled pale and medium gray, segments darker from first through fifth. Abdomen: dorsal surface shining pale yellowish white, ventral surface same ground color with some medium to dark-gray scales medially. Wing length: 6.2–6.8 mm. Male: as in text figures 18 *e*, *g*, *l*, *o*; abdominal sterna membranous; eighth abdominal tergum moderately heavily sclerotized on posterior part and with a pair of posteriorly directed lobes; posterior part of eighth abdominal sternum a heavily sclerotized, C-shaped band; valvae nearly symmetrical, broad, right valva with pair of short lobes from base, lobes varying in size relative to one another; left valva with a very broad-based lobe, constricted at $\frac{2}{5}$ length and tapering to very narrow apex; aedoeagus curved, tapering on distal $\frac{1}{2}$ to very slender apex; right brachium a long, stout spine, somewhat downcurved at apex, lacking a basal projection. Female: as in text figure 28 *e*; ostium bursae near left margin of eighth abdominal tergum, opening as two, slender teardrop shaped structures; ductus bursae lightly sclerotized at base then membranous to corpus bursae, a short lobe near ostium bursae; corpus bursae membranous with an accessory pouch from base; ductus seminalis arising from pouch; signa lacking but pair of reticulate structures on opposite walls present; posterior margin of eighth abdominal sternum nearly straight; eighth abdominal tergum slightly convex on posterior margin, anterior margin heavily sclerotized with pair of anteriorly directed lobes medially, these bounded by apophyses anteriores.

The immature stages are unknown.

TYPES: Holotype: ♀. Quartzite, Yuma Co., Arizona; 4 April 1973; J. H. Baker; USNM slide 6093. USNM. Paratypes: 12 ♂♂, 7 ♀♀. Same locality as for holotype; 9 April, 23 May 1973 (5 ♂♂). Baboquivari Mts., Pima Co., Arizona; 1–15 May 1924; O. C. Poling (1 ♂). Baboquivaria [*sic*] Mts., Pima Co., Arizona (2 ♀♀). Ajo, Pima Co., Arizona; 1–30 March 1923; O. C. Poling (5 ♂♂, 1 ♀). Sells P.O., Indian Oasis, Pima Co., Arizona; 1–15 April 1923; O. C. Poling (1 ♀). En route Dewey to Salome, Arizona; 8–15 May (1 ♂). Redington, Arizona (2 ♀♀). Inyo Co., California; 1–15 June 1922; O. C. Poling (1 ♀). Albuquerque, New Mexico; 19 July 1903; Oslar. Paratypes to USNM.

*Venis* is very closely related to *episcia* on male and female characters but can be separated as indicated under *episcia*. In addition veins $M_3$ and $Cu_1$ in the hindwing usually are stalked in *venis*, connate to short stalked in *episcia*. *Venis* varies somewhat in forewing maculation, mainly in the amount of dark scales.

### *Anoncia piperata* Hodges

PL. 4, FIG. 34. TEXT FIGS. 18*f*, *k*, *n*; 28 *c*.

*Anoncia piperata* Hodges, 1962, *Ent. Americana*, **42** (n.s.): 106, figs. 78, 192.
Type-locality: Smoky Valley, 6200′, Tulare Co., California. [USNM]

Although *piperata* is extremely closely related to *episcia* and *venis* on genital characters, the moths are distinct by color pattern. The oblique, dark band running from the posterior margin of the forewing toward the costal margin at ½ the length of the cell is unique in the genus. Additionally, most of the forewing pattern consists of dark lines running parallel to the veins. The most distinctive feature of *piperata* is the heavily sclerotized pouch from the base of the ductus bursae. Wing length varies from 8.0 to 9.4 mm. Moths vary in the amount of dark-tipped scales present on the head, thorax, legs and forewings.

The immature stages are unknown.

*Piperata* has been collected from Prescott and Redington, Arizona west to Siskiyou and Kern Counties, California. Most specimens have been collected in San Bernardino and Kern Counties. Moths have been taken at light from 8 April through 16 July.

### *Anoncia mosa* Hodges, NEW SPECIES

PL. 4, FIG. 22. TEXT FIGS. 18*j*, *m*, *p*; 28 *b*.

*Anoncia mosa* Hodges.
Type-locality: 2 mi W Paso Robles, San Louis Obispo Co., California. [UCB]

Upper surface as figured. Head: tongue mottled dark gray brown and pale gray basally, somewhat paler and more nearly unicolorous distally; maxillary palpus dark brown and pale gray; frons, vertex and occiput mainly pale grayish white, many scales tipped with dark gray to brown; labial palpus with first and second segments mainly dark brown, many pale scale bases, apex of second segment white, inner surface of second segment mainly white at base, mottled dark brown and white medially and white at apex, third segment mainly white with some dark-brown scales on lateral surface and at apex; antenna mainly dark brown on first segment and base of shaft, ventral surface of scape white, scales of pecten pale gray, nearly white on basal ½ and at apex, brown between. Foreleg: mainly dark brown mottled with pale-gray scales; apex of tibia with several white scales; apices of first four tarsal segments with several white scales. Midleg: much as for foreleg but with more, pale-gray based scales, apex of each tarsal segment pale. Hindleg: coxa and trochanter yellowish gray, nearly white; femur yellowish gray, nearly white on dorsal surface, mottled yellowish gray to white and dark brown on ventral ⅔; tibia pale with many gray or medium-gray scales, dorsal scales pale yellowish gray to white, a concolorous streak at base of first pair of spurs, shorter spurs dark brown contrasting with tibia; tarsal segments mottled dark gray and pale gray basally, apex of each segment pale. Wing length: 6.0–7.7 mm. Male genitalia: as in text figures 18 *j*, *m*, *p*; seventh abdominal tergum often with short projection on posterolateral margins; eighth abdominal tergum with pair of slender, submedial projections; eighth abdominal sternum a heavily sclerotized, C-shaped band; valvae nearly symmetrical; base of right valva with pair of lobes, medial lobe longer than costal lobe; base of left valva with heavily sclerotized long lobe, base of lobe tapering in short distance to slender part; aedoeagus with distal ¼ tapering to slender apex; right brachium heavily sclerotized, distal ⅔ evenly tapering to narrow apex, with a slight basal projection. Female: as in text figure 28 *b*; posterior margin of eighth abdominal sternum broadly rounded, a lightly sclerotized, free band extending anteriorly from posterior margin; ostium bursae on left side of eighth abdominal tergum, surrounded by heavily sclerotized, semicircular plate; base of ductus bursae heavily sclerotized, a short lobe at base, rest membranous; corpus bursae with an accessory pouch at base, ductus seminalis arising from pouch; signa absent but paired structures with numerous, inwardly directed bulges present; eighth abdominal tergum deeply excavated on left side but excavation not reaching midline, anterior margin with two, somewhat rounded lobes; apophyses anteriores asymmetrical, left apophysis arising from nearer middle of margin than right apophysis.

The immature stages are unknown.

TYPES: Holotype: ♂. California, San Louis Obispo Co., 2 mi W Paso Robles; 25 April 1968;

at light; J. A. Chemsak & J. A. Powell; USNM slide 4036. UCB. Paratypes: 43 ♂♂, 15 ♀♀. Same locality as for holotype; 25, 28 April 1968 (5 ♂♂, 3 ♀♀). 3 mi SW Atascadero, San Luis Obispo Co., California; 26 April 1968; at light; J. Doyen (1 ♂). 4 mi NE Lemoncove, Tulare Co., California; 13 May 1963; J. Powell (1 ♂). Fairview, Tulare Co., California; 30 April 1964; at light; J. Doyen (1 ♂). 5 mi S Sage, Riverside Co., California; 16 April 1965; A. Toschi (1 ♂). Chino Canyon, nr. Palm Sprgs., Riverside Co., California; 19, 20 April 1960; black light; J. Powell (15 ♂♂, 5 ♀♀). 14 mi NE Mojave, Kern Co., California; 13 April 1962; C. MacNeill, D. Rentz & R. Brown (4 ♂♂). Piute Mts., 4350′, Rancheria Creek, T29S, R33E, Sec. 23, Kern Co., California; 1–3 June 1973; J. P. & K. E. Donahue (5 ♂♂). Bishop, California; 14 June 1937; E. C. Johnston (1 ♂, 1 ♀). San Bernardino, California (1 ♂). 2 mi NW Valyermo, Los Angeles Co., California; 1 May 1968; J. Powell (1 ♀). Mt. Palomar, San Diego Co., California; 18 July 1963; J. Powell (1 ♀). En route Dewey to Salome, Arizona; April–7 May (6 ♂♂, 3 ♀♀). Redington, Arizona (2 ♂♂, 1 ♀). Paratypes to CAS; LACM; UCB; USNM.

*Mosa* is very closely related to *piperata*, *smogops* and *psentia*. Adults can be recognized by the presence of a dark, gray-brown transverse fascia at ⅓ the wing length, a well-developed spot or dash in the cell, a spot at the end of the cell and a dash at ⅔ to ¾ the length of the fold. The male genitalia of *mosa* and *smogops* are very similar, but the large process at the base of the left valva has a slight shoulder in *mosa*, none in *smogops*. The shape of the eighth abdominal sternum distinguishes females of *mosa* from these three species. *Mosa* varies in coloration from yellow brown to gray brown. The dusting of dark-brown scales on the distal ⅓ of the forewing is less on specimens from Arizona and eastern California than on others. A series of specimens from the Piute Mountains has a relatively suffused forewing color pattern.

### *Anoncia smogops* Hodges, NEW SPECIES
PL. 6, FIGS. 12, 13. TEXT FIGS. 18 *d, h, i*; 28 *a*.

*Anoncia smogops* Hodges.
Type-locality: Baboquivari Mts., Pima Co., Arizona. [USNM]

Upper surface as figured. Head: base of tongue mainly pale gray, nearly white with very few darker gray scales; maxillary palpus mainly pale gray with some brown-tipped scales toward apex; frons, vertex and occiput mainly pale gray with numerous brown-tipped scales on lower part of frons in front of eye; labial palpus with first segment mainly pale gray with few medium-gray scales, lateral surface of second segment mainly dark with scales having pale bases, apex white, inner surface white at base, mottled white and brown medially, apex white, third segment white at base and apex, brown medially; antenna with scape mottled pale gray and dark brown dorsally, white ventrally, scales of pecten pale gray on basal ½ and at apex, brown between, shaft with alternating rows of pale-gray and brown scales on basal ⅓ becoming unicolorous yellow gray. Foreleg: mottled pale gray and brown, apex of tibia and apices of first four tarsal segments white. Midleg: much as for foreleg, but all surfaces with more pale scales. Hindleg: coxa, trochanter and much of femur pale yellowish white shining with blue or purple reflections; tibia mainly pale gray to white, dorsal scales yellowish gray, spurs mottled pale gray and brown; tarsus with first segment mainly pale gray with numerous medium-gray scales, apex pale, second through fifth segments becoming increasingly darker, apices of second, third and fourth segments pale. Wing length: 5.5–7.3 mm. Forewing and hindwing with large area of brown scales on undersurface basally, hindwing with comparable brown scales on dorsal surface in female. Male: as in text figures 18 *d, h, i*; eighth abdominal tergum with heavily sclerotized margin and a pair of short, posteriorly directed lobes; eighth abdominal sternum a heavily sclerotized, C-shaped band; valvae symmetrical, right valva with pair of lobes from base, costal lobe usually shorter than medial lobe, left valva with strong lobe from base tapering from base to apex, lacking shoulder; aedoeagus tapering from ⅝ length to moderately stout apex; right brachium somewhat J-shaped with short basal projection. Female: as in text figure 28 *a*; posterior margin of eighth abdominal sternum nearly straight; ostium bursae at middle of eighth abdominal tergum; eighth abdominal tergum deeply emarginate on left side, excavation extending more than ½ distance across segment; ductus bursae membranous; corpus bursae with broad-based accessory pouch at base; ductus seminalis arising from accessory pouch; signa absent; apophyses anteriores slightly asymmetrical, arising from middle ⅓ of anterior margin of eighth abdominal tergum then directed laterally.

The immature stages are unknown.

TYPES: Holotype: ♀. Baboquivari Mts., Pima Co., Arizona; 1–15 May 1924; O. C. Poling; USNM slide 4181. USNM. Paratypes: 7 ♂♂, 7 ♀♀. Brown's Canyon, Baboquivari Mts., Pima Co., Arizona; 1–15 May 1923; O. C. Poling (1 ♀). Mohave Co., Arizona; 1–15 May (1 ♂, 1 ♀). En route Dewey to Salome, Arizona; 8–15 May (2 ♀♀). Olancha, Inyo Co., California; 8–30 June (5 ♂♂, 1 ♀). Victorville, California; 19 May 1939; C. M. Dammers (1 ♀). San Diego, California; 24–30 May (1 ♂). Mason Valley, San Diego Co., California; 20 May 1939; Lloyd M. Martin (1 ♀). Paratypes to LACM; USNM.

*Smogops* is closely allied with *piperata*, *mosa* and *psentia* as indicated under *mosa*. The indistinct forewing pattern separates *smogops* from *mosa*. The sex scaling on the forewings and hindwings of the female is unique in the genus. The most noticeable variation is the number of dark-brown scales on the head, thorax and forewings.

### *Anoncia psentia* Hodges, NEW SPECIES
PL. 6, FIG. 19. TEXT FIG. 28 *d*.

*Anoncia psentia* Hodges.
Type-locality: 14 mi NE Mojave, Kern Co., California. [CAS]

Upper surface as figured. Head: base of tongue mottled pale gray and dark brown; maxillary palpus mainly very pale gray with some dark-tipped scales at apex; frons, vertex and occiput mainly pale gray to white with several dark-brown tipped scales, vertex and occiput paler than frons; labial palpus mainly white, lateral and antero-ventral surfaces of second segment dark with brown-tipped scales, third segment with some gray-tipped scales toward apex; antenna with scape mottled pale grayish white and brown dorsally, ventral surface yellow gray to white, scales of pecten unicolorous pale gray to white, shaft mottled pale gray and brown basally becoming yellowish gray and darker gray distally. Foreleg: mottled pale gray to white and brown; apex of coxa and all of trochanter pale; tibia white on epiphysis and at apex; apices of all tarsal segments pale. Midleg: coxa, trochanter and femur mainly pale yellowish white to white with some dark-tipped scales; tibia with more numerous dark-tipped scales, a white streak at $\frac{3}{5}$ length and apex white; tarsus with first two segments mainly pale with some dark-tipped scales, distal three segments with more numerous brown-tipped scales at bases. Hindleg: coxa, trochanter and femur mainly pale yellowish gray to white; tibia and tarsus with more numerous gray scales, last two tarsal segments with many brown scales, particularly on ventral surface. Wing length: 7.0–8.0 mm. Male: no specimens available. Female: as in text figure 28 *d*; eighth abdominal sternum with posterior margin nearly straight; ostium bursae on left side of eighth abdominal tergum; eighth abdominal tergum excavated on right side around ostium bursae to approximately $\frac{2}{5}$ width of segment; ductus bursae membranous throughout, a short lobe from near base; corpus bursae membranous with an accessory pouch from base; ductus seminalis arising from this pouch; paired signa present, each an irregular plate with numerous inwardly directed points; posterior margin of seventh abdominal tergum heavily sclerotized with one or two, short, heavily sclerotized lobes; apophyses anteriores arising from $\frac{1}{2}$ distance from middle to lateral margins; left apophysis directed toward lateral margin more than right apophysis.

The immature stages are unknown.

TYPES: Holotype: ♀. 14 mi NE Mojave, Kern Co., California; 13 April 1962; C. MacNeill, D. Rentz & R. Brown; USNM slide 4107. CAS. Paratypes: 2 ♀♀. Same data as for holotype. Paratypes to CAS; USNM.

*Psentia* is closely allied with *piperata*, *mosa* and *smogops* but can be separated from all three species by the straight posterior margin of the eighth abdominal sternum, the forewing coloration and pattern, the short lobe on the ductus bursae, and the position of the ostium bursae and degree of emargination of the eighth abdominal tergum. Specimens in the short type series are nearly uniform in coloration and pattern.

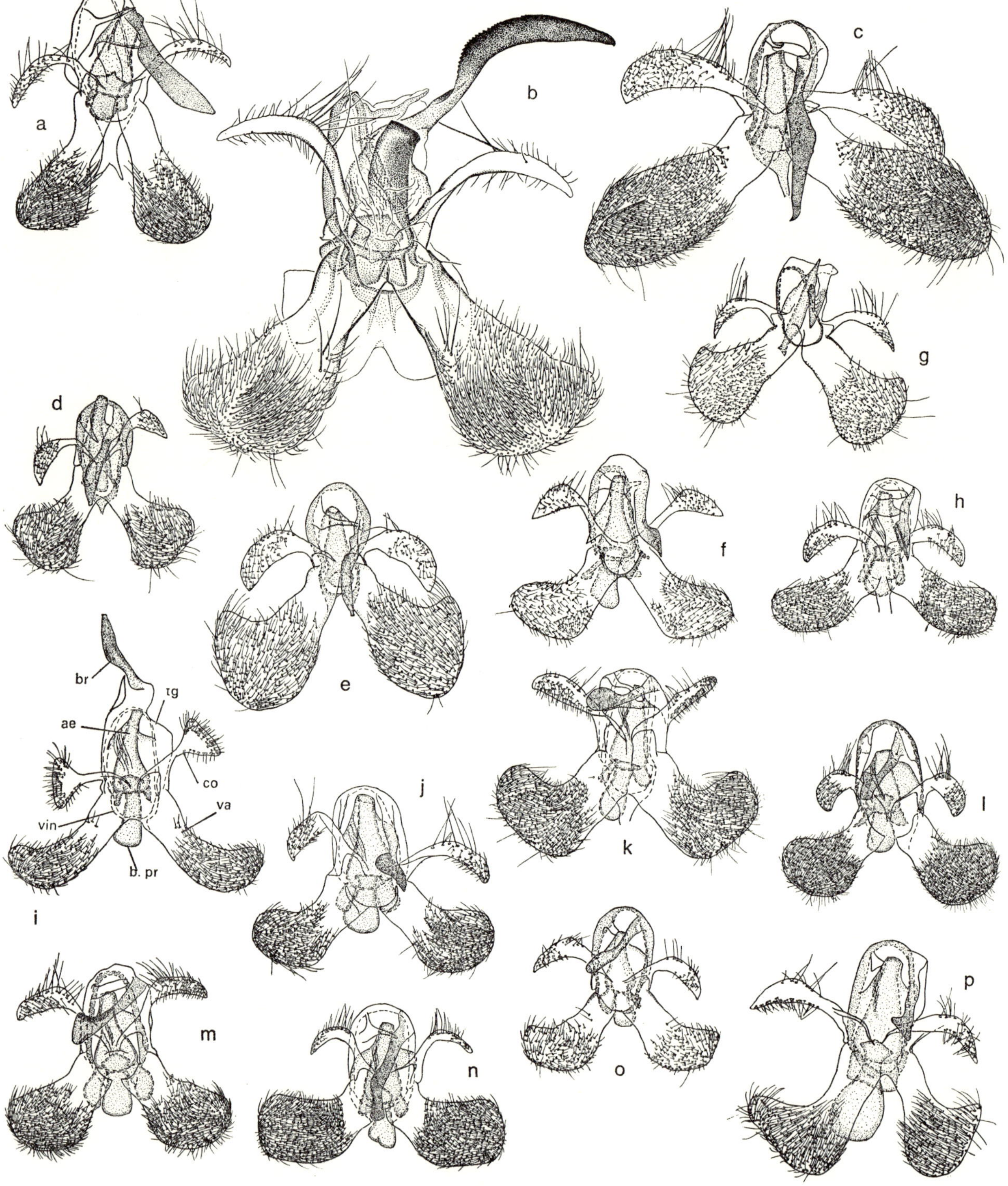


FIGURE 5: MALE GENITALIA OF *COSMOPTERIX* SPECIES

*a. C. nitens.*
*b. C. sinelinea.*
*c. C. molybdina.*
*d. C. pulchrimella.*
*e. C. bendidia.*
*f. C. montisella.*
*g. C. attenuatella.*
*h. C. magophila.*
*i. C. clandestinella.*
*j. C. gracilens.*
*k. C. lespedezae.*
*l. C. dicacula.*
*m. C. delicatella.*
*n. C. opulenta.*
*o. C. quadrilineella.*
*p. C. dapifera.*

Figures of varying scales.

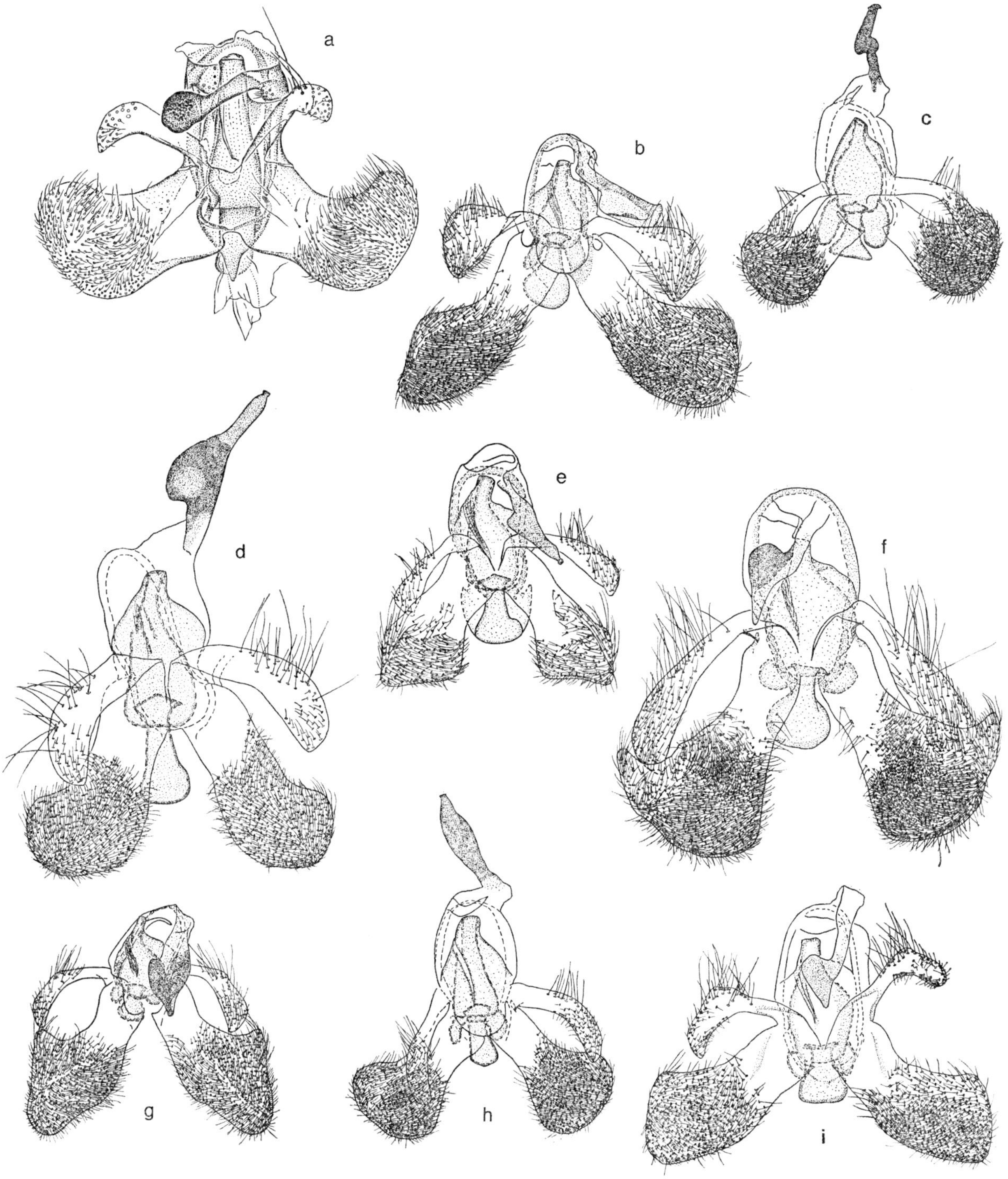


FIGURE 6: MALE GENITALIA OF *COSMOPTERIX* SPECIES

*a. C. chisosensis.* *d. C. gemmiferella.* *g. C. damnosa.*
*b. C. minutella.* *e. C. inopis.* *h. C. bacata.*
*c. C. abdita.* *f. C. scirpicola.* *i. C. clemensella.*

Figures of varying scales.

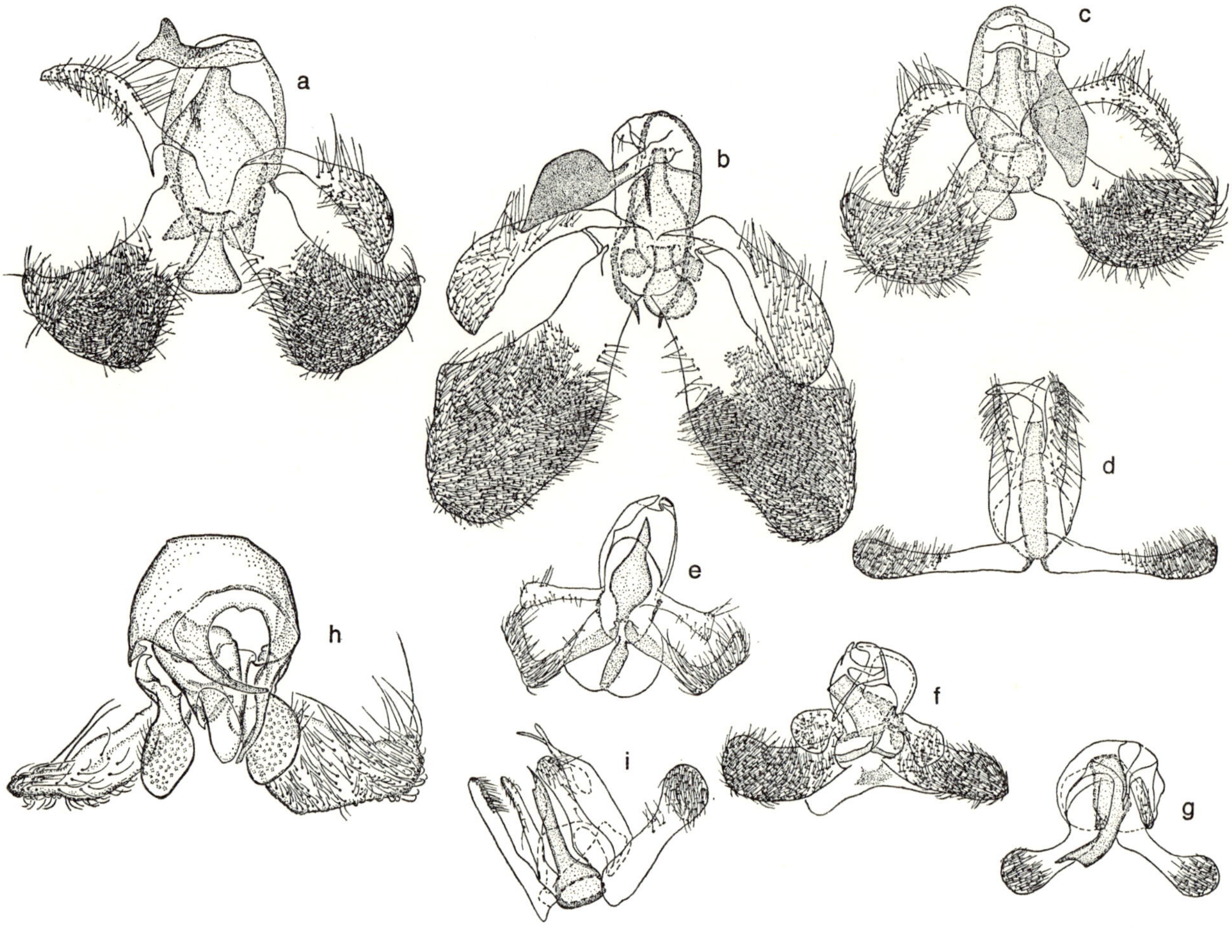


FIGURE 7: MALE GENITALIA OF COSMOPTERIGINAE SPECIES

*a. Cosmopterix ebriola.*
*b. Cosmopterix fernaldella.*
*c. Cosmopterix floridanella.*
*d. Pebops ipomoeae.*
*e. Tanygona lignicolorella.*
*f. Eralea albalineella.*
*g. Melanocinclis nigrilineella.*
*h. Eralea abludius.*
*i. Melanocinclis lineigera.*

Figures of varying scales.

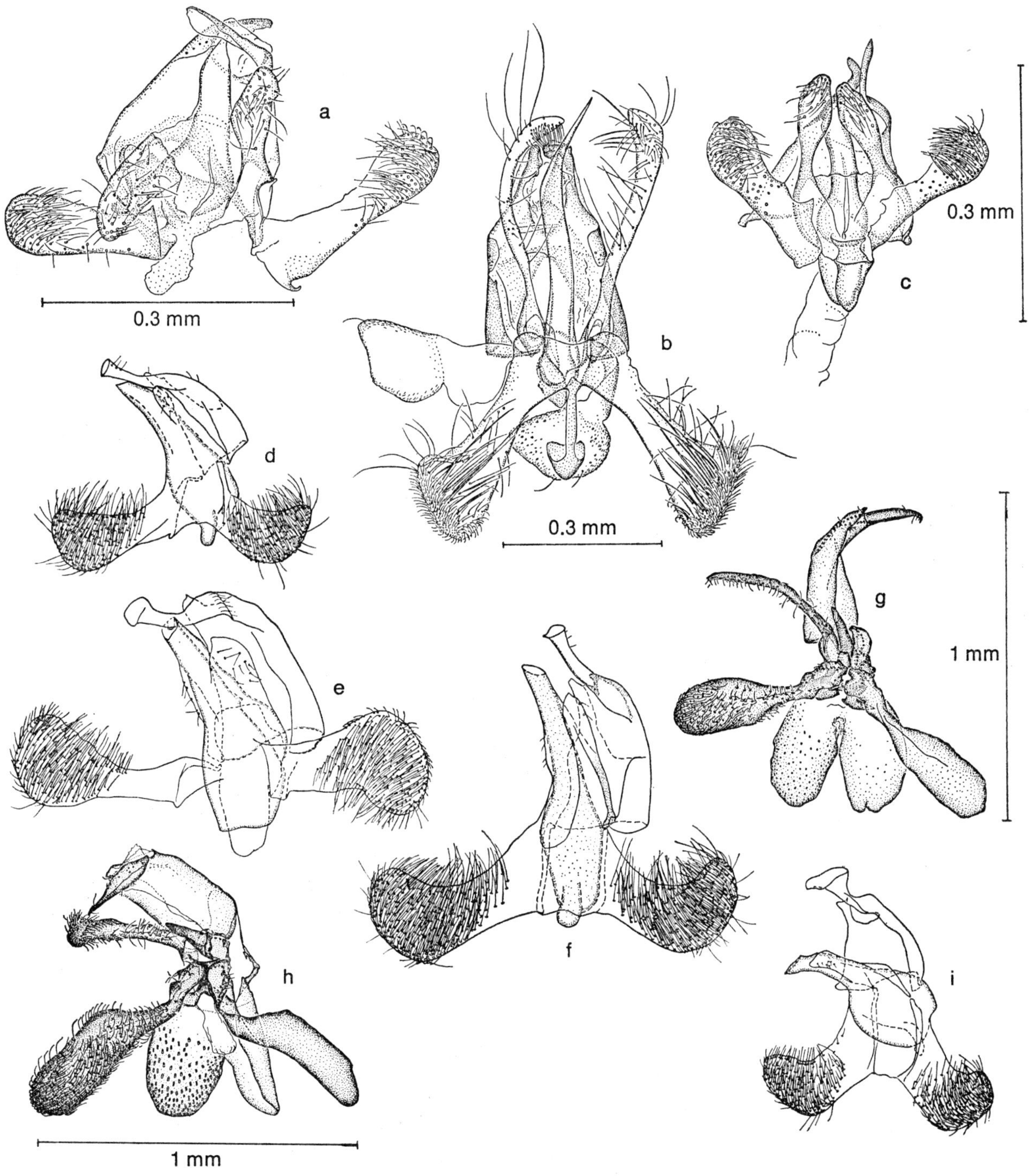


FIGURE 8: MALE GENITALIA OF COSMOPTERIGINAE SPECIES

*a. Melanocinclis sparsa.*
*b. Melanocinclis vibex.*
*c. Melanocinclis gnoma.*
*d. Stagmatophora sexnotella.*
*e. Stagmatophora wyattella.*
*f. Stagmatophora enchrysa.*
*g. Pyroderces badia.*
*h. Pyroderces rileyi.*
*i. Stagmatophora iridella.*

Not to scale: *d, e, f, i.*

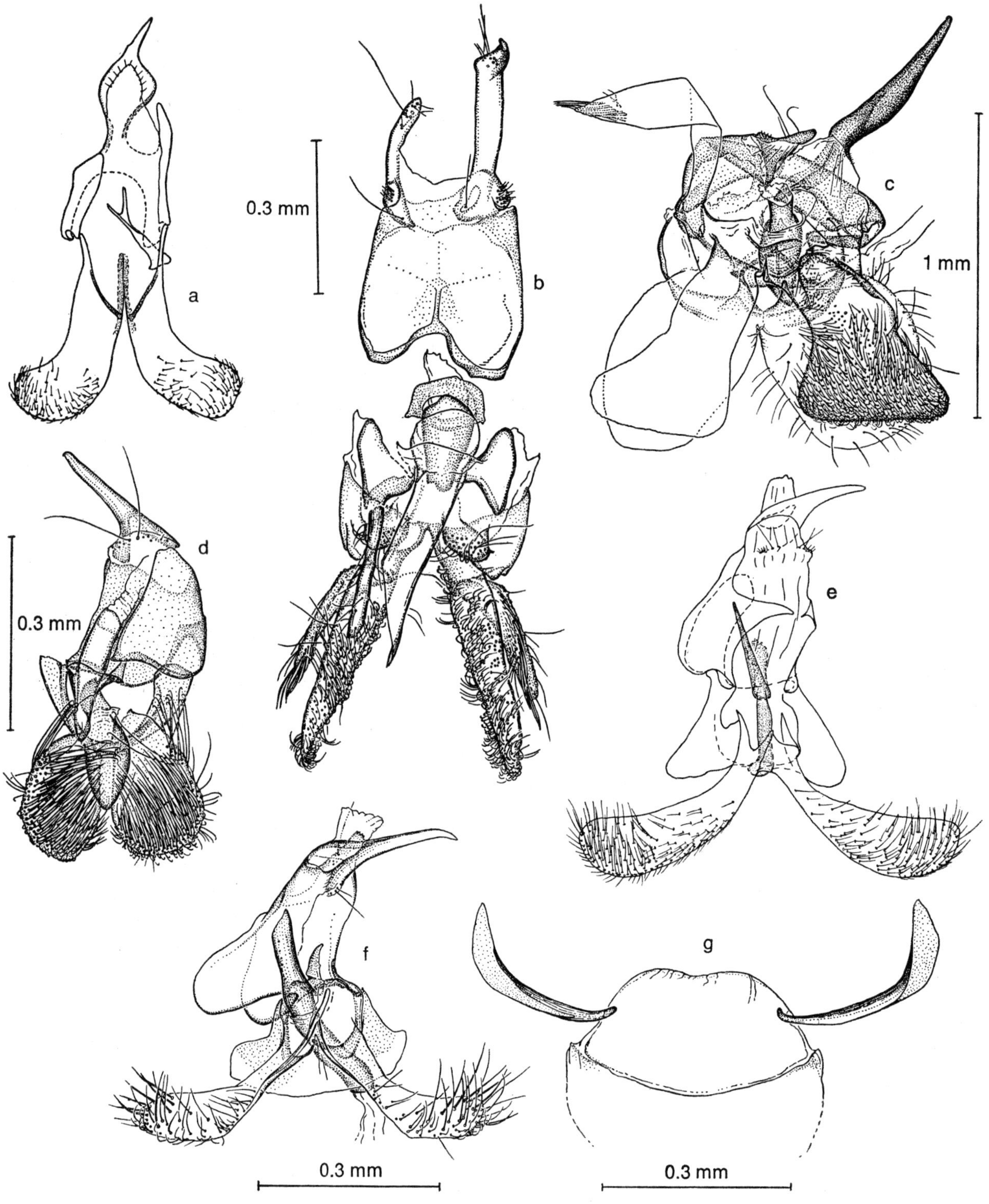


FIGURE 9: MALE GENITAL FEATURES OF COSMOPTERIGINAE SPECIES

*a. Teladoma helianthi.* *b. Pyroderces albistrigella.* *c. Limnaecia phragmitella.* *d. Teladoma nebula.* *e. Teladoma astigmatica.* *f. Teladoma tonia.*
*g.* Seventh and eighth abdominal sterna, *Teladoma tonia.*

Not to scale: *a, e.*

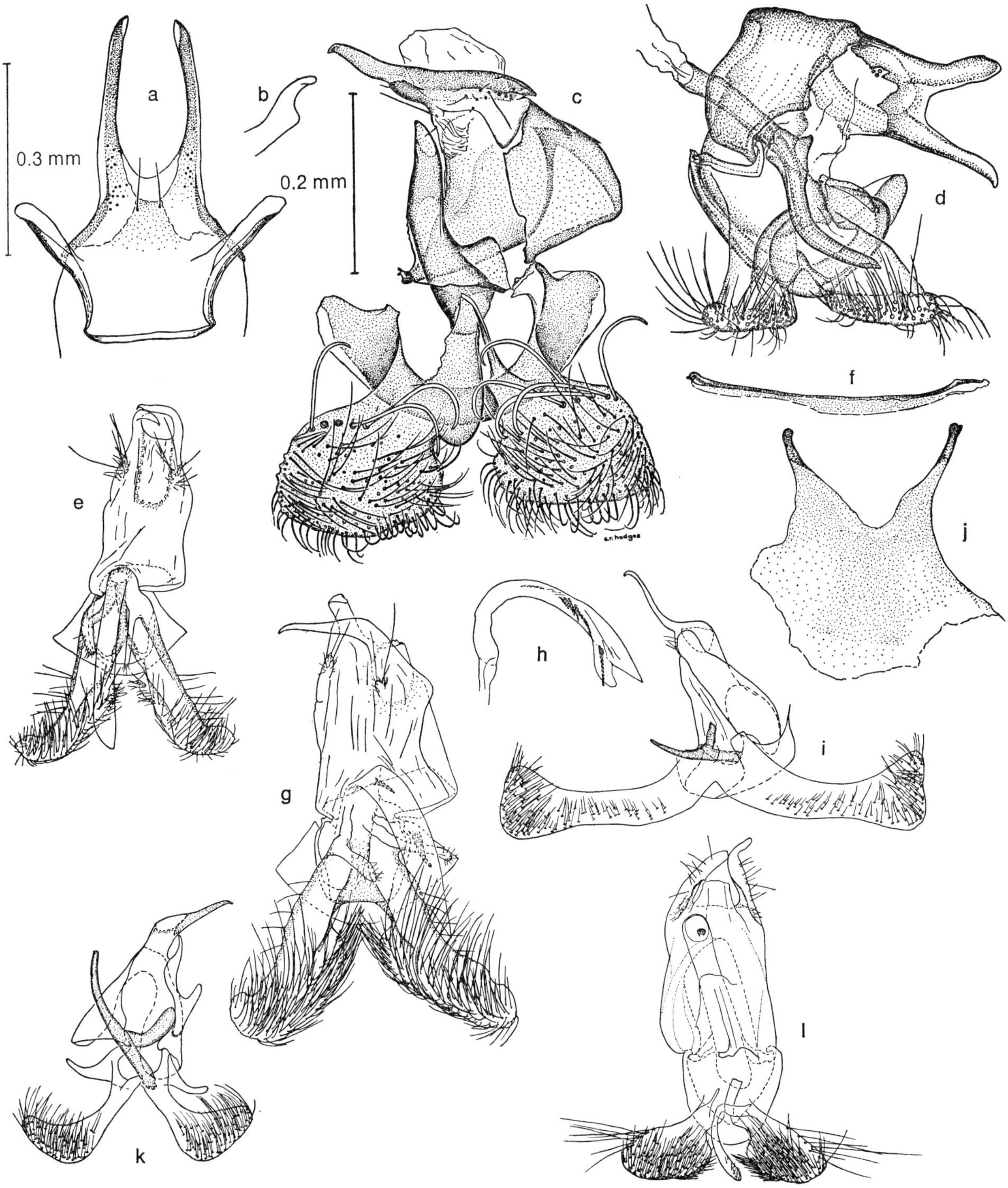


FIGURE 10: MALE GENITAL FEATURES OF *TELADOMA* AND *TRICLONELLA* SPECIES

*a.* Seventh and eighth abdominal terga, *Tel. tonia.*
*b.* Apex of uncus, *Tel. habra.*
*c.* Genitalia, *Tel. habra.*
*d.* Genitalia, *Tel. exigua.*
*e.* Genitalia (aedoeagus omitted), *Tr. xuthocelis.*
*f.* Eighth abdominal sternum, *Tel. exigua.*
*g.* Genitalia, *Tr. pergandeella.*
*h.* Aedoeagus, *Tr. xuthocelis.*
*i.* Genitalia, *Tel. incana.*
*j.* Eighth abdominal tergum, *Tel. exigua.*
*k.* Genitalia, *Tel. murina.*
*l.* Genitalia, *Tr. antidectis.*

Different scales for *a, d, f, j*; and *c*. Not to scale: *b, e, g, h, i, k, l.*

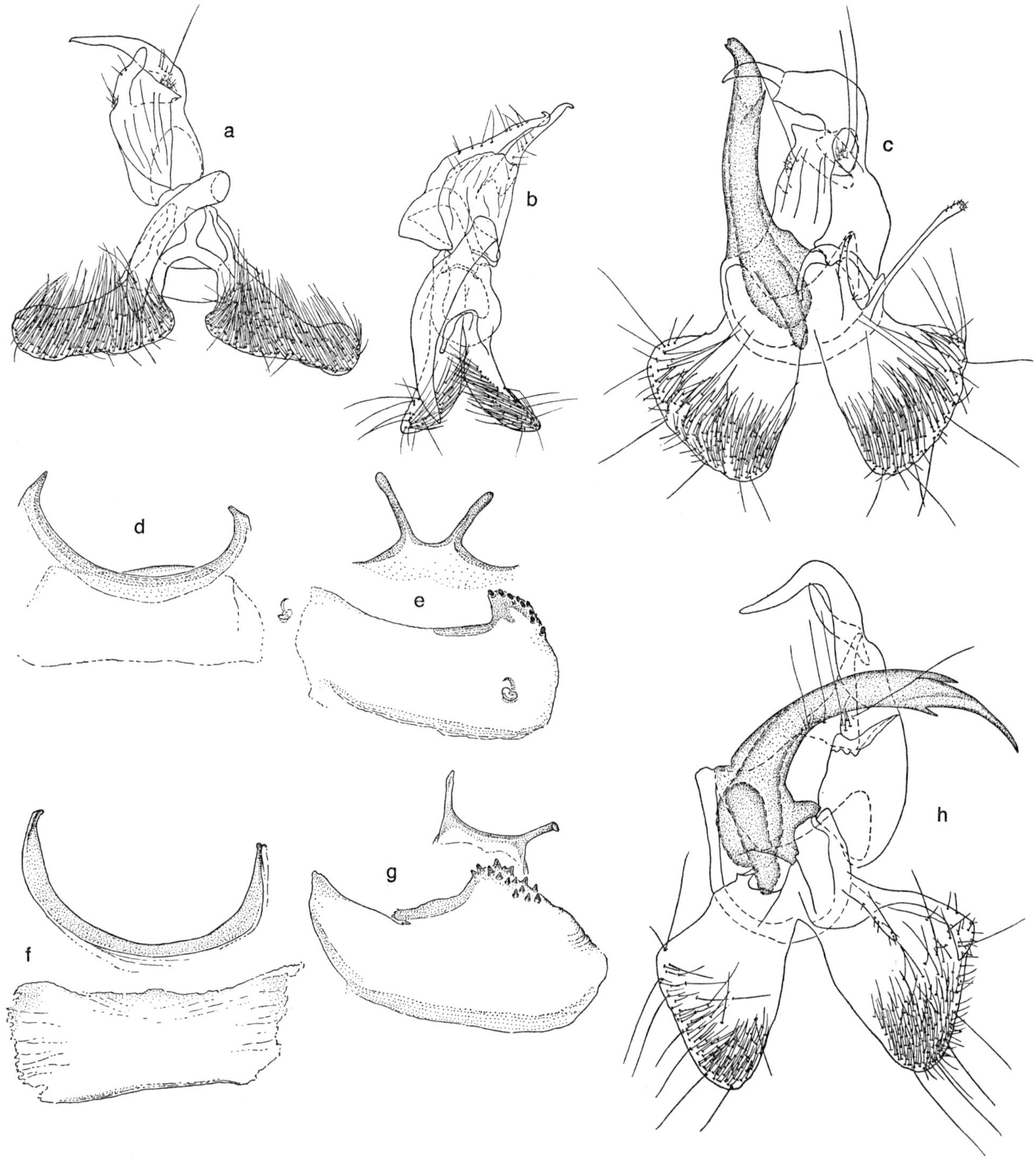


FIGURE 11: MALE GENITAL FEATURES OF *TRICLONELLA* AND *ANONCIA* SPECIES

*a*. Genitalia, *T. determinatella*.
*b*. Genitalia, *T. bicoloripennis*.
*c*. Genitalia, *A. conia*.
*d*. Seventh and eighth abdominal terga, *A. conia*.
*e*. Seventh and eighth abdominal sterna, *A. conia*.
*f*. Seventh and eighth abdominal terga, *A. brunneipes*.
*g*. Seventh and eighth abdominal sterna, *A. brunneipes*.
*h*. Genitalia, *A. brunneipes*.

Figures of varying scales.

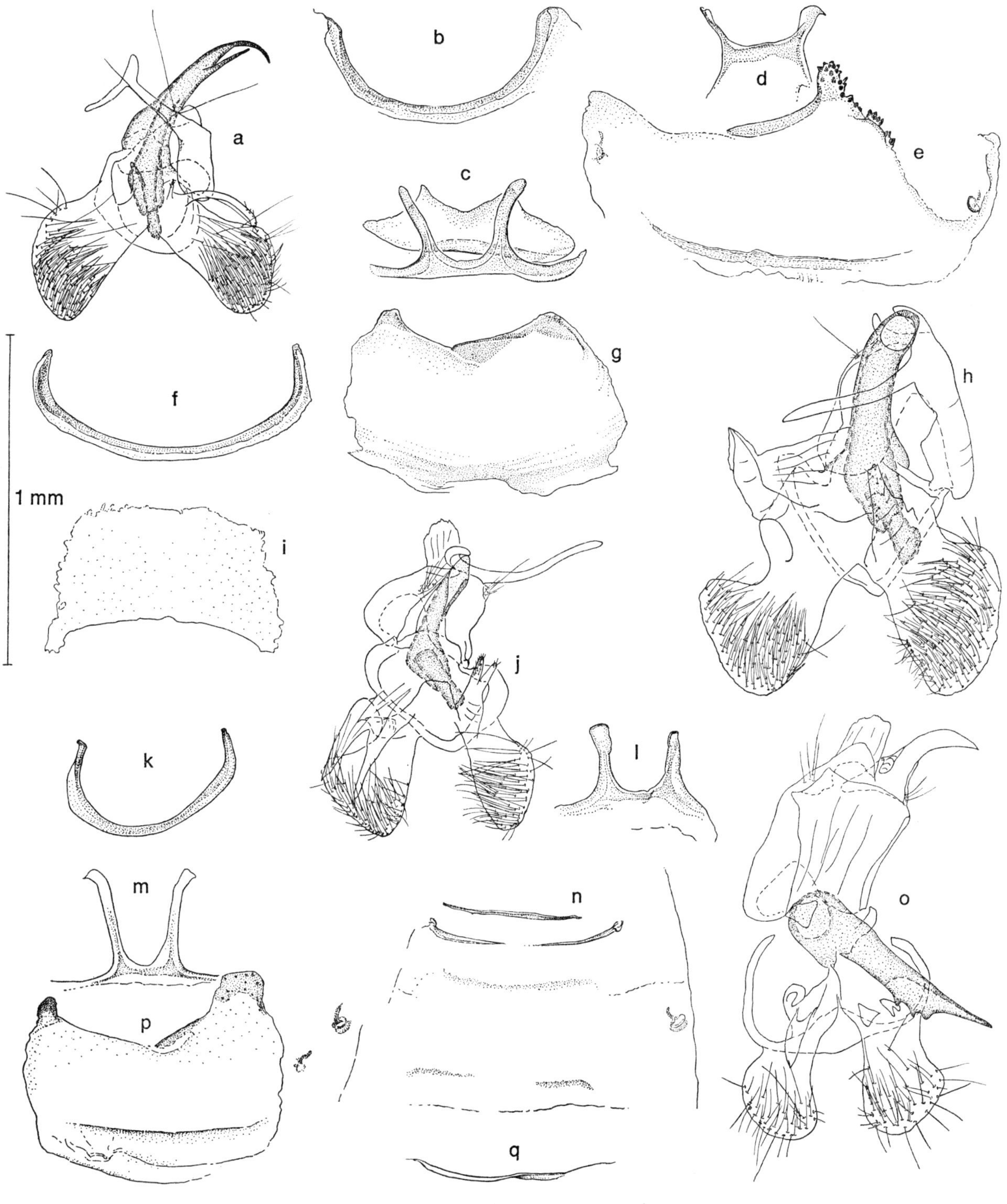


FIGURE 12: MALE GENITAL FEATURES OF *ANONCIA* SPECIES

*a*. Genitalia, *A. sphacelina*.
*b*. Eighth abdominal tergum, *A. sphacelina*.
*c*. Eighth abdominal sternum, *A. fasciata*.
*d*. Eighth abdominal sternum. *A. sphacelina*.
*e*. Seventh abdominal sternum, *A. sphacelina*.
*f*. Eighth abdominal tergum, *A. fasciata*.
*g*. Seventh abdominal sternum, *A. fasciata*.
*h*. Genitalia, *A. alboligula*.
*i*. Seventh abdominal tergum, *A. fasciata*.
*j*. Genitalia, *A. fasciata*.
*k*. Eighth abdominal tergum, *A. alboligula*.
*l*. Eighth abdominal sternum, *A. fasciata*.
*m*. Eighth abdominal sternum, *A. alboligula*.
*n*. Seventh and eighth abdominal sterna, *A. glacialis*.
*o*. Genitalia, *A. glacialis*.
*p*. Seventh abdominal sternum, *A. alboligula*.
*q*. Eighth abdominal tergum, *A. glacialis*.

Same scale for *c*, *f*, *g*, *i*, *k*, *l*, *m*, *n*, *p* and *q*. Not to scale: *a*, *b*, *d*, *e*, *h*, *j*, *o*.

FIGURE 13: MALE GENITAL FEATURES OF *ANONCIA* SPECIES

*a.* Genitalia, *A. bitoqua.*
*b.* Eighth abdominal tergum, *A. bitoqua.*
*c.* Eighth abdominal sternum, *A. bitoqua.*
*d.* Seventh abdominal sternum, *A. bitoqua.*
*e.* Genitalia, *A. porriginosa.*
*f.* Eighth abdominal tergum, *A. flegax.*
*g.* Genitalia, *A. flegax.*
*h.* Seventh and eighth abdominal sterna, *A. porriginosa.*
*i.* Seventh and eighth abdominal terga, *A. porriginosa.*
*j.* Seventh and eighth abdominal terga, *A. aciculata.*
*k.* Genitalia, *A. aciculata.*
*l.* Eighth abdominal sternum, *A. flegax.*
*m.* Seventh abdominal sternum, *A. flegax.*

All same scale except *j* (as shown). Not to scale: *e.*

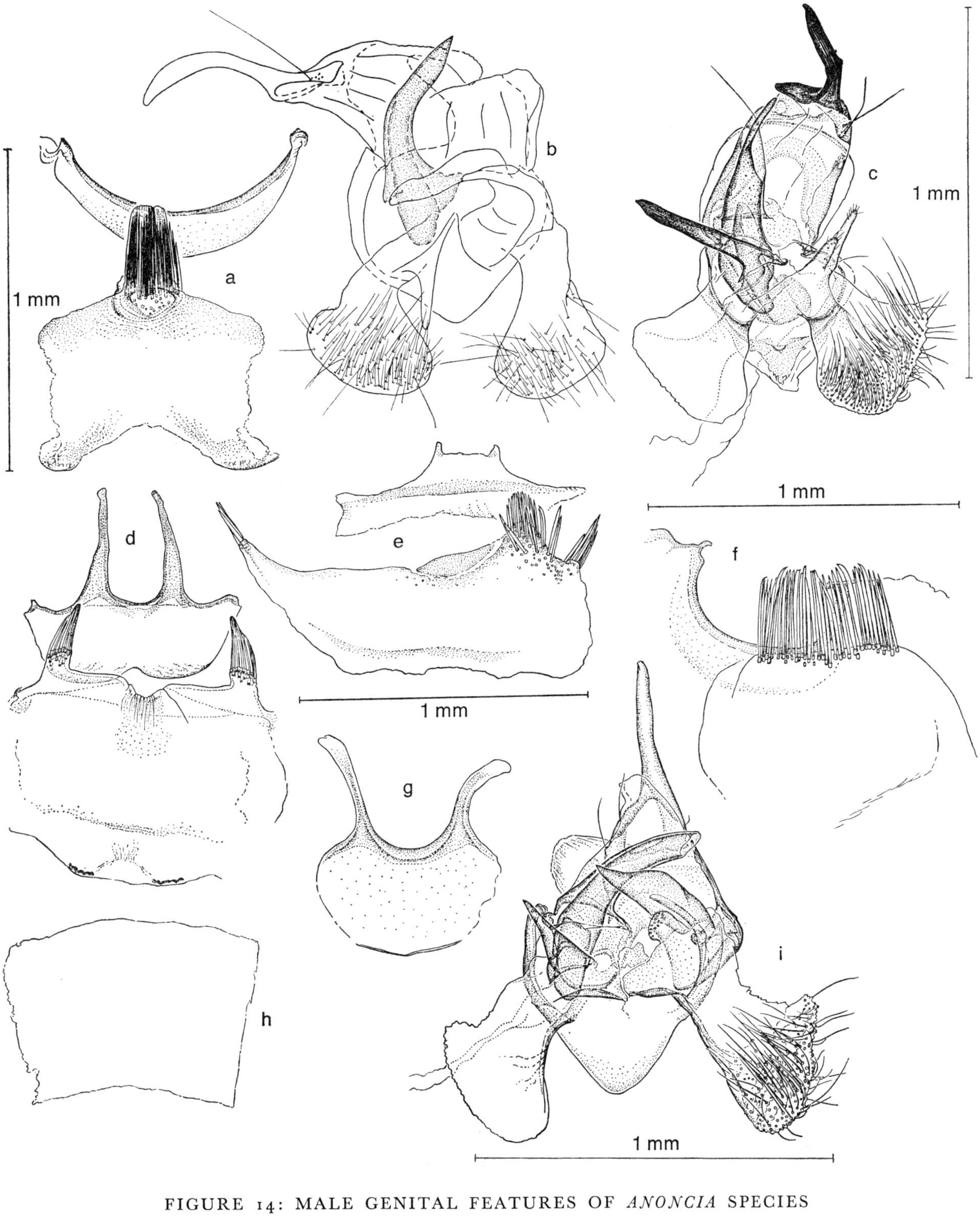


FIGURE 14: MALE GENITAL FEATURES OF *ANONCIA* SPECIES

*a.* Seventh and eighth abdominal terga, *A. mones*.
*b.* Genitalia, *A. callida*.
*c.* Genitalia, *A. mones*.
*d.* Seventh and eighth abdominal sterna, *A. mones*.
*e.* Seventh and eighth abdominal sterna, *A. aciculata*.
*f.* Seventh and eighth abdominal terga, *A. naclia*.
*g.* Eighth abdominal sternum, *A. naclia*.
*h.* Seventh abdominal tergum, *A. callida*.
*i.* Genitalia, *A. naclia*.

Different scales for *a, d, f, g, h*; *c, i*; and *e*. Not to scale: *b*.

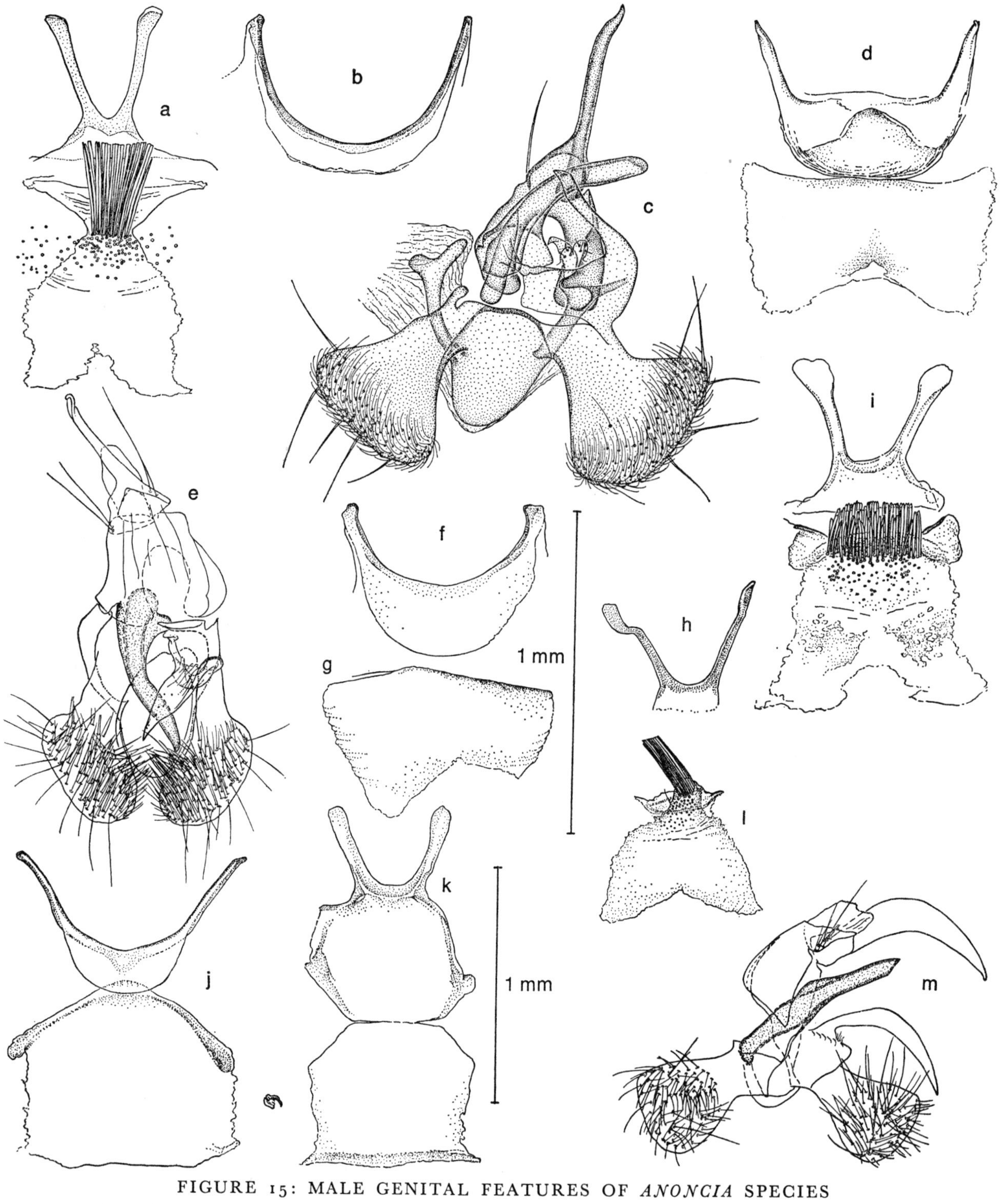


FIGURE 15: MALE GENITAL FEATURES OF *ANONCIA* SPECIES

*a.* Seventh and eighth abdominal sterna, *A. callida.*
*b.* Eighth abdominal tergum, *A. callida.*
*c.* Genitalia, *A. nocticola.*
*d.* Seventh and eighth abdominal terga, *A. nocticola.*
*e.* Genitalia, *A. nebritis.*
*f.* Eighth abdominal tergum, *A. nebritis.*
*g.* Seventh abdominal tergum, *A. nebritis.*
*h.* Eighth abdominal sternum, *A. nebritis.*
*i.* Seventh and eighth abdominal sterna, *A. nocticola.*
*j.* Seventh and eighth abdominal terga, *A. leucoritis.*
*k.* Seventh and eighth abdominal sterna, *A. leucoritis.*
*l.* Seventh abdominal sternum, *A. nebritis.*
*m.* Genitalia, *A. leucoritis.*

All same scale except *j* and *k* (as shown). Not to scale: *c*, *e*, *m*.

FIGURE 16: MALE GENITAL FEATURES OF *ANONCIA* SPECIES

*a*. Genitalia, *A. longa*.
*b*. Seventh and eighth abdominal terga, *A. longa*.
*c*. Seventh and eighth abdominal sterna, *A. longa*.
*d*. Genitalia, *A. furvicosta*.
*e*. Seventh and eighth abdominal sterna, *A. longa*.
*f*. Seventh and eighth abdominal terga, *A. furvicosta*.
*g*. Seventh and eighth abdominal terga, *A. longa*.
*h*. Seventh and eighth abdominal sterna, *A. furvicosta*.

All same scale except *a* and *d* (not to scale).

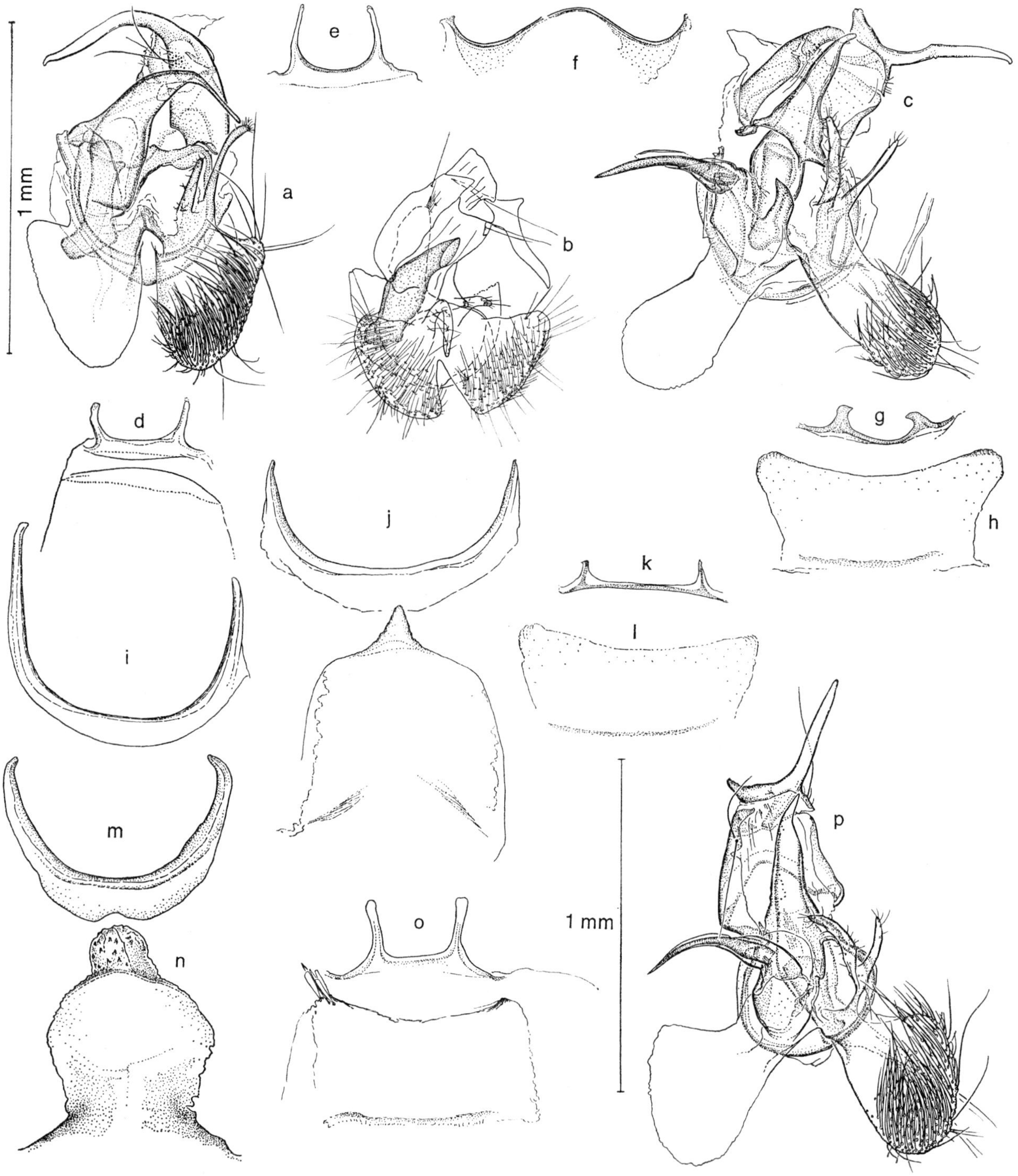


FIGURE 17: MALE GENITAL FEATURES OF *ANONCIA* SPECIES

*a.* Genitalia, *A. orites.*
*b.* Genitalia, *A. diveni.*
*c.* Genitalia, *A. loexya.*
*d.* Seventh and eighth abdominal sterna, *A. orites.*
*e.* Eighth abdominal sternum, *A. diveni.*
*f.* Eighth abdominal tergum, *A. diveni.*
*g.* Eighth abdominal sternum, *A. loexya.*
*h.* Seventh abdominal sternum, *A. loexya.*
*i.* Eighth abdominal tergum, *A. orites.*
*j.* Seventh and eighth abdominal terga, *A. noscres.*
*k.* Eighth abdominal sternum, *A. loexya.*
*l.* Seventh abdominal sternum, *A. loexya.*
*m.* Eighth abdominal tergum, *A. loexya.*
*n.* Seventh abdominal tergum, *A. loexya.*
*o.* Seventh and eighth abdominal sterna, *A. noscres.*
*p.* Genitalia, *A. noscres.*

All same scale except *b* (not to scale).

FIGURE 18: MALE GENITAL FEATURES OF *ANONCIA* SPECIES

*a.* Genitalia, *A. episcia.*
*b.* Eighth abdominal tergum, *A. episcia.*
*c.* Eighth abdominal sternum, *A. episcia.*
*d.* Genitalia, *A. smogops.*
*e.* Genitalia, *A. venis.*
*f.* Genitalia, *A. piperata.*
*g.* Seventh and eighth abdominal terga, *A. venis.*
*h.* Eighth abdominal sternum, *A. smogops.*
*i.* Eighth abdominal tergum, *A. smogops.*
*j.* Eighth abdominal sternum, *A. mosa.*
*k.* Eighth abdominal sternum, *A. piperata.*
*l.* Eighth abdominal sternum, *A. venis.*
*m.* Genitalia, *A. mosa.*
*n.* Eighth abdominal tergum, *A. piperata.*
*o.* Seventh abdominal sternum, *A. venis.*
*p.* Eighth abdominal tergum, *A. mosa.*

All same scale except *a* and *f* (not to scale).

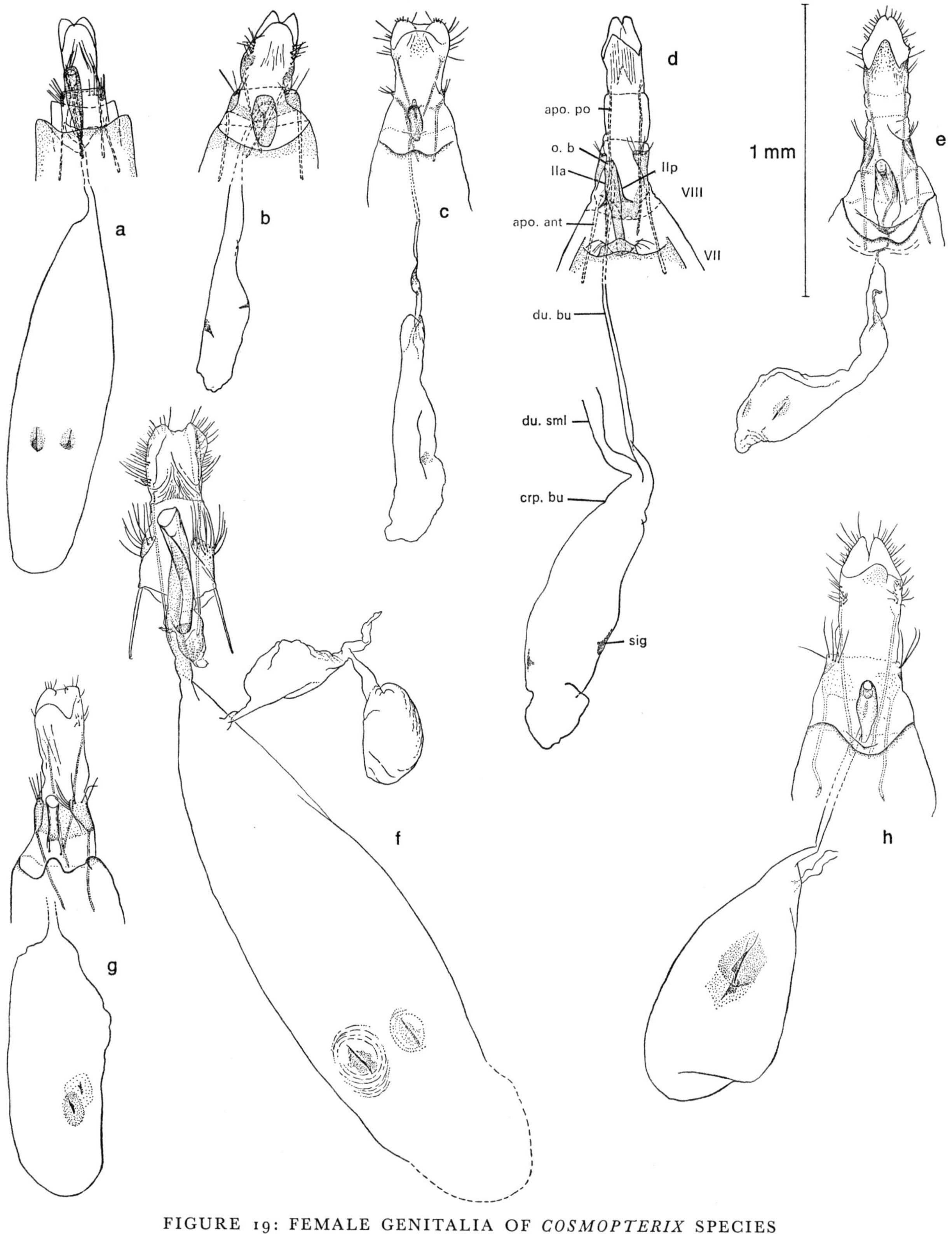


FIGURE 19: FEMALE GENITALIA OF *COSMOPTERIX* SPECIES

*a. C. nitens.* *b. C. pulchrimella.* *c. C. attenuatella.* *d. C. molybdina.* *e. C. bendidia.* *f. C. sinelinea.* *g. C. clandestinella.* *h. C. montisella.*

Same scale for *e*, *f*, *g* and *h*. Not to scale: *a*, *b*, *c*, *d*.

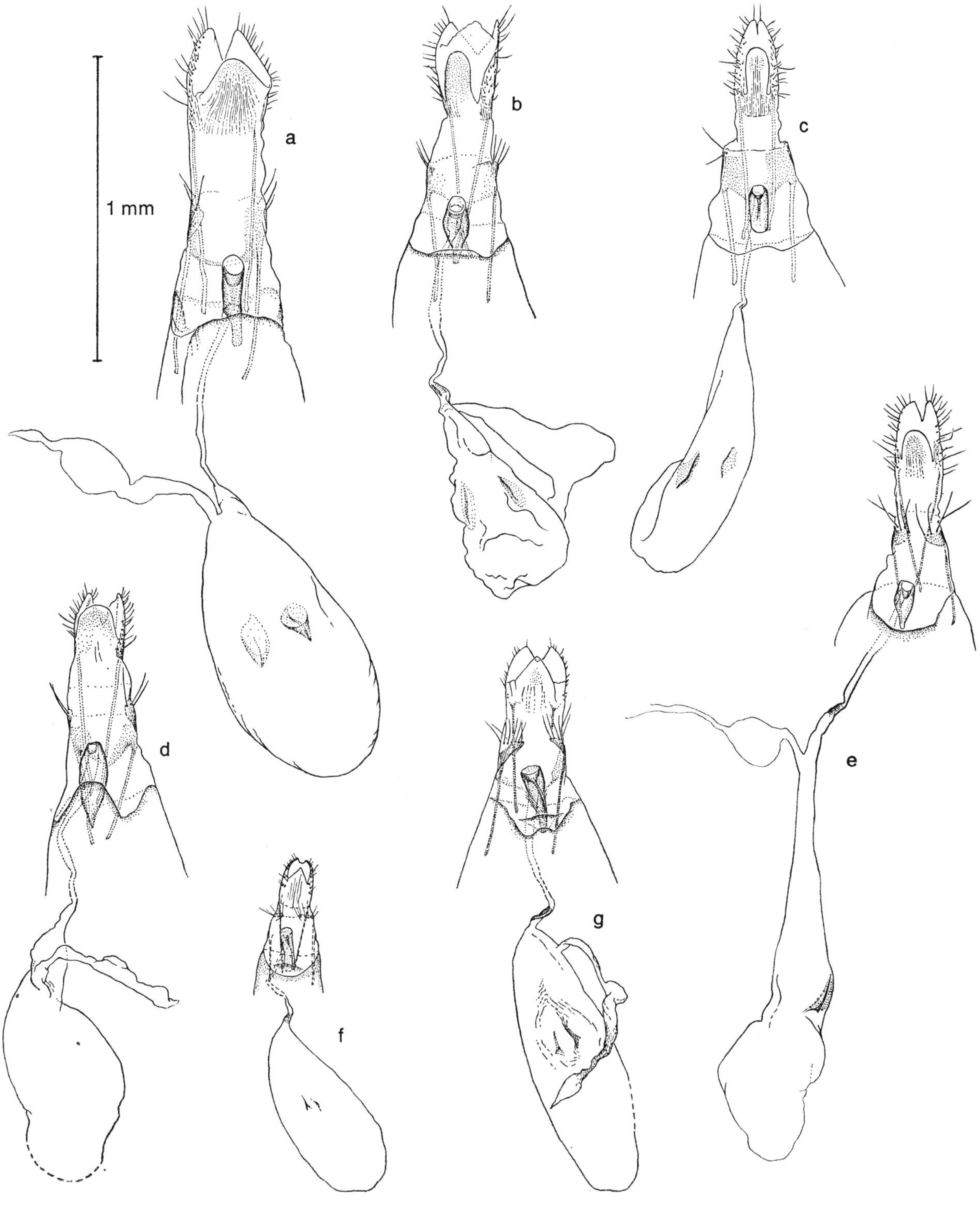


FIGURE 20: FEMALE GENITALIA OF *COSMOPTERIX* SPECIES

*a. C. dapifera.* *b. C. dicacula.* *c. C. quadrilineella.* *d. C. lespedezae.* *e. C. opulenta.* *f. C. inopis.* *g. C. abdita.*

All same scale except *f* (not to scale).

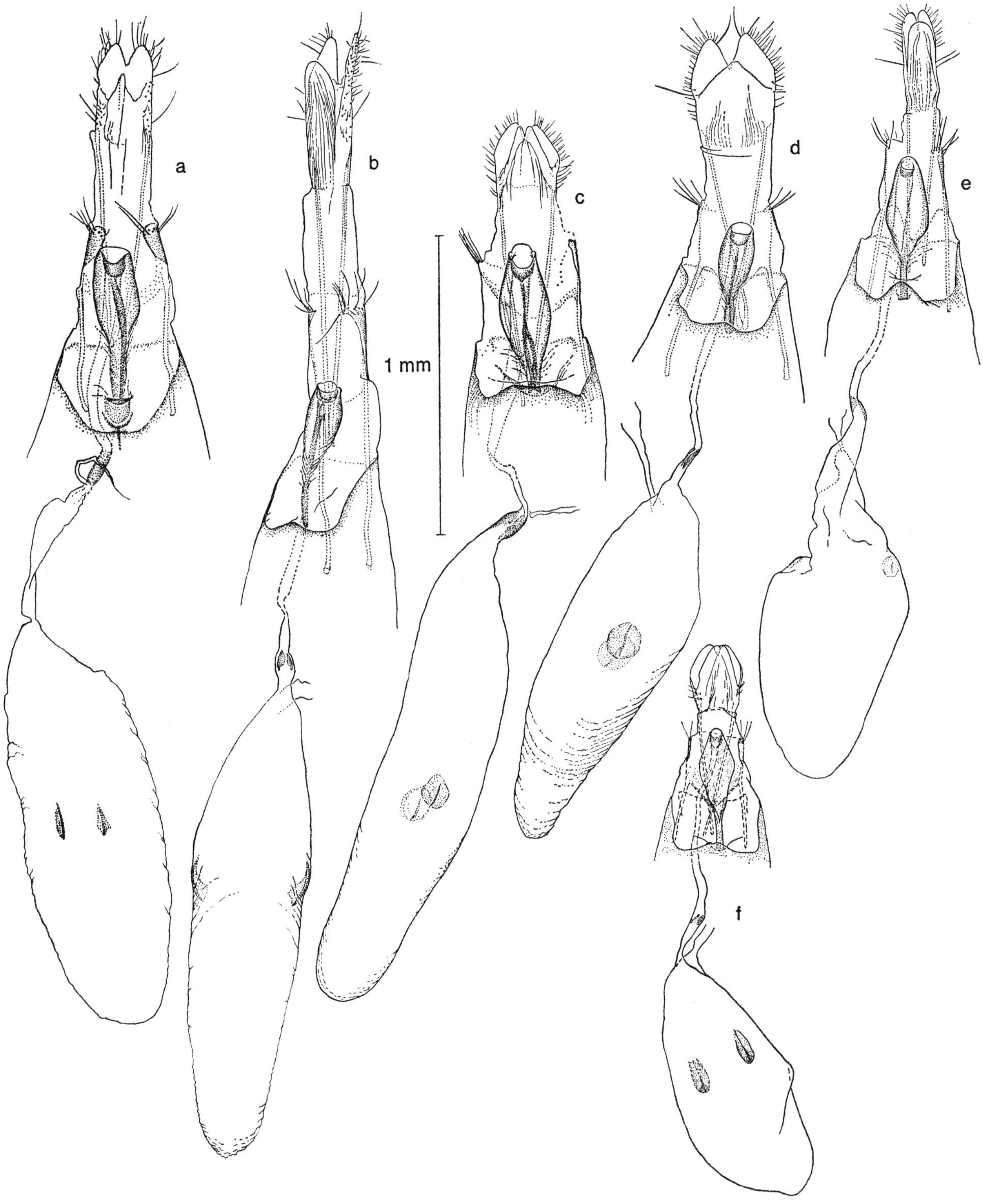


FIGURE 21: FEMALE GENITALIA OF *COSMOPTERIX* SPECIES

*a. C. gemmiferella.*
*b. C. scirpicola.*
*c. C. clemensella.*
*d. C. ebriola.*
*e. C. fernaldella.*
*f. C. fernaldella.*

All same scale except *f* (not to scale).

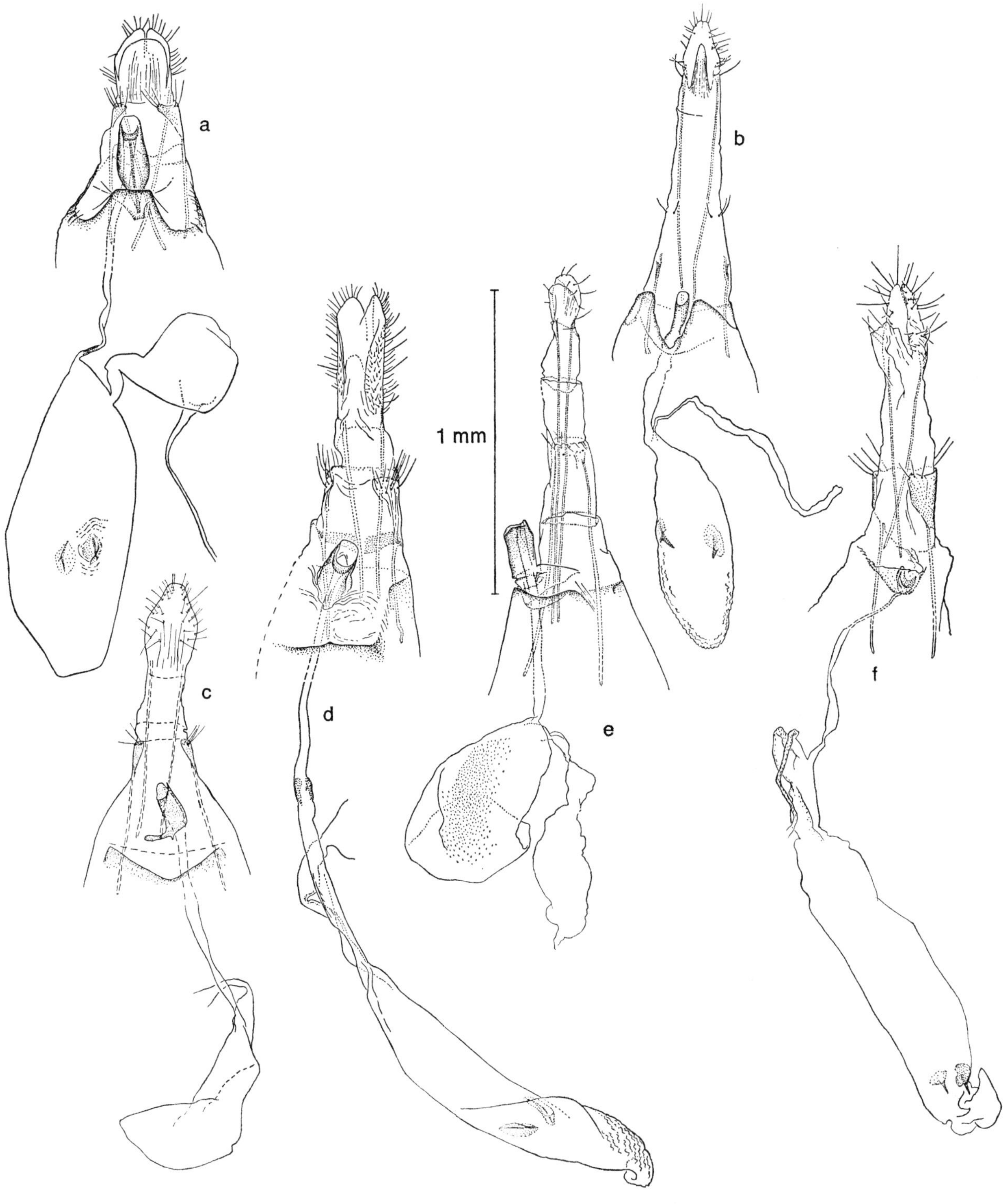


FIGURE 22: FEMALE GENITALIA OF COSMOPTERIGINAE SPECIES

*a. Cosmopterix floridanella.*
*b. Pebops ipomoeae.*
*c. Melanocinclis lineigera.*
*d. Cosmopterix facunda.*
*e. Tanygona lignicolorella.*
*f. Eralea albalineella.*

All same scale except *c* (not to scale).

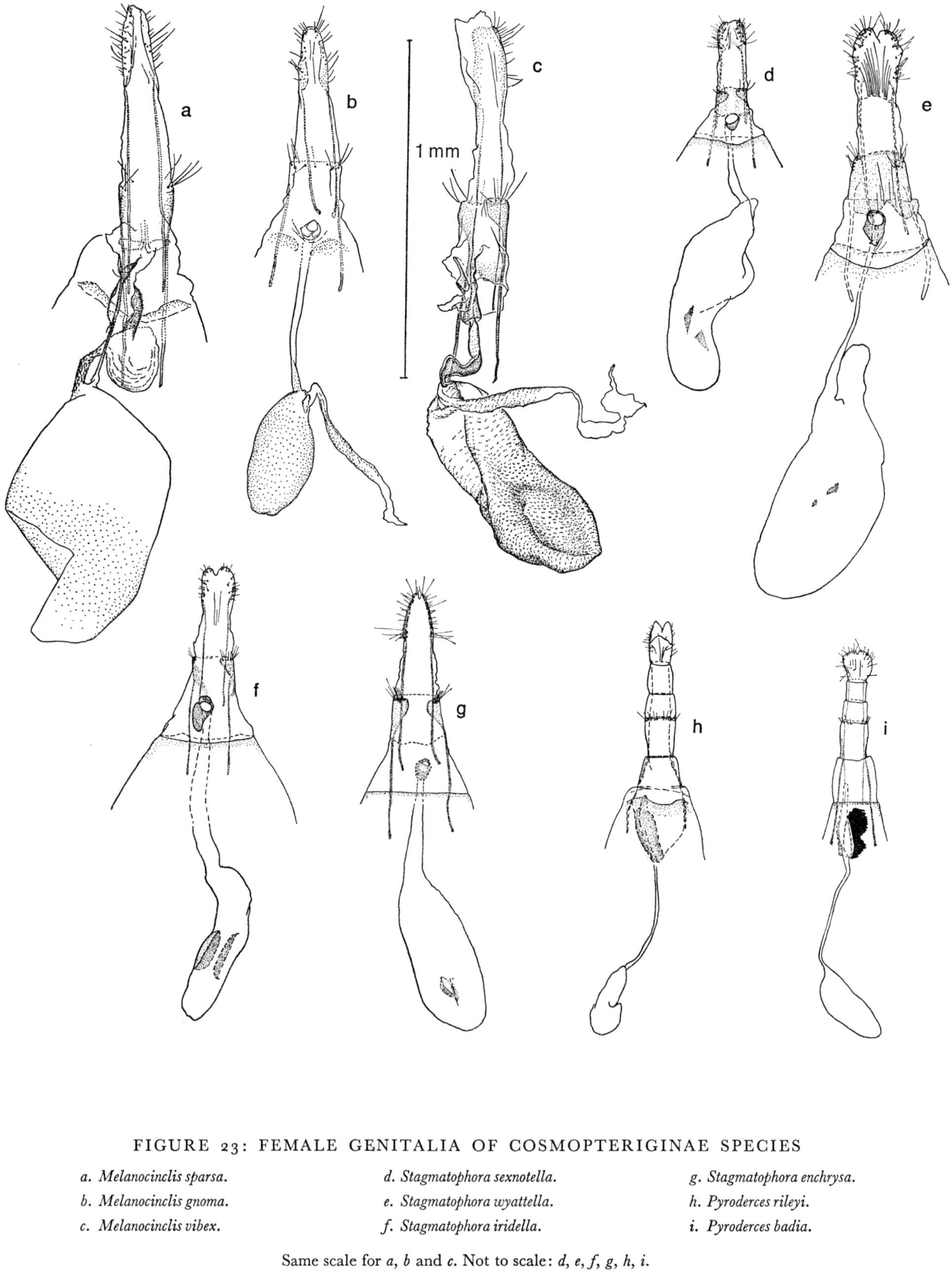


FIGURE 23: FEMALE GENITALIA OF COSMOPTERIGINAE SPECIES

*a. Melanocinclis sparsa.*
*b. Melanocinclis gnoma.*
*c. Melanocinclis vibex.*
*d. Stagmatophora sexnotella.*
*e. Stagmatophora wyattella.*
*f. Stagmatophora iridella.*
*g. Stagmatophora enchrysa.*
*h. Pyroderces rileyi.*
*i. Pyroderces badia.*

Same scale for *a*, *b* and *c*. Not to scale: *d*, *e*, *f*, *g*, *h*, *i*.

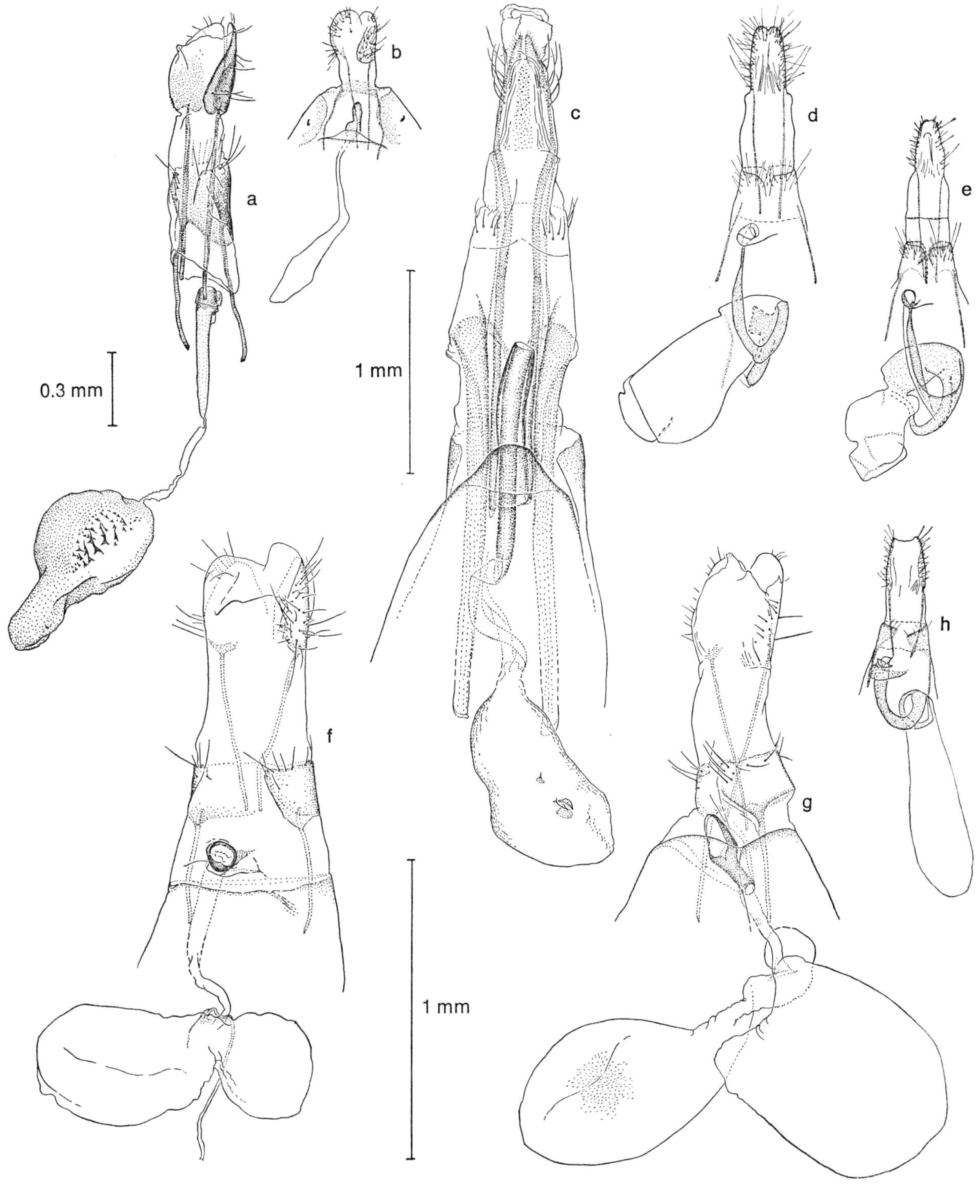


FIGURE 24: FEMALE GENITALIA OF COSMOPTERIGINAE SPECIES

*a. Pyroderces albistrigella.*
*b. Teladoma helianthi.*
*c. Limnaecia phragmitella.*
*d. Triclonella xuthocelis.*
*e. Triclonella pergandeella.*
*f. Triclonella determinatella.*
*g. Triclonella bicoloripennis.*
*h. Triclonella antidectis.*

Different scales for *a*; *c*; and *f*, *g*. Not to scale: *b*, *d*, *e*, *h*.

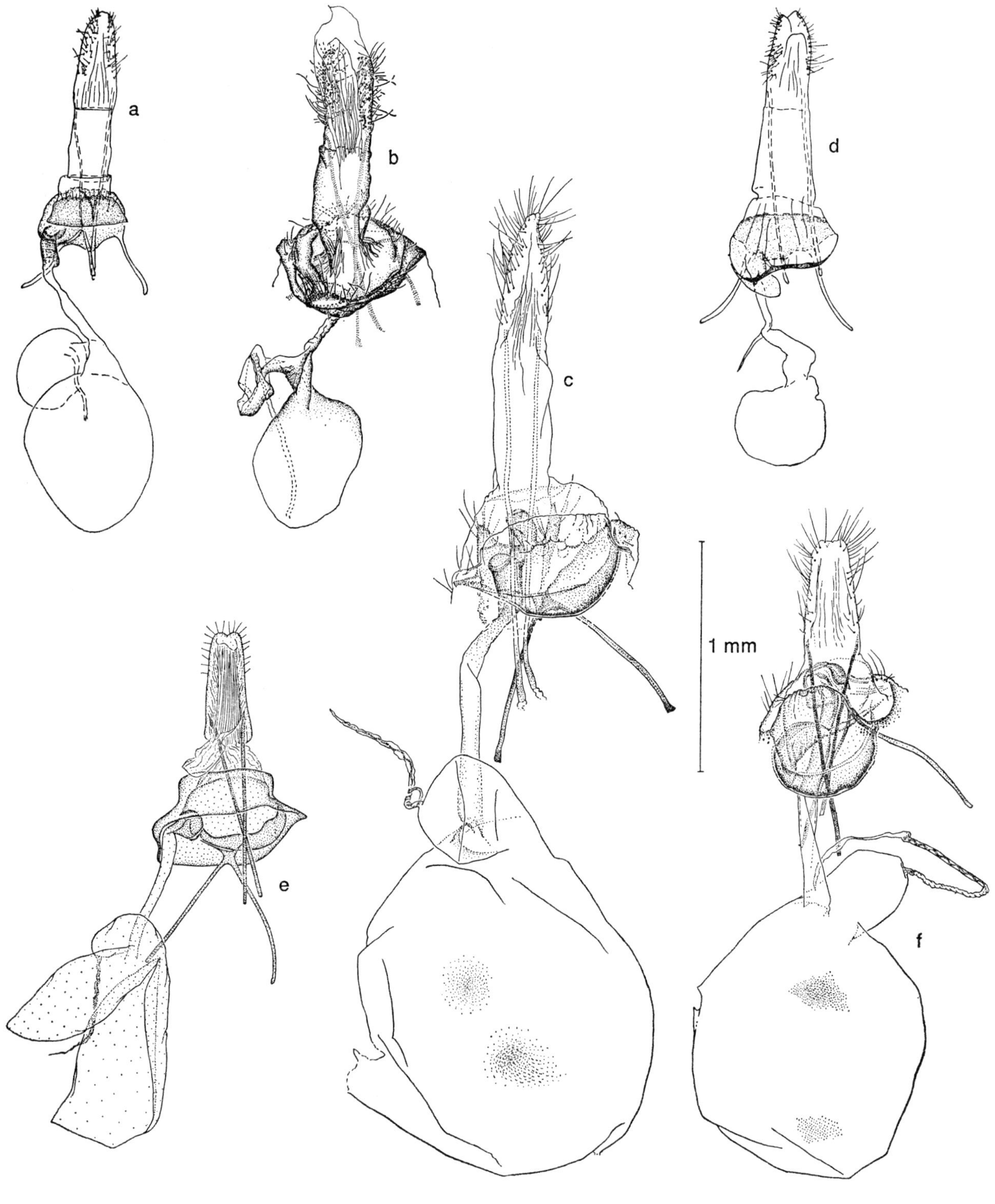


FIGURE 25: FEMALE GENITALIA OF *ANONCIA* SPECIES

*a. A. conia.* *b. A. brunneipes.* *c. A. fasciata.* *d. A. sphacelina.* *e. A. alboligula.* *f. A. slales.*

Same scale for *c* and *f*. Not to scale: *a, b, d, e.*

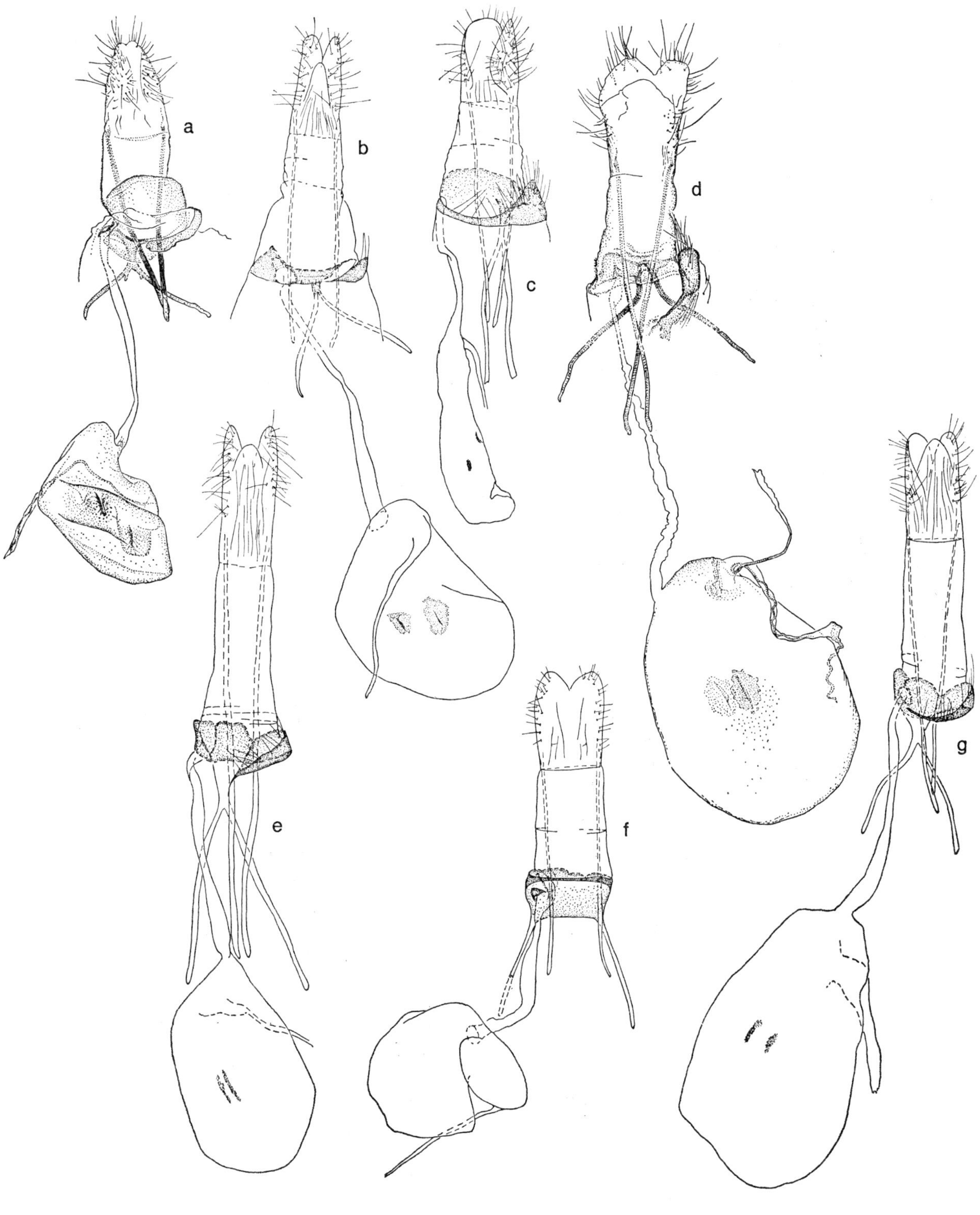


FIGURE 26: FEMALE GENITALIA OF *ANONCIA* SPECIES

*a. A. nocticola.* *b. A. nebritis.* *c. A. leucoritis.* *d. A. psepsa.* *e. A. longa.* *f. A. diveni.* *g. A. furvicosta.*

Figures of varying scales.

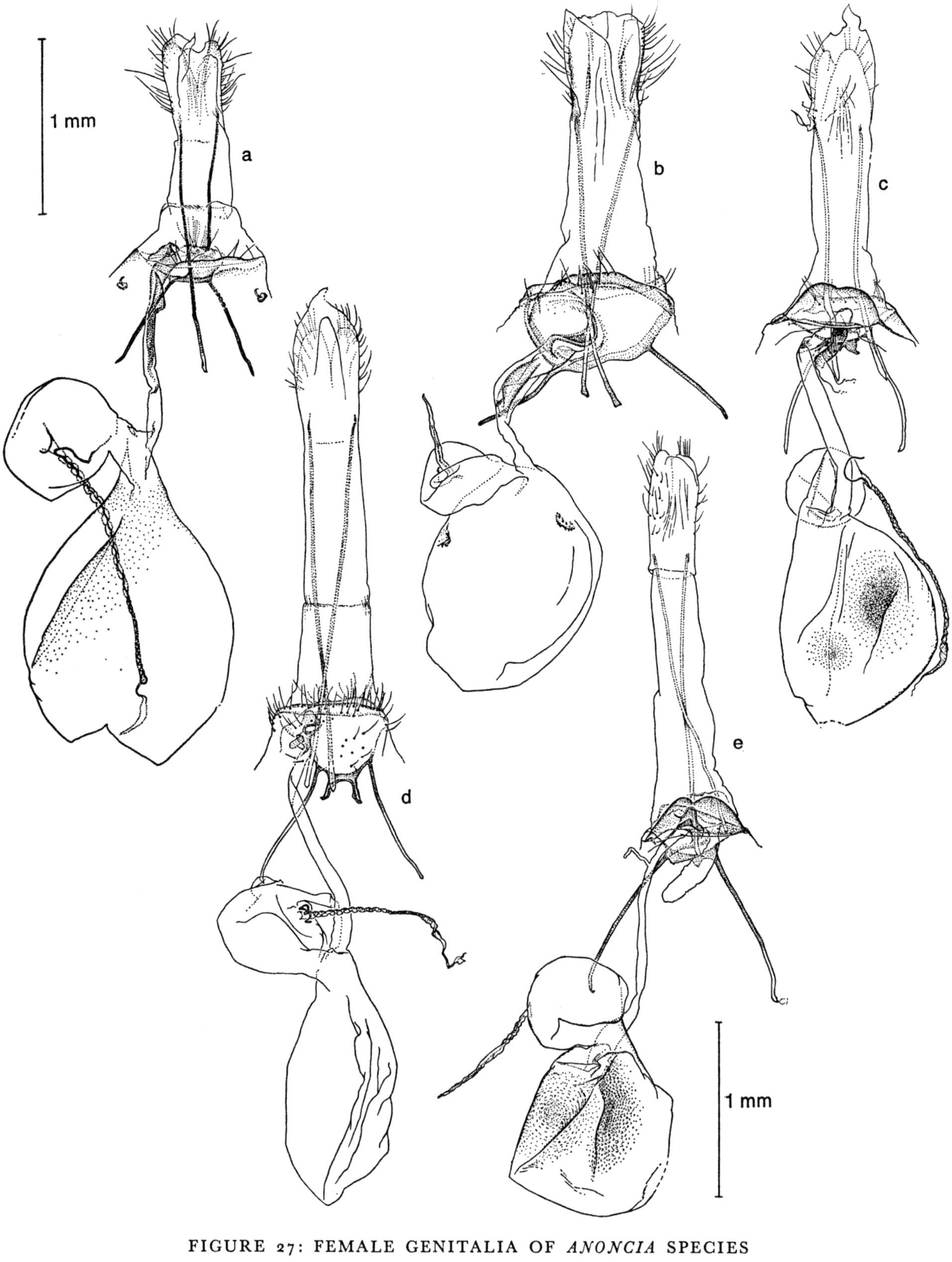


FIGURE 27: FEMALE GENITALIA OF *ANONCIA* SPECIES

*a. A. orites.* *b. A. fregeis.* *c. A. noscres.* *d. A. episcia.* *e. A. loexya.*

All same scale.

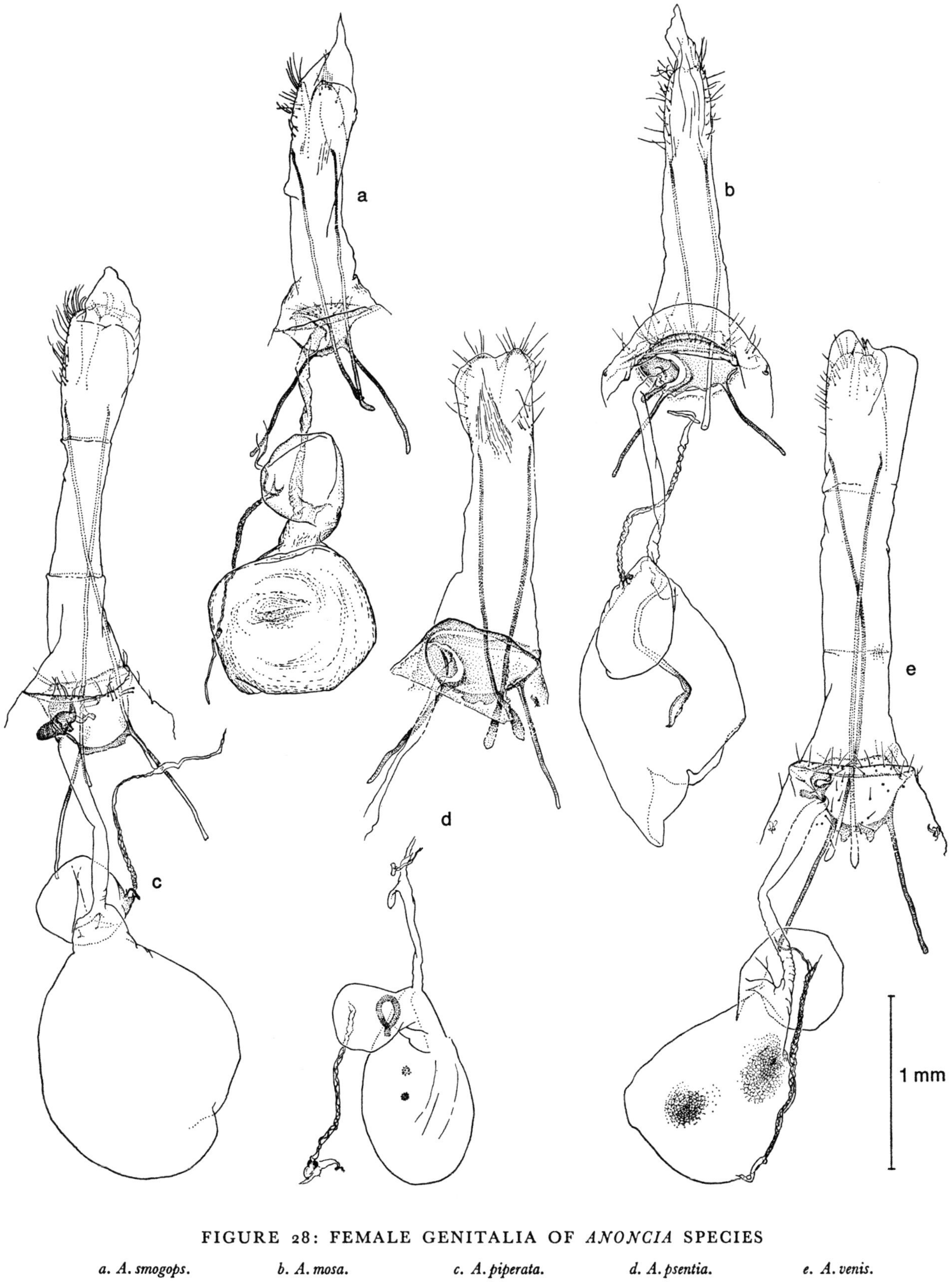


FIGURE 28: FEMALE GENITALIA OF *ANONCIA* SPECIES

*a. A. smogops.* *b. A. mosa.* *c. A. piperata.* *d. A. psentia.* *e. A. venis.*

All same scale.

SUBFAMILY

## **Chrysopeleiinae** Mosher

Type-genus: *Chrysopeleia* Chambers, 1874.

Chrysopeleiidae Mosher, 1916, *Bull. Illinois State Laboratory Nat. Hist.*, **12**: 104.

Walshiidae Hodges, 1962, *Jour. Lep. Soc.*, **15**: 82.

Type-genus: *Walshia* Clemens, 1860.

NOTE—An incorrect interpretation of the Code of Zoological Nomenclature led me to propose Walshiidae as the valid name for the Chrysopeleiinae. Even though the species used to define Chrysopeleiinae, *Stilbosis ostryaeella*, is not a member of the genus *Chrysopeleia*, the family group name based on *Chrysopeleia* is valid.

Published information does not permit a reasonable estimate of numerical size or geographic distribution of the Chrysopeleiinae. *Cholotis* Meyrick occurs in Australia; six genera and 30 species occur in the Palearctic Region from the Canaries east to Pakistan, Afghanistan and the Seychelles; 14 species are in the Ethiopian Region; and no species are known from Oceania. Small size and inconspicuousness have contributed to this subfamily being poorly collected and studied. Before 1961, 18 species were recognized in North America north of Mexico. Subsequently, 46 species have been described, and 22 new species are added in this Fascicle. Judging from numerous undetermined specimens I have seen, the Chrysopeleiinae are well developed in the Neotropical Region.

Chrysopeleiine adults are narrow winged. The forewing usually is narrowly lanceolate with the apex acute, and the hindwing is narrowly lanceolate to nearly linear with the apex acute. Although the major characters for defining the subfamily are in the male genitalia, most species can be placed to subfamily by appearance. Study of the colored plates reveals that many species are black or very dark gray and smooth scaled, medium to dark gray with patches of raised scales or shades of brown and black with patches of raised scales. Other families with slender-winged species such as the Blastobasidae, Agonoxenidae, Coleophoridae, Momphidae, Oecophoridae and Scythridae have the forewing with vein 1A present; vein 1A is absent in the Chrysopeleiinae. The male genitalia have the aedoeagus ankylosed and often massive relative to the rest of the genitalia; the vinculum and tegumen usually are short; the gnathos is absent; and the uncus is present or absent.

Characters of the Chrysopeleiinae are as follows: head smooth scaled; tongue scaled basally; labial palpus recurved, usually smooth scaled and somewhat stout; maxillary palpus short, folded at base of tongue; antenna $\frac{1}{2}$ to $\frac{3}{4}$ length of forewing, pecten of one or two setae or absent; ocelli present or absent. Forewing lanceolate, apex narrowly rounded to acute, raised scales present or absent, cell open or closed, 11 or 12 veins present, $Cu_2$ present or absent, $M_1$ separate or stalked with $R_{4+5}$, $M_2$ usually separate, sometimes stalked with $R_{4+5}$ and $M_1$, 1A absent, 2A and 3A forked basally. Hindwing narrowly lanceolate to nearly linear, apex acute, cell usually open, six to eight veins present, $M_1$ stalked with $R_s$ or separate, $M_2$ and $M_3$ usually present, 2A often present. Abdomen without stout dorsal setae; male often with eighth segment having sclerotized bands or plates. Male genitalia symmetrical or asymmetrical; valva single lobed or compound with two or three lobes, sometimes with fracture zones, sometimes much reduced; tegumen and vinculum usually reduced, short; aedoeagus ankylosed with saccus, often massive; juxta absent; transtilla usually absent; subscaphium rarely present; gnathos absent; socii absent; uncus present or absent. Female: ostium bursae variously placed on seventh abdominal sternum, often with heavily sclerotized plates preceding and succeeding it, lamellae vaginales often with finely netted surface, modified scales often associated with folds around ostium bursae; signa usually present, paired, occasionally single; apophyses usually slender, sometimes massive, apophyses anteriores usually joined by sclerotized band.

Larva: characterization of the larva is based on *Stilbosis ostryaeella*, *S. tesquella*, *Walshia amorphella* and *Sorhagenia nimbosa*. Characters are drawn from an unpublished chart made by Heinrich and Capps, formerly of the U.S. Department of Agriculture, and terminology is that proposed by Hinton (1946). Head smooth, flattened or not; submental pit lacking; ocelli 6; adfrontal ridge meeting longitudinal ridge of frons at $\frac{3}{4}$ or more distance to vertex; seta P2 posterior to P1; L1 below level of P1; A3 approximately equidistant between A2 and L1 (closer to A2 in *Stilbosis tesquella*); A1, A2, A3 and L1 not in a line. Body with smooth skin (granulose in *W. amorphella*). Thorax without secondary setae; prothorax with D2 lower than XD1 (approximately level with XD1 in *S. tesquella*); SV1 and SV2 present; coxae not touching; meso- and metathorax with SV1 present; metathorax with coxae less than twice width apart (twice width apart in *S. ostryaeella*). Abdominal segments one to six with D2 remote from D1 (approximate to D1 in *S. ostryaeella*), L1 approximate to L2, each on separate pinaculum; eighth abdominal segment with SD1 dorsoanterad of spiracle, spiracle normally positioned, L3 ventrocaudad of L1 and L2, D1 and D2 not much nearer each other than on seventh segment and not on a pinaculum; ninth abdominal segment with D1 and D2 closer to each other than on eighth abdominal segment and not on a single pinaculum, D1 closer to SD1 than to D2 in *S. tesquella* and *Sorhagenia nimbosa*, D1 approximately equidistant between D2 and SD1 in *Stilbosis ostryaeella* and *W. amorphella*, D1, D2 and SD1 approximately in a line, SD1 not closer to L1 and L2 than to D1 and D2, L3 absent; subventral group with two setae on segment one, two or three setae on segment two, three setae on segments three to six, two setae on segment seven, and one seta on segments eight and nine; anal fork absent; anal legs without secondary setae. Crochets uniserial and uniordinal, in a complete circle (weaker outwardly in *S. tesquella*) and continuous on anal legs.

Pupa (based on *Walshia*, *Stilbosis*, *Periploca*, *Sorhagenia* and *Ithome*): frontoclypeal suture complete or incomplete; mandible and/or eye piece defined or not; appendages slightly elevated and firmly appressed to body, extending to posterior part of fifth abdominal segment in *Walshia*, to posterior part of seventh abdominal segment in *Ithome*, *Stilbosis* and *Sorhagenia* and nearly to the anal opening in *Periploca*; cremaster absent; abdominal spiracles cylindrical and slightly elevated; abdominal segments apparently immobile, males of *Sorhagenia* and *Ithome* movable between segments five and six and six and seven.

Extreme similarity in appearance among many species of *Periploca*, *Siskiwitia*, *Stilbosis*, *Walshia*, *Perimede*, *Sorhagenia* and *Ithome* makes examination of genital characters necessary to identify specimens.

## KEY TO GENERA OF CHRYSOPELEIINAE

1. Forewing with patches of raised scales . . . . . . . . . 2
— Forewing without patches of raised scales . . . . . . 7

2. Forewing with $M_1$ stalked with $R_{4+5}$ . . . . . . . . . . 3
— Forewing with $M_1$ separate from $R_{4+5}$ . . . . . . . . . 5

3. Male with aedoeagus massive; valvae reduced, asymmetrical (text figure 36 *f*) . . . . . . . . *Chrysopeleia* p. 122
— Not as above . . . . . . . . . . 4

4. Male with uncus weakly developed, a small rounded lobe, lightly sclerotized; aedoeagus reduced (text figure 40 *g*) . . . . . . . . *Sorhagenia* p. 131
— Male with well-developed uncus; aedoeagus developed, heavily sclerotized (text figure 34 *e*) . . . . . . . . *Stilbosis* p. 114

5. Hindwing with $R_s$ and $M_1$ distant basally (text figure 3 *e*), cell closed or nearly closed; male with uncus; female with nonpiercing ovipositor . . . . . . . . 6
— Hindwing with $R_s$ and $M_1$ approximate basally (text figure 4 *b*), cell open; male without uncus; female with piercing ovipositor, apophyses massive . . . . . . . . *Obithome* p. 138

6. Forewing with $Cu_2$ developed from cell, hindwing with cell closed, male with lightly sclerotized uncus (text figure 37 *d*) . . . . . . *Walshia* p. 122
— Forewing with $Cu_2$ present at margin, hindwing with cell open, male with short, heavily sclerotized uncus (text figure 38 *c*) . . . . . . . . *Nepotula* p. 126

7. Forewing with $M_1$ stalked with $R_{4+5}$ (text figure 2 *g*) . . . . . . . . 8
— Forewing with $M_1$ separate from $R_{4+5}$ (text figure 3 *h*) . . . . . . . . 12

8. Forewing with $M_2$ stalked with $R_{4+5}+M_1$ (text figure 2 *g*) . . . . . . . . 9
— Forewing with $M_2$ separate from $R_{4+5}+M_1$ (text figure 3 *j*) . . . . . . . . 10

9. Male with uncus, forewing with $M_2$ arising before $R_4$ . . . . . . . . *Siskiwitia* p. 111
— Male without uncus, forewing with $M_2$ arising beyond $R_4$ . . . . . . . . *Periploca* p. 100

10. Antenna with pecten present; forewing with $Cu_1$ and $Cu_2$ slightly downcurved from end of cell (text figure 3 *j*) . . . . . . . . *Ithome* p. 134

— Antenna with pecten absent, forewing with $Cu_1$ and $Cu_2$ not downcurved from end of cell . . . . . . . . . . . . . . . . . . . . . . . . . . . . . . . . . 11

11. Forewing with $Cu_1$ developed from cell to margin; male with uncus very short, divided apically (text figure 33 *d*) . . . . . . . *Synploca* p. 113

— Forewing with $Cu_1$ developed at margin; male with uncus very heavily sclerotized, triangular (text figure 33 *b*) . . . . . . . . . . . . . *Pristen* p. 113

12. Hindwing with cell open, male with uncus divided apically, female with apophyses anteriores joined by sclerotized band . . . . *Perimede* p. 128

— Hindwing with cell closed, male without uncus, female with apophyses anteriores not joined by sclerotized band . . . . . . . . . . . . *Afeda* p. 127

GENUS
## *Periploca* Braun

*Periploca* Braun, 1919, *Ent. News*, **30**: 261.
Type-species: *Periploca purpuriella* Braun, 1919, now considered to be a junior synonym of *Elachista* ? *orichalcella* Clemens, 1864.
Original designation.

*Periploca* is a genus of 27 small to very small moths. All species are from the United States and Canada, but further collecting and study of the smaller moths south of our area probably will extend the geographic range of the genus and increase the number of species. *Periploca* species generally are somberly colored, immaculate moths with dark-gray, dark-bronze or black forewings (exceptionally shining yellow), and the forewings lack raised scales. These two characters will separate adults from other chrysopeleiids except for *Pristen* and *Synploca* species. *Periploca* species lack an uncus, whereas *Pristen* and *Synploca* species have an uncus.

Characters of *Periploca* are as follows: labial palpus with second and third segments usually subequal in length, sometimes second segment twice length of third segment; antenna simple, pecten of one or two hairlike scales; ocelli present or absent. Forewing lanceolate, apex acute; cell open; 11 or 12 veins present ($Cu_2$ absent); $R_4$, $R_5$, $M_1$ and $M_2$ stalked. Hindwing narrowly lanceolate to linear, apex acute; cell open; eight veins present; $R_s$ and $M_1$ stalked. Male with relatively simple genitalia; valvae symmetrical or asymmetrical; vinculum sometimes produced on caudal margin; aedoeagus large relative to entire genitalia, sometimes armed with teeth or pointed projections; manica heavily sclerotized, joining saccal region of vinculum; uncus and gnathos absent; sometimes with paired, eversible, lateral scent glands between second and third abdominal segments. Female with ostium bursae on seventh abdominal sternum; ductus bursae membranous or sclerotized basally, coiled or straight; corpus bursae membranous; signa present and paired or absent; apophyses anteriores and posteriores subequal in length.

Many *Periploca* larvae are associated with junipers as needle miners, stem miners, fruit borers and gall makers. One species is a stem gall maker on *Ceanothus* species, and another bores in the thorns of honey locust and locust. *Periploca nigra* is a pest of ornamental junipers in California, and *P. mimula* is a pest of Chinese juniper in Tennessee.

Adults are very similar superficially; thus, characters of the male and female genitalia must be used to separate them.

### KEY TO SPECIES OF *PERIPLOCA*

1. Males . . . . . . . . . . . . . . . . . . . . . . . . . . . . . . . . . . 2

— Females . . . . . . . . . . . . . . . . . . . . . . . . . . . . . . . 27

2. Aedoeagus concave dorsally (text figure 29 *a*) . . . . . . . . . . . . . . . . . . . . . . . . . . . . . . . . . 3

— Aedoeagus convex or sinuous dorsally . . . . . . . . . 5

3. Valva with saccular lobe (text figure 29 *a*) . . . . . . . . . . . . . . . . . . . . . . . . . . . . . . . *orichalcella* p. 103

— Valva without saccular lobe (text figure 29 *g*) . . . . . . . . . . . . . . . . . . . . . . . . . . . . . . . . . 4

4. Valva broadly triangular, costal margin slightly convex (text figure 29 *g*) . . . . . . *ceanothiella* p. 103

— Valva not triangular, costal margin sinuous (text figure 29 *h*) . . . . . . . . . . . . . . . . . . . . . . . *arsa* p. 103

5. Valvae asymmetrical (text figure 31 *g*) . . . . *gulosa* p. 109

— Valvae symmetrical . . . . . . . . . . . . . . . . . . . . . . . 6

6. Valva with long, cylindrical extension from inner surface (text figure 32 *e*) . . . . . . . . . . . *labes* p. 111

— Valva unarmed . . . . . . . . . . . . . . . . . . . . . . . . . . 7

7. Vinculum evenly rounded or produced posteriorly (text figures 32 *a,f*) . . . . . . . . . . . . . . 8
— Vinculum emarginate posteriorly (text figure 32 *b*) . . . . . . . . . . . . . . . . . . . . . . . . . . . . 20

8. Vinculum produced medially (text figure 32 *a*) . . . . . . . . . . . . . . . . . . . . . . . . . . . . . . . . . . 9
— Vinculum evenly rounded, not produced medially . . . . . . . . . . . . . . . . . . . . . . . . . . . . . . . 11

9. Aedoeagus armed dorsally (text figure 32 *f*) . . . 10
— Aedoeagus unarmed (text figure 32 *a*) . . . . . *fessa* p. 109

10. Aedoeagus with numerous small and large dorsal teeth, apex truncated (text figure 32 *f*) . . . . . . . . . . . . . . . . . . . . . . . . . . . . . . . *dentella* p. 111
— Aedoeagus with three or four large dorsal teeth, without small teeth, distal part tapering to rounded apex (text figure 32 *h*) . . . . . . . . . . . . . . . . . . . . . . . . . . . . . . . . . *serrulata* p. 110

11. Aedoeagus armed externally (text figure 31 *a*) . . . . . . . . . . . . . . . . . . . . . . . . . . . . . . . . . . 12
— Aedoeagus unarmed . . . . . . . . . . . . . . . . . . . . . . 16

12. Valva with costal margin deeply excavated before apex, apex of valva narrowly acute (text figure 31 *a*) . . . . . . . . . . . . . . . . . . . . . . . . . 13
— Valva with costal margin sinuous or rounded, apex broadly acute . . . . . . . . . . . . . . . 15

13. Aedoeagus C-shaped (text figure 30 *b*) . . . . . . . . 14
— Aedoeagus sinuous (text figure 31 *a*) . . . . . . *tridens* p. 105

14. Valva with distal part having parallel dorsal and ventral margins (text figure 30 *b*) . . . . *teres* p. 105
— Valva with distal part having converging dorsal and ventral margins (text figure 29 *e*) . . . . . . . . . . . . . . . . . . . . . . . . . . . . . . . . . *hostiata* p. 105

15. Valva with sinuous costal margin, one or two stout setae from costal margin (text figure 30 *a*) . . . . . . . . . . . . . . . . . . . . . . . . . *opinatrix* p. 106
— Valva with costal margin rounded; setae on valva generally uniform in size, none extremely stout (text figure 32 *g*) . . . . . . . . *dipapha* p. 111

16. Aedoeagus with apex upturned (text figure 30 *d*) . . . . . . . . . . . . . . . . . . . . . . . . . . . . . . . . . . 17
— Aedoeagus with apex straight (text figure 29 *b*) . . . . . . . . . . . . . . . . . . . . . . . . . . . . . . . . . . 19

17. Valva with costal margin strongly indented before apex, apex narrowly acute (text figures 29 *d,f*) . . . . . . . . . . . . . . . . . . . . . . . . . . . 18
— Valva with costal margin evenly convex or slightly concave before apex, apex more broadly acute (text figure 30 *d*) . . . . . . . . . *hortatrix* p. 106

18. Aedoeagus very slender apically, length about three times maximum height of valva (text figure 29 *d*) . . . . . . . . . . . . . . . . *gleditschiaeella* p. 104
— Aedoeagus broad to apex, length nearly two times maximum height of valva (text figure 29 *f*) . . . . . . . . . . . . . . . . . . . . . . . . *intermedia* p. 104

19. Valva with costal margin strongly indented before apex, apex narrowly acute (text figure 29 *c*) . . . . . . . . . . . . . . . . . . . . . . . . . . *repanda* p. 104
— Valva with costal margin even, apex broadly acute (text figure 30 *c*) . . . . . . . . . . *atrata* p. 106

20. Aedoeagus unarmed . . . . . . . . . . . . . . . . . . . *facula* p. 109
— Aedoeagus armed . . . . . . . . . . . . . . . . . . . . . . . . 21

21. Aedoeagus armed dorsally (text figure 32 *d*) . . . . . . . . . . . . . . . . . . . . . . . . . . . . . . . . . . . *funebris* p. 110
— Aedoeagus armed laterally (text figure 31 *e*) . . . 22

22. Aedoeagus armed on right side (text figure 31 *e*) . . . . . . . . . . . . . . . . . . . . . . . . . . . . . . . . . . 23
— Aedoeagus armed on left side (text figure 31 *c*) . . . . . . . . . . . . . . . . . . . . . . . . . . . . . . . . . . 26

23. Aedoeagus armed with at least a long projection (text figure 31 *e*) . . . . . . . . . . . . . . . . . . . . 24
— Aedoeagus armed with several, short spines (text figure 31 *d*) . . . . . . . . . . . . . . . . . . . . . . . . . 25

24. Aedoeagus with spine on right side at right angle, distal part of aedoeagus nearly straight (text figure 31 *e*) . . . . . . . . . . . . . . . . *nigra* p. 108
— Aedoeagus with spine on right side parallel with long axis of aedoeagus, distal part

upturned (text figure 32 *c*) . . . . . . . . . . . . . *juniperi* p. 109

25. Aedoeagus with five or six readily visible projections, apex sharply constricted and turned to left (text figure 31 *d*) . . . . . . . . . . . *devia* p. 108

— Aedoeagus with numerous, very small projections, apex slightly twisted to right (text figure 31 *f*) . . . . . . . . . . . . . . . . . . . . . . . . . . . . *cata* p. 107

26. Aedoeagus with a long projection preceded by a few short ones (text figure 31 *c*) . . . . . . . . *laeta* p. 107

— Aedoeagus with several, short, asymmetrical projections (text figure 31 *b*) . . . . . . . . . . . *mimula* p. 107

27. Ductus bursae not heavily sclerotized before ostium bursae (text figure 43 *e*) . . . . . . . . . . . . . . 28

— Ductus bursae heavily sclerotized before ostium bursae (text figure 46 *d*) . . . . . . . . . . . . . 35

28. Sixth abdominal sternum with posterior margin invaginated medially (text figure 43 *c*) . . . . . . . . . . . . . . . . . . . . . . . . . . . . *orichalcella* p. 103

— Sixth abdominal sternum with posterior margin even (text figure 43 *e*) . . . . . . . . . . . . . . . 29

29. Ductus bursae nearly straight, or broadly curved, slender or broad (text figure 43 *e*) . . . . . 30

— Ductus bursae twisted several times, slender (text figure 44 *c*) . . . . . . . . . . . . . . . . . . . . . . . . . 31

30. Genital plate with posterior margin curved, upturned part of lateral margin angled before ½ length (text figure 43 *e*); ductus bursae less than twice length of corpus bursae . . . . . . . . . . . . . . . . . . . . . . . . . . *ceanothiella* p. 103

— Genital plate with posterior margin straight, lateral margin straight (text figure 43 *d*); ductus bursae more than twice length of corpus bursae . . . . . . . . . . . . . . . . . . . . . . . . . . *arsa* p. 103

31. Lamella antevaginalis with a fingerlike invagination anteriorly (text figure 44 *c*) . . . . . . 32

— Lamella antevaginalis without such invagination . . . . . . . . . . . . . . . . . . . . . . . . . . . . 33

32. Width of signum more than ½ basal width of corpus bursae, diameter of invagination on lamella antevaginalis about equal to diameter of ductus bursae (text figure 44 *c*) . . . . . . . . . . . . . . . . . . . . . . . . . . . . . . . . . . . . *opinatrix* p. 106

— Width of signum less than ⅓ basal width of corpus bursae, diameter of invagination on lamella antevaginalis less than ½ diameter of ductus bursae (text figure 44 *a*) . . . . . . *hortatrix* p. 106

33. Length of apophyses posteriores more than five times maximum width of ninth abdominal segment . . . . . . . . . . . . . . . . *gleditschiaeella* p. 104

— Length of apophyses posteriores less than twice maximum width of ninth abdominal segment . . . . . . . . . . . . . . . . . . . . . . . . . . . . . . . . 34

34. Ostium bursae on anterior margin of seventh abdominal sternum, lamella postvaginalis a broadly curving script "X", signum less than ¼ width of corpus bursae (text figure 45 *a*) . . . . . . . . . . . . . . . . . . . . . . . *atrata* p. 106

— Ostium bursae at ⅓ on seventh sternum, lamella postvaginalis a very narrow, open ellipse, signum about ⅓ width of corpus bursae (text figure 46 *b*) . . . . . . . . . . . . . . . . . *labes* p. 111

35. Posterior margin of eighth abdominal segment with numerous setae (text figure 46 *d*) . . . 36

— Posterior margin of eighth abdominal segment with few setae (text figure 45 *e*) . . . . . . . . . 37

36. Ductus bursae with heavily sclerotized part hook shaped (text figure 46 *d*) . . . . . . . . . . *juniperi* p. 109

— Ductus bursae with heavily sclerotized part angulate slightly beyond middle (text figure 46 *a*) . . . . . . . . . . . . . . . . . . . . . . . . . *funebris* p. 110

37. Seventh abdominal sternum with straight anterior margin (text figure 45 *e*) . . . . . . . . . . . *cata* p. 107

— Seventh abdominal sternum with lobes, invaginations or productions on anterior margin (text figures 45 *c*, 46 *c*) . . . . . . . . . . . . . . 38

38. Seventh abdominal sternum with medially produced anterior margin (text figure 45 *c*) . . *laeta* p. 107

— Seventh abdominal sternum medially emarginate or slightly concave . . . . . . . . . . . . . . . . . . 39

39. Seventh abdominal sternum with anterior margin emarginate medially (text figure 45 *b*)....................................40
— Seventh abdominal sternum with anterior margin slightly convex medially, flanked by a pair of lobes (text figure 46 *c*)..............41

40. Lamella postvaginalis well developed, ½ times as long as wide (text figure 45 *b*)......*soror* p. 108
— Lamella postvaginalis very short, ⅙ times as long as wide (text figure 45 *d*).............*devia* p. 108

41. Eighth abdominal tergum usually with a transverse row of setae (text figure 46 *c*)....*nigra* p. 108
— Eighth abdominal tergum usually without a row of setae (text figure 44 *d*)..........*mimula* p. 107

### *Periploca orichalcella* (Clemens)

PL. 5, FIG. 1. TEXT FIGS. 2 *g*; 29 *a*; 43 *c*, *g* (McD. 7693, 9011).

*Elachista ? orichalcella* Clemens, 1864, *Proc. Ent. Soc. Philadelphia*, **2**: 430.
Type-locality: Virginia. [ANSP]

*Elachista* (?) *concolorella* Chambers, 1875, *Can. Ent.*, **7**: 55.
Type-locality: Bosque Co., Texas. [MCZ]

*Periploca purpuriella* Braun, 1919, *Ent. News*, **30**: 261.
Type-locality: Fredalba, California. [ANSP]

*Orichalcella*, *arsa*, *ceanothiella* and *gleditschiaeella* have the head, thorax and forewings clothed with scales that are metallic bronze medially and nonreflecting dark gray to black marginally. The small size (wing length: 2.5–3.6 mm) will separate *orichalcella* from *ceanothiella* and *gleditschiaeella*, and the unicolorous, dark-gray antenna will separate *orichalcella* from *arsa* which has the antennae pale yellow distally. Males of *orichalcella* have a pair of lateral, eversible scent glands between the second and third abdominal segments. These bear numerous scales apically. California specimens tend to be larger (wing length: 3.1–3.6 mm) and often have less reflective forewings than do those from the rest of the range. Specimens from southern Arizona and western Texas have a wing length of 2.5 to 3.2 mm and consistently are very shiny. No genital differences have been found to separate the two groups.

The immature stages are unknown.

*Orichalcella* occurs in Virginia, central Illinois and from western Texas to California. Moths have been collected from 26 May to 6 June in western Texas, 7 July to 3 August in southern Arizona and July, August, September and March in California.

### *Periploca arsa* Hodges, NEW SPECIES

PL. 6, FIG. 16. TEXT FIGS. 29 *h*; 43 *d*, *h*.

*Periploca arsa* Hodges.
Type-locality: Key Largo Key, Monroe Co., Florida. [USNM]

Upper surface as figured. Head: tongue, maxillary palpus, labial palpus and lower part of frons dark, shining yellowish bronze, upper part of frons paler, shining yellow to bronze; vertex, occiput and base of antenna dark, shining bronze; ocelli absent; antenna pale yellowish gray distally. Legs: mainly shining bronze; coxae paler than rest of segments; hindtarsus with some pale scales at apices of segments. Abdomen greasy, dark yellowish brown dorsally and ventrally. Male with pair of lateral, eversible scent glands between second and third abdominal segments. Male: as in text figure 29 *h*; valva with costal margin sinuous, apex turned up slightly, a small distal lobe on saccular margin; aedoeagus concave dorsally. Female: as in text figures 43 *d*, *h*; ostium bursae at apex of parallel-margined, horseshoe-shaped sclerite, nearly at middle of seventh sternum; seventh sternum with a folded, sclerotized band on anterior margin; ductus bursae slender, approximately twice length of corpus bursae; corpus bursae with somewhat bubbled walls; signa paired, thorn shaped.

The immature stages are unknown.

TYPES: Holotype: ♂. Key Largo Key, Monroe Co., Florida; 17 Oct. 1968; Mrs. Spencer Kemp; USNM slide 4881. USNM. Paratypes: 6 ♂♂, 4 ♀♀. Same locality as for holotype; 1–25 Oct. 1968, 1971. Paratypes to CPK; USNM.

*Arsa* is similar to *orichalcella* but can be separated as indicated in the key. Additionally, *arsa* is somewhat larger; the forewing bends down just before the apex; and the antenna is pale yellow apically. Contrasting characters apply for *orichalcella*.

### *Periploca ceanothiella* (Cosens)

PL. 4, FIG. 45. TEXT FIGS. 29 *g*; 43 *b*, *e* (McD. 7714).

*Stagmatophora ceanothiella* Cosens, 1908, *Can. Ent.*, **40**: 107.
Type-locality: Toronto, Canada.

*Ceanothiella* is very similar to *gleditschiaeella* in size and shining scales but usually can be separated by the third segment of the labial palpus being ½ the length of the second segment, whereas in *gleditschiaeella* the second and third segments are nearly equal in length. Genital differences are as given in the key. Ocelli are absent. Wing length varies from 3.8 to 6.2 mm. Although the third segment of the labial palpus normally is much shorter than the second segment, rarely it is nearly as long as the second segment.

Cosens (1908: 107) reared *ceanothiella* from Jerseytea, *Ceanothus americanus* L. The spindle-shaped gall is normally terminal on the stem. Overwintering occurs in the larval stage.

*Ceanothiella* has been collected rarely in the East in southern Ontario and central New York. It occurs in the Black Hills of South Dakota, northern Arizona and southwestern Oregon south to Los Angeles, California. Moths are present from May through July.

### *Periploca gleditschiaeella* (Chambers)

PL. 4, FIG. 47. TEXT FIGS. 29 *d*; 44 *b* (McD. 7719).

*Laverna* (?) (*Anybia*?) *gleditschiaeella* Chambers, 1876, *Can. Ent.*, **8**: 135.
Type-locality: Kentucky. [MCZ]

*Gleditschiaeella* is superficially very similar to *ceanothiella* but can be separated as indicated in the key and under *ceanothiella*. Ocelli are present. The male genitalia of *gleditschiaeella* are nearest those of *intermedia* and *repanda*. Wing length varies from 4.0 to 6.2 mm. Worn specimens are much paler and somewhat more yellow than fresh specimens.

The larva bores in the thorns of honey locust, *Gleditschia triacanthos* L.

*Gleditschiaeella* has a limited range from western Maryland to southern Ohio and northern Kentucky. Specimens have been collected from 20 May to 7 June.

### *Periploca intermedia* Hodges, NEW SPECIES

PL. 4, FIG. 46. TEXT FIG. 29 *f*.

*Periploca intermedia* Hodges.
Type-locality: Devil's Den State Park, Washington Co., Arkansas. [USNM]

Upper surface as figured. Head: base of tongue with medium yellowish-gray scales; maxillary palpus paler yellowish gray; labial palpus mainly dark gray, inner surface of first and base of second segment yellowish gray, second and third segments subequal in length; frons and vertex shining yellow to yellowish brown; upper part of vertex and occiput dark gray, nearly black and somewhat shining; antenna with scape and base of shaft dark gray, nearly black, shaft becoming pale yellowish to yellowish gray at ⅛ length; ocelli present. Legs mainly dark gray, nearly black, somewhat shining, apices of tarsal segments somewhat paler than rest of segments, hindtarsus mottled pale gray and dark gray. Wing length: 4.2–5.0 mm. Male genitalia: as in text figure 29 *f*; valva with costal margin strongly sinuous, excavated at ⅔ length, apex narrowly acute, saccular margin sinuous; aedoeagus slightly longer than length of valva, sinuous, apex slightly upturned. Female: no specimens available.

The immature stages are unknown.

TYPES: Holotype: ♂. Devil's Den State Park, Washington Co., Arkansas; 11 July 1966; R. W. Hodges; USNM slide 4272. USNM. Paratypes: 5 ♂♂. Same locality as for holotype; 12 June–22 July. Paratypes to University of Arkansas; USNM.

Variation in the short type series is restricted to wing length and degree of wear of the dark scales on the forewings. The two specimens collected on 12 June are very pale, almost yellowish; those collected in early July are dark gray and perhaps represent a second brood. *Intermedia* is allied with *gleditschiaeella* and *repanda* on genital characters but can be separated from them as indicated in the key.

### *Periploca repanda* Hodges, NEW SPECIES

PL. 6, FIG. 17. TEXT FIGS. 29 *b*, *c*.

*Periploca repanda* Hodges.
Type-locality: Nugent Mt., Chisos Mts., Big Bend National Park, Brewster Co., Texas. [USNM]

Upper surface as figured. Head: scales of tongue and frons shining pale yellow; maxillary palpus mottled pale yellow and gray; labial palpus dark gray and black laterally, first and second segments pale yellowish gray becoming darker on third segment mesally, second and third segments subequal in length; vertex and occiput dark gray with shining yellow reflections; antenna with scape dark gray dorsally, pale yellow ventrally, on anterior margin and on distal margin, base of shaft dark gray, rest mainly pale yellowish gray; ocelli present. Foreleg: coxa yellow to yellowish gray; femur darker yellowish gray, apex paler than rest of segment; tibia and tarsus dark gray, apex of tibia pale gray nearly white, apex of fifth tarsal segment with some pale-gray scales. Midleg: coxa and trochanter shining yellow; femur mainly yellow mottled with some gray scales; tibia mainly medium to dark gray, scale bases pale, apex pale yellow; tarsus dark gray with some paler gray scales at apices of first and fifth segments. Hindleg: coxa, trochanter and femur shining yellow with some gray scales on femur; tibia and tarsus mainly dark gray, a few pale scales at base of first pair of tibial spurs and at apex of tibia; tarsal segments with pale scales at apices and base of first segment. Wing length: 4.0–4.3 mm. Male genitalia: as in text figures 29 *b*, *c*; valva with costal margin nearly right angled at $\frac{2}{3}$ length, apex slightly upturned and acute, angled but not strongly sinuous; aedoeagus at least $1\frac{1}{2}$ times length of valva, slightly expanded on distal $\frac{1}{6}$, apex acute. Female: no specimens available.

The immature stages are unknown.

TYPES: Holotype: ♂. Texas, Brewster Co., [Big Bend National Park], Chisos Mts., Nugent Mt.; 4 June 1973; R. W. Hodges; USNM slide 4724. USNM. Paratypes: 2 ♂♂. Phoenix Mts., Arizona; 21 Sept. 1924; L. J. Bottimer (1 ♂). California, Riverside Co., 5 mi NW Palm Sprs.; 20 Sept. 1961; J. A. Chemsak (1 ♂). Paratypes to UCB; USNM.

Little variation is shown in the short type series. Specimens of *repanda* can be separated from those of *intermedia* and *gleditschiaeella* as indicated in the key.

### *Periploca hostiata* Hodges

PL. 5, FIG. 2. TEXT FIG. 29 *e*.

*Periploca hostiata* Hodges, 1969, *Smithsonian Contrib. Zool.*, **18**: 4, fig. 7.
Type-locality: Fort Simcoe, Yakima Co., Washington. [USNM]

*Hostiata* is a shining, gray-black species which is similar to *gleditschiaeella*, *intermedia* and *repanda* in genital characters but can be separated from them as indicated in the key. Ocelli are present. The two specimens have wing lengths of 4.0 and 4.2 mm.

The immature stages are unknown.

*Hostiata* has been collected at Rustler Park, Chiricahua Mountains, Arizona on 27 July and at the type-locality on 31 July.

### *Periploca teres* Hodges, NEW SPECIES

PL. 6, FIG. 18. TEXT FIG. 30 *b*.

*Periploca teres* Hodges.
Type-locality: Rustler Park, 8500′, Chiricahua Mts., Arizona. [UCB]

Upper surface as figured. Head: tongue, maxillary palpus and frons shining pale yellow; dark gray with shining yellow reflections; labial palpus with first and second segments pale yellow, third segment dark gray laterally with many pale-based scales and mainly pale yellowish gray medially; antenna mainly shining pale yellow, dorsal surface of scape with many dark-gray scales; ocelli present. Foreleg: coxa and trochanter shining yellow to yellowish gray; femur slightly darker with pale apex; tibia mottled pale yellowish gray and dark gray, apex off white; tarsal segments mottled pale and dark gray, apex of first segment off white and apices of fourth and fifth segments broadly off white. Midleg: coxa, trochanter and femur shining yellowish white; tibia mottled pale and dark gray, mainly dark, apex off white; tarsus mainly dark gray, apices of first and second segments with off-white scales. Hindleg: coxa, trochanter and femur shining yellow to yellowish white; tibia mottled medium gray and pale yellowish gray, an oblique yellowish-gray streak at base of first pair of spurs and another just before terminal spurs; tarsus mottled medium and pale gray, apices of first four segments off white. Wing length: 3.5 mm. Male genitalia: as in text figure 30 *b*; valva with costal margin strongly excavated just beyond $\frac{1}{2}$ length, distal part ending in slender, parallel-margined lobe; aedoeagus C-shaped, about twice length of valva, a series of short, pointed projections on right side of $\frac{3}{5}$ length. Female: no specimens available.

The immature stages are unknown.

TYPE: Holotype: ♂. Rustler Park, 8500′, Chiricahua Mts., Arizona; 20 Aug. 1972; at light; J. Powell; SBA slide 184. UCB.

The male genitalia ally *teres* with *hostiata*; differences are as given in the key. *Teres* is known only from the type specimen.

### *Periploca tridens* Hodges, NEW SPECIES

PL. 6, FIG. 20. TEXT FIG. 31 *a*.

*Periploca tridens* Hodges.
Type-locality: Twain Harte, Tuolumne Co., California. [UCB]

Upper surface as figured. Head: tongue, maxillary palpus, frons and vertex shining pale yellow; occiput medium to dark gray with shining yellow reflections; labial palpus shining pale yellow, second and third segments with dark scales toward apex [many scales absent and possibly darker on fresh specimen]; antenna with scape dark gray, nearly black dorsally, mottled dark gray and pale yellow ventrally, shaft dark at base becoming pale yellowish gray shortly beyond base. Ocelli present. Foreleg: coxa pale yellow mottled with gray; femur, tibia and tarsus mainly dark gray, apex of tibia and apices of first and fifth tarsal segments pale yellowish gray. Midleg: coxa and trochanter mottled pale yellow and dark gray; mainly dark gray mottled with pale yellow, apex pale; tibia mainly dark gray, apex yellowish white; tarsus mottled dark gray with some pale, yellowish-gray based scales, apices of first, second and fifth segments pale yellowish gray. Hindleg: coxa and trochanter mainly shining yellowish gray; femur mottled yellowish gray and darker gray; tibia medium gray with pale-based scales, apex pale yellowish gray; tarsus pale yellowish gray, base and apex of first segment and apices of second through fifth segments pale yellowish gray to off white. Wing length: 4.9 mm. Male genitalia: as in text figure 31 *a*; valva with costal margin sinuous, sharply downcurved at $\frac{2}{3}$ length, apex acute, saccular margin sinuous; aedoeagus strongly sinuous, apex slightly downcurved, with three, moderately long, conical projections at $\frac{3}{5}$ length. Female: no specimens available.

The immature stages are unknown.

TYPE: Holotype: ♂. Twain Harte, Tuolumne Co., California; 20 June 1959; M. Lundgren; SBA slide 205. UCB.

*Tridens* is related to *hortatrix* on genital characters but can be separated as indicated in the key.

### *Periploca hortatrix* Hodges

PL. 5, FIG. 3. TEXT FIGS. 30 *d*; 44 *a*.

*Periploca hortatrix* Hodges, 1969, *Smithsonian Contrib. Zool.*, **18**: 4, figs. 1, 10, 28, 29.

Type-locality: Devil's Den State Park, Washington Co., Arkansas. [USNM]

*Hortatrix* is a shining gray-black species that is most nearly allied to *atrata*, *opinatrix* and *tridens*. The male of *hortatrix* differs from that of *atrata* by having the apex of the valva acute and lacking a triangular projection on the saccular margin. It differs from those of *opinatrix* and *tridens* by having the aedoeagus unarmed. The female of *hortatrix* differs from that of *opinatrix* in having few uniserial setae at the posterior margin of the eighth abdominal segment and in having the fingerlike invagination of the lamella antevaginalis slender, nearly twice as long as the opening. In *opinatrix* the setae are numerous and biserial, and the fingerlike process is broad and blunt. The female genitalia of *atrata* lack a fingerlike process. Wing length varies from 3.2 to 4.0 mm, and males are slightly larger than females.

The immature stages are unknown.

*Hortatrix* has been collected at light in central Illinois and northwest Arkansas from late May through early June.

### *Periploca opinatrix* Hodges

PL. 5, FIG. 4. TEXT FIGS. 30 *a*; 44 *c*.

*Periploca opinatrix* Hodges, 1969, *Smithsonian Contrib. Zool.*, **18**: 4, figs. 8, 9, 27.

Type-locality: 6 mi NW Newcastle, Wyoming. [USNM]

*Opinatrix* is a somewhat shining, dark, gray-black species. Males can be separated from those of other species of *Periploca* by having the aedoeagus armed with four conical projections at $\frac{3}{4}$ its length, the posterior margin of the vinculum emarginate and the apex of the valva acute. Females of *opinatrix* are nearest those of *hortatrix* but can be separated as indicated under *hortatrix*. Ocelli are present. Wing length varies from 3.8 to 4.5 mm. Specimens vary in the amount of dark scales remaining on the forewings.

The immature stages are unknown.

The short type series was collected in eastern Wyoming at light from 13 to 20 July.

### *Periploca atrata* Hodges (Juniper Cone Moth)

PL. 4, FIG. 48. TEXT FIGS. 30 *c*; 45 *a*.

*Periploca atrata* Hodges, 1962, *Pan-Pacific Ent.*, **38**: 90, figs. 11, 19.

Type-locality: Madera Canyon, 4880', Santa Rita Mts., Santa Cruz Co., Arizona. [CU]

*Atrata* is a dark, gray-black species. The antennae are dark basally, becoming greasy yellowish gray distally. Ocelli are present. Wing length varies from 3.8 to 6.0 mm. Major differences among specimens result from degree of wear of the dark scales on the head, thorax and forewings; fresh specimens are very dark; worn specimens are very pale with shining yellowish-gray reflections. Genital characters ally males with *hortatrix* and females with *labes*. Differences are as given in the key.

Powell (1963) describes larval and pupal habits of *atrata*, which feeds on the fruit of *Juniperus pachyphloea* Torrey and *Juniperus californica* Carr. Pupation occurs in late January and February, and adults emerge in April and May. An ichneumonid, *Pristomerus* species, was reared from unsegregated juniper berries infested by *atrata*.

*Atrata* occurs in the mountains of northern and southern Arizona west to Contra Costa County, California. Field-collected specimens are recorded from 11 July to 4 August in Arizona and 5 to 19 April in California.

### *Periploca mimula* Hodges

PL. 1, FIG. 62; PL. 5, FIG. 5. TEXT FIGS. 31 *b*; 44 *d*.

*Periploca mimula* Hodges, 1962, *Pan-Pacific Ent.*, **38**: 92, figs. 5, 15.
Type-locality: Madera Canyon, 4880′, Santa Rita Mts., Santa Cruz Co., Arizona. [CU]

*Mimula* is a dark, gray-black species with wing length of 3.3 to 5.0 mm. Ocelli are present. On characters of the male genitalia it is near *laeta* from which it differs mainly by lacking a long spine on the left side of the aedoeagus. Females of *mimula* and *nigra* are nearly inseparable on maculation or genital characters. Some females of *nigra* have a transverse, dorsal row of six to eight setae on the ninth abdominal segment, whereas this area has no setae in most examined specimens of *mimula*. Features of the genital plate and the ductus bursae are somewhat variable and so far have been useless to distinguish between the two.

Charles R. Patrick (*in litt.*) has reared *mimula* from cells in the cambium layer of *Juniperus chinensis* variety *keteleeri* Corman in Haywood County, Tennessee. Two generations occur per year. First instar larvae were found on 25 February; mature larvae were found on 5 May. Pupation occurred on 10 June, and adults were recovered on 18 June. First instar larvae of the second generation were present on 5 August, second instar larvae on 15 August, and adults emerged from mid-September. Eggs are laid singly in crevices on the bark. The larva forms a cell and does not girdle branches or stems. Plants over three feet tall were preferred. Much red-brown frass is associated with the cell, and some plant sap exudes from the entrance of the cell.

*Mimula* has been collected from Tennessee, Arkansas and Washington south to Texas, New Mexico and southern Arizona. Adults have been collected from 4 June through September.

### *Periploca laeta* Hodges

PL. 5, FIG. 6. TEXT FIGS. 31 *c*; 45 *c*.

*Periploca laeta* Hodges, 1962, *Pan-Pacific Ent.*, **38**: 92, figs. 13, 18.
Type-locality: Monticello, Florida. [CU]

*Laeta* is a dark, gray-black species with wing length varying from 4.0 to 4.2 mm. Ocelli are present. The antennal scape and first four or five segments of the shaft are nearly black; then the shaft gradually becomes greasy yellowish gray. Males can be separated from those of other species of *Periploca* by having the left side of the aedoeagus armed with a long projection preceded by a few short ones. The heavily sclerotized basal part of the ductus bursae in combination with the produced anterior margin of the seventh sternum will separate females of *laeta* from those of other species of *Periploca*.

The larva feeds beneath the bark on stems of *Juniperus* species.

*Laeta* occurs from Washington, D.C. south to northern Florida and west to southern Arizona. Adults have been collected on 1 June and 22 July. Reared specimens from Florida emerged on 15 and 18 February.

### *Periploca cata* Hodges

PL. 5, FIGS. 7, 8. TEXT FIGS. 31 *f*; 45 *e*.

*Periploca cata* Hodges, 1962, *Pan-Pacific Ent.*, **38**: 93, figs. 10, 17, 24.
Type-locality: Putnam Co., Illinois. [CU]

*Cata* is a shining, dark, gray-black species with wing length varying from 3.5 to 5.3 mm. The antenna is dark gray basally becoming greasy, yellow gray distally. Ocelli are present. The shape of the valva varies on the costal and distal margins. Males can be characterized by having the vinculum emarginate posteriorly and the right side of the aedoeagus bearing numerous, very small projections. Females have the base of the ductus bursae sclerotized, a small number of setae on the posterior margin of the eighth abdominal segment and the anterior margin of the seventh sternum simple.

Larvae feed in cedar apple galls, *Gymnosporangium* species, on junipers, and pupation occurs within the galls.

*Cata* has been reared in central Illinois and southeastern Iowa with adults emerging from early May through early June. A series of adults was collected in northwest Arkansas from mid-May through late June.

### *Periploca devia* Hodges

PL. 5, FIG. 9. TEXT FIGS. 31 *d*; 45 *d*.

*Periploca devia* Hodges, 1969, *Smithsonian Contrib. Zool.*, **18**: 5, figs. 11, 30–32.
Type-locality: Fort Valley, 7350′, 7½ mi NW Flagstaff, Coconino Co., Arizona. [USNM]

*Devia* is a shining, dark, gray-black moth with wing length of 4.0 to 4.8 mm. Ocelli are present. The male genitalia of *devia* are nearest those of *cata* in configuration but differ in having the aedoeagus armed with five to eight easily visible projections and the apex constricted and curved to the left and extending but a short distance beyond the posterior margin of the vinculum. The aedoeagus of *cata* has a larger number of very fine projections, and the apex is slightly pointed to the right. The female genitalia are nearest those of *laeta*, differing in having the anterior margin of the seventh abdominal sternum incurved medially and the basal, sclerotized part of the ductus bursae heavily sclerotized immediately before the ostium bursae, then less heavily sclerotized.

The immature stages are unknown.

*Devia* occurs from the Flagstaff area in northern Arizona and the Chiricahua Mountains in southern Arizona west to San Bernardino and Contra Costa County, California. Specimens have been collected from 22 June to 25 September.

### *Periploca soror* Hodges, NEW SPECIES

PL. 6, FIG. 21. TEXT FIG. 45 *b*.

*Periploca soror* Hodges.
Type-locality: Green Gulch, 5500′, Chisos Mts., Big Bend National Park, Brewster Co., Texas. [USNM]

Upper surface as figured. Head: tongue, maxillary palpus and frons pale yellow to yellowish gray; vertex and occiput dark gray; labial palpus mainly dark gray with some scale bases paler gray; antenna with scape mainly dark gray, nearly black, dorsally, pale yellowish gray on anterior margin and at apex, shaft dark gray basally becoming yellowish gray at ⅓ length; ocelli present. Foreleg: coxa and trochanter yellowish gray; femur, tibia and tarsus dark gray, nearly black with some pale-based scales, apices of tibia and tarsal segments pale. Midleg: coxa and trochanter yellowish gray to pale yellow; femur dark gray; tibia dark gray with pale apex; tarsus dark gray, nearly black with pale scales at apices of first, second and fifth segments. Hindleg: coxa and trochanter shining, pale yellowish to yellowish gray; femur, tibia and tarsus medium gray with pale scale bases, tibia with dorsal tuft pale on distal ½ and with pale scales at apex, tarsus with pale scales at base and apex of first segment and apices of other segments. Wing length: 4.0 mm. Male genitalia: no specimens available. Female: as in text figure 45 *b*; ostium bursae at ¼ length of seventh abdominal sternum; a semicircular, sclerotized plate posterad of ostium bursae; anterior margin of seventh abdominal sternum with narrow, more heavily sclerotized band; base of ductus bursae straight, heavily sclerotized, followed by a slight constriction then twisted three or four times; corpus bursae with paired signa, each with an invaginated, inwardly directed plate.

The immature stages are unknown.

TYPES: Holotype: ♀. Texas: Brewster Co., [Big Bend National Park], Chisos Mts., Green Gulch, 5500′; 6 June 1973; R. W. Hodges; USNM slide 4727. USNM.

*Soror* is nearest *devia* but can be separated as indicated in the key.

### *Periploca nigra* Hodges (Juniper Twig Girdler)

PL. 5, FIGS. 10, 11. TEXT FIGS. 31 *e*; 46 *c*.

*Periploca nigra* Hodges, 1962, *Pan-Pacific Ent.*, **38**: 94, fig. 12.
Type-locality: Sacramento, California. [USNM]

*Nigra* is a shining, dark, gray-black species with wing length of 3.0 to 4.5 mm. Ocelli are present. Males of *nigra* can be separated from those of other species of *Periploca* by having the aedoeagus armed with one long and one short projection on the right side. As was pointed out under *mimula*, the females of *nigra* and *mimula* are nearly inseparable.

Koehler and Tauber (1964) reported that the larvae girdle twigs of *Juniperus chinensis* L., *J. horizontalis* Moench., *J. sabina* L. and *J. virginiana* L. In the San Francisco, California area there is one generation per year. Larvae hatch in late July and remain active until the following March. Pupation occurs in the larval galleries, and adults emerge from mid-May through July. They reported a braconid parasite, *Microchelonus* species, on the immature stages.

*Nigra* occurs from New York south to Virginia and Louisiana then west to Sacramento and San Diego, California. Most specimens are from California. Adults have been collected from June through August in the East, late May and early June in Arkansas, 28 March to 21 April in Louisiana and 3 May to 7 August in California.

### *Periploca fessa* Hodges

PL. 5, FIG. 12. TEXT FIG. 32 *a*.

*Periploca fessa* Hodges, 1962, *Pan-Pacific Ent.*, **38**: 95, fig. 4.
Type-locality: Siesta Key, Sarasota Co., Florida. [CU]

*Fessa* is a dark, gray-black species with wing length varying from 3.2 to 4.0 mm. Ocelli are present. The base of the antenna is dark becoming greasy yellowish gray distally. Males can be recognized by the following combination of genital characters: valvae symmetrical, vinculum produced posteriorly with apex of projection narrowly rounded and apex of aedoeagus directed to the left. Females are unknown.

The immature stages are unknown.

*Fessa* has been collected at light at six widely separated localities: Washington, D.C. in May and on 6 June, west-central Florida in early April, Baton Rouge, Louisiana on 1 and 16 April, northwest Arkansas from 27 May to 22 July, Fort Worth, Texas on 29 September and eastern Wyoming on 13 July.

### *Periploca gulosa* Hodges

PL. 5, FIG. 13. TEXT FIG. 31 *g*.

*Periploca gulosa* Hodges, 1962, *Pan-Pacific Ent.*, **38**: 95, fig. 2.
Type-locality: Madera Canyon, 4880′, Santa Rita Mts., Santa Cruz Co., Arizona. [CU]

*Gulosa* is a shining, dark gray-black species with wing length of 4.0 mm. Ocelli are present. The antenna is dark basally becoming greasy yellow gray distally. Only males of *gulosa* have been collected, and they can be recognized by the asymmetrical valvae, the sinuous aedoeagus and the posteriorly produced vinculum. This is the only species of *Periploca* with asymmetrical valvae.

The immature stages are unknown.

Adults have been collected at light in the Santa Rita and Huachuca Mountains of southern Arizona from mid-July through mid-August.

### *Periploca facula* Hodges

PL. 6, FIG. 22. TEXT FIG. 32 *b*.

*Periploca facula* Hodges, 1962, *Pan-Pacific Ent.*, **38**: 96 (in part), fig. 7.
Type-locality: Madera Canyon, 4880′, Santa Rita Mts., Santa Cruz Co., Arizona. [CU]

The forewings of *facula* are pale brown overlaid with burnished gray black, and the wing length varies from 4.2 to 4.4 mm. Ocelli are present. The scape and base of the shaft of the antenna are mottled gray, and the apex of the scape and distal $\frac{3}{4}$ of the shaft are yellow gray to yellowish white. The male genitalia are characterized by having the posterior margin of the vinculum emarginate medially and by having an unarmed aedoeagus. Females are unknown. In earlier papers (1962: 96; 1969: 6) I confused *juniperi* with *facula*. Males of these two species are very similar but separable as indicated in the key. Females that had been associated with *facula* prove to be those of *juniperi* on the basis of reared material.

The immature stages are unknown.

*Facula* has been collected at the type-locality at light in early October.

### *Periploca juniperi* Hodges, NEW SPECIES

PL. 5, FIG. 14. TEXT FIGS. 32 *c*; 46 *d*.

*Periploca juniperi* Hodges.
Type-locality: 6 mi NW Newcastle, Wyoming. [USNM]

Upper surface as figured. Head: tongue dark gray at base becoming pale yellowish to yellowish gray beyond base; frons mainly dark gray with overlay of pale yellow scales; vertex and occiput dark gray with yellowish reflections; labial palpus uniformly dark gray, nearly black; antenna dark gray, nearly black, extreme apex slightly paler gray; ocelli present. Foreleg: mainly dark gray, nearly black becoming greasy; apex of tibia with some pale yellowish-gray scales. Mid- and hindleg: much as for foreleg but generally paler. Wing length: 3.8–4.8 mm. Male genitalia: as in text figure 32 *c*; valva relatively short, apex broadly rounded, saccular margin generally concave; aedoeagus at least four times as long as valva, a long, slender spine on right side running parallel to aedoeagus, distal part with much reduced cross section and upturned, apex acute. Female: as in text figure 46 *d*; ostium bursae near anterior margin of seventh abdominal sternum, a folded and produced, lightly sclerotized area laterad and ventrad of ostium bursae; base of ductus bursae heavily sclerotized, somewhat J-shaped and twisted to left, rest of ductus bursae membranous and coiled three or four times; corpus bursae with paired signa; eighth abdominal segment with numerous slender scales on posterior margin.

The type series was reared from cedar apple galls, *Gymnosporangium* species, on juniper. Other specimens have rearing records of juniper only.

TYPES: Holotype: ♀. Wyoming, 6 mi NW Newcastle; 23 July 1965; R. W. Hodges; issued 3 Aug. 1965; reared from *Juniperus* sp. USNM. Paratypes: 12 ♂♂, 11 ♀♀. Same locality and data as for holotype; issued 3–18 July (4 ♂♂, 5 ♀♀). Same locality as for holotype; 13 July 1965 (1 ♂). Fort Valley, 7350′, 7½ mi NW Flagstaff, Coconino Co., Arizona; 22 June 1961; Ronald W. Hodges (2 ♀♀). Eureka, California; "8–6"; H. S. Barber (1 ♀). Buck Cr. Rgr. Sta., 5150′, Modoc Co., California; 5 June 1970; *Juniperus* sp.; Dietz and Powell (7 ♂♂, 1 ♀). Jacumba, California; *Juniperus californica*; reared 7 Dec. 1915; F. P. Keen (2 ♀♀). Paratypes to UCB; USNM.

*Juniperi* is very similar to *facula* on male genitalia but differs by the aedoeagus bearing a long spine on the right side and the much shorter valvae.

## *Periploca funebris* Hodges

PL. 5, FIG. 15. TEXT FIGS. 32 *d*; 46 *a*.

*Periploca funebris* Hodges, 1962, *Pan-Pacific Ent.*, **38**: 96, figs. 8, 22.
Type-locality: Madera Canyon, 4880′, Santa Rita Mts., Santa Cruz Co., Arizona. [CU]

*Funebris* is a dark, gray-black species with wing length varying from 3.8 to 4.0 mm. The antenna is dark gray, almost black basally becoming gray distally. Ocelli are present. Males have the vinculum produced posteriorly and deeply emarginate medially and have three or four teeth on the dorsal surface of the aedoeagus on the distal ⅓. The female genitalia are nearest those of *facula* but differ in that the heavily sclerotized, basal part of the ductus bursae is bent slightly beyond the middle in *funebris*, hook shaped in *facula*.

One specimen has been reared from the cone of oneseed juniper, *Juniperus monosperma* Engelmann.

*Funebris* has been collected in the Santa Rita Mountains of southern Arizona from early July through early October and central Colorado in late July.

## *Periploca serrulata* Hodges, NEW SPECIES

PL. 6, FIG. 23. TEXT FIG. 32 *h*.

*Periploca serrulata* Hodges.
Type-locality: Johnson's Pass, 5 mi W Clover, Tooele Co., Utah. [UCB]

Upper surface as figured. Head: tongue with dark-gray scales basally having shining yellow reflections; labial palpus dark gray, nearly black with shining yellow reflections, third segment slightly longer than second segment; frons shining pale yellowish gray; vertex and occiput dark gray, nearly black with shining yellow reflections; ocelli present; antenna dark gray, nearly black on scape and base of shaft becoming yellowish gray at ⅓ length of shaft to apex. Under surface of fore- and hindwing pale yellowish gray, costal margin of forewing darker gray. Foreleg dark gray, nearly black. Midleg dark gray, nearly black with some pale-gray scales at apices of tarsal segments. Hindleg mottled gray and pale yellowish gray. Wing length: 3.7 mm. Male genitalia: as in text figure 32 *h*; valva with costal margin strongly excavated at ⅖ length then tapering concave-up to acute apex; aedoeagus C-shaped, ending in relatively blunt apex, six or seven dorsal, triangular projections on distal ¼; vinculum strongly and narrowly produced posteriorly, posterior margin rounded medially. Female: no specimens available.

The moth was reared from cecidomyiid (?) galls on juniper.

TYPE: Holotype: ♂. Utah: Johnson's Pass, 5 mi W Clover, Tooele Co.; 13 Aug. 1973; em. 20 Jan. 1974; reared from cecid? galls on *Juniperus*; J. Powell; SBA slide 186. UCB.

The male genitalia of *serrulata* are very similar to those of *funebris* but differ in the unemarginate posterior margin of the vinculum and the broadly triangular spines on the dorsal surface of the aedoeagus. In *funebris* the vinculum is emarginate posteriorly, and the dorsal projections on the aedoeagus are very slender.

*Periploca dentella* Hodges, NEW SPECIES
TEXT FIG. 32*f*.

*Periploca dentella* Hodges.
Type-locality: Yosemite National Park, California. [USNM]

Head: mainly pale yellowish gray; labial palpus with lateral surface darker gray; ocelli present. Foreleg: mottled gray and pale yellowish gray, epiphysis and apices of tarsal segments paler yellowish gray. Midleg much as for foreleg but generally paler. Hindleg mainly pale yellowish gray. Wing length: 4.1 mm. Male genitalia: as in text figure 32 *f*; vinculum strongly produced on posterior margin, rounded medially; valva about $\frac{1}{2}$ length of aedoeagus, very slender on distal $\frac{1}{2}$, costal margin downcurved at $\frac{1}{3}$ length and apex acute; aedoeagus with truncated apex, dorsal surface with numerous teeth on distal $\frac{2}{5}$, some teeth large, most small. Female: no specimens available.

Larvae were found on California incensecedar, *Libocedrus decurrens* Torrey.

TYPE: Holotype: ♂. Yosemite Natl. Park; 23 May 1941; E. Ernest; reared from *Libocedrus decurrens*; USNM slide 4406. USNM.

The male genitalia of *dentella* are similar to those of *funebris* and *serrulata*. They differ from those of *funebris* in the rounded, posterior margin of the vinculum and from those of *serrulata* by the numerous small and large teeth on the dorsal surface of the aedoeagus.

*Periploca dipapha* Hodges
PL. 5, FIG. 16. TEXT FIG. 32 *g*.

*Periploca dipapha* Hodges, 1969, *Smithsonian Contrib. Zool.*, **18**: 6, figs. 12, 13.
Type-locality: Madera Canyon, 4880′, Santa Rita Mts., Santa Cruz Co., Arizona. [CU]

The two specimens of *dipapha* differ in coloration, probably a result of wear. The fresh specimen is shining dark gray to black; frons, vertex and occiput heavily overlaid with shining pale-yellowish scales; inner surface of labial palpus paler than very dark-gray lateral surface; antenna dark gray basally becoming pale yellow at $\frac{1}{5}$ length of shaft; foreleg mainly dark gray with some pale-gray scales at apex of tibia; mid- and hindleg much paler yellowish gray with tibia and tarsus darker than coxa, trochanter and femur. The worn specimen is mainly shining yellowish brown and pale yellow. Ocelli are present. Wing length varies from 3.8 to 4.0 mm. *Dipapha* can be separated from other species as indicated in the key.

The immature stages are unknown.

*Dipapha* has been collected in the southern Guadalupe Mountains in western Texas on 22 May and in the Santa Rita Mountains in southern Arizona on 27 July.

*Periploca labes* Hodges
PL. 5, FIG. 17. TEXT FIGS. 32 *e*; 46 *b*.

*Periploca labes* Hodges, 1969, *Smithsonian Contrib. Zool.*, **18**: 6, figs. 14, 33.
Type-locality: Vail Lake Road, 6500′, 9½ mi SE Flagstaff, Coconino Co., Arizona. [USNM]

*Labes* is a pale-yellow species and is thereby immediately distinguishable from the other species of *Periploca*. Ocelli are present. Wing length varies from 3.2 to 3.4 mm. In males the long projection arising from the inner surface of the valva is unique to this species. The presence of crescent-shaped signa, membranous ductus bursae, ostium bursae lying posterad of the anterior margin of the seventh sternum, seventh sternum simple and a single row of setae on the eighth segment, in combination, characterize females of *labes*.

The immature stages are unknown.

*Labes* has been collected south of Flagstaff, Arizona on 11 July and in the Santa Rita Mountains, Arizona on 17 July.

GENUS
*Siskiwitia* Hodges

*Siskiwitia* Hodges, 1969, *Smithsonian Contrib. Zool.*, **18**: 10.
Type-species: *Siskiwitia alticolans* Hodges, 1969. Original designation.

*Siskiwitia*, with three species, is endemic to North America and known only from Arkansas, Florida and southern Arizona. The species are very small, and the forewings are dark brown with shining yellow reflections. One species has a transverse, white band at $\frac{2}{3}$ the wing length. *Siskiwitia* is nearest *Periploca*, but the species usually can be separated by the small size and shining forewings. The presence of the uncus and a long costal lobe on the valva separate all species of *Siskiwitia* from *Periploca* species.

Characters of *Siskiwitia* are as follows: labial palpus slightly recurved, second segment usually slightly longer than third segment; antenna $\frac{2}{3}$

length of forewing, basal segment with pecten consisting of a single scale; ocelli present. Forewing lanceolate, apex acute; $Cu_1$ or $Cu_2$ absent; $M_1$ and $M_2$ either stalked or separate from $R_{4+5}$. Hindwing nearly linear, anal veins absent or represented by fold; $R_s$ and $M_1$ long stalked; cell open; $M_2$, $M_3$, $Cu_1$ and $Cu_2$ very weak to well developed. Male genitalia symmetrical; valva roughly triangular with heavily sclerotized costal lobe arising from near end of costa; aedoeagus curved dorsally, vesica without cornuti; uncus a heavily sclerotized, acute lobe. Female unknown.

The immature stages are unknown.

## KEY TO SPECIES OF *SISKIWITIA*

1. Forewing with transverse, white band at $\frac{2}{3}$ length .......... *alticolans* this page

— Forewing without transverse white band, uniformly dark .......... 2

2. Valva with costal lobe tapering from base to apex, uncus very heavily sclerotized throughout length, aedoeagus broadly curved dorsally at $\frac{1}{3}$ length (text figure 33 *c*), Arkansas .......... *latebra* this page

— Valva with costal lobe narrow near base, becoming broad distally, then tapering to acute apex; uncus very heavily sclerotized distally; aedoeagus more strongly curved dorsally at $\frac{1}{3}$ to $\frac{1}{2}$ length (text figure 33 *e*); Florida .......... *falcata* this page

### *Siskiwitia alticolans* Hodges

PL. 5, FIG. 43. TEXT FIGS. 2 *f*; 33 *a*.

*Siskiwitia alticolans* Hodges, 1969, *Smithsonian Contrib. Zool.*, **18**: 10, figs. 5, 19.
Type-locality: Bog Springs Campground, Madera Canyon, 5100′, Santa Rita Mts., Santa Cruz Co., Arizona. [USNM]

*Alticolans* is unique within the Chrysopeleiinae in having the dark-brown forewing overlaid with shining yellow to gold scales and in having a transverse white band at $\frac{2}{3}$ the length. Wing length varies from 2.5 to 2.8 mm. Male genital characters are as illustrated.

The immature stages are unknown.

*Alticolans* has been collected from 10 to 31 July in Madera Canyon at elevations of 4880 feet and 5100 feet.

### *Siskiwitia latebra* Hodges, NEW SPECIES

PL. 5, FIG. 44. TEXT FIG. 33 *c*.

*Siskiwitia latebra* Hodges.
Type-locality: Devil's Den State Park, Washington Co., Arkansas. [USNM]

Upper surface as figured. Head, thorax and forewing dark brown with shining yellow-gold reflections; tongue and mesal surface of labial palpus medium yellowish gray, lateral surface of labial palpus darker; labial palpus with second segment longer than third segment, slightly recurved, often appearing almost decumbent. Foreleg: coxa and trochanter mainly shining yellowish gray, femur slightly darker, tibia and tarsus yet darker. Midleg: much as for foreleg but with many pale-based scales on tibia and tarsus. Hindleg: much as for foreleg, apex of tibia and bases of tarsal segments pale yellowish gray. Wing length: 2.4–2.8 mm. Male genitalia: as in text figure 33 *c*; valva with costal process broad basally, gradually tapering to acute apex, somewhat angulate at $\frac{1}{2}$ length; uncus heavily sclerotized from base to apex and more heavily sclerotized on distal $\frac{1}{3}$.

TYPES: Holotype: ♂. Devil's Den State Park, Washington Co., Arkansas; 11 July 1966; R. W. Hodges. USNM. Paratypes: 27♂♂. Same locality as for holotype; 24 May–20 July (25 ♂♂). Hartford, Sebastian Co., Arkansas; 9 June 1966; R. W. Hodges (2 ♂♂). Paratypes to University of Arkansas; USNM.

Collection dates for the type series indicate two broods, one in late May and early June, a second from late June into July.

*Latebra* can be separated from its congeners as indicated in the key.

### *Siskiwitia falcata* Hodges, NEW SPECIES

PL. 6, FIG. 24. TEXT FIG. 33 *e*.

*Siskiwitia falcata* Hodges.
Type-locality: Archbold Biological Station, L. Placid, Florida. [USNM]

Upper surface as figured. Head, dorsal surface of thorax and upper surface of forewing dark brown with shining yellow to gold reflections; labial palpus with lateral surface brown, mesal surface yellowish gray becoming darker on third segment, second and third segments nearly equal in length. Foreleg: coxa, trochanter and femur gray with some yellowish reflections, apex of coxa paler than rest of coxa; tibia and tarsus dark gray. Midleg: coxa and trochanter shining yellowish gray; femur darker with shining yellowish reflections; tibia and tarsus mainly dark gray, apex of tibia and apices of first four tarsal segments with medium-gray scales. Hindleg: coxa yellowish gray with shining purple and yellow reflections; femur much the same; tibia and tarsus dark gray, apex of tibia and apices of first four tarsal segments pale gray. Wing length: 2.0–2.3 mm. Male genitalia: as in text figure 33 *e*; valva with costal lobe broad at base, narrow medially, then becoming broad before apex; aedoeagus upcurved before ½ length; uncus heavily sclerotized on distal ½, lightly sclerotized basally.

TYPES: Holotype: ♂. Florida, L. Placid, Archbold Biological Station; 16–22 May 1964; R. W. Hodges; USNM slide 7825. USNM. Paratype: 1 ♂. Florida, Fisheating Creek, Palmdale; 7–10 May 1964; R. W. Hodges. Paratype to USNM.

The forewing venation differs from that of *alticolans* by having only $M_1$ stalked with $R_{4+5}$. Two veins are absent from the cell; these are either $M_2$, $M_3$, $Cu_1$ or $Cu_2$. *Falcata* is very similar to *latebra* but can be separated as indicated in the key.

GENUS
## *Pristen* Hodges, NEW NAME

Gender: Feminine.
Type-species: *Neoploca corusca* Hodges, 1964. Original designation as type-species of *Neoploca* Hodges.

*Neoploca* Hodges, 1964, *Proc. U.S. Natl. Mus.*, **115**: 294, figs. 7, 11, 59.
Type-species: *Neoploca corusca* Hodges, 1964. Original designation.

NOTE—*Neoploca* Hodges, 1964, is preoccupied by *Neoploca* Matsumura, 1927, in the Insecta.

*Pristen* is a monobasic genus whose single species is known from the Santa Rita Mountains in southern Arizona. *Pristen* is close to *Periploca* and *Siskiwitia* but can be separated from the former by the male genitalia with an uncus and the relatively broad tegumen; males of *Periploca* lack an uncus and have a very narrow tegumen. Males of *Pristen* differ from those of *Siskiwitia* by lacking a costal lobe on the valva and having a coiled aedoeagus; those of *Siskiwitia* have a costal lobe and have the aedoeagus straight laterally. Females of *Pristen* differ from those of *Periploca* by having the ostium bursae on a small, sclerotized plate posterad of the sclerotized region of the seventh abdominal sternum. In *Periploca* species the ostium bursae is on the sclerotized part of the seventh abdominal sternum or on the anterior margin of the seventh abdominal sternum. Females of *Siskiwitia* are unknown.

Characters of *Pristen* are as follows: labial palpus recurved, second segment slightly longer than third segment; ocelli present; antenna ⅝ to ⅔ length of forewing, basal segment lacking pecten. Forewing lanceolate, apex acute; 12 veins present; $R_4$, $R_5$ and $M_1$ stalked; $Cu_2$ present at margin. Hindwing narrowly lanceolate, apex acute; cell open; eight veins present; $R_s$ and $M_1$ stalked; $M_2$, $M_3$, $Cu_1$ and $Cu_2$ apparently arising from common stem of cubitus. Male genitalia symmetrical; valva somewhat triangular; aedoeagus coiled, vesica without cornuti; gnathos absent; uncus present, broadly triangular. Female with ostium bursae well defined on small, sclerotized plate posterad of sclerotized part of seventh abdominal sternum; bursa copulatrix membranous, signa paired; apophyses slender, apophyses anteriores joined by sclerotized band; ovipositor nonpiercing.

The immature stages are unknown.

### *Pristen corusca* (Hodges), NEW COMBINATION
PL. 4, FIG. 49. TEXT FIGS. 3 *a*; 33 *b*; 47 *a*.

*Neoploca corusca* Hodges, 1964, *Proc. U.S. Natl. Mus.*, **115**: 294, figs. 19, 37, 59.
Type-locality: Madera Canyon, 4880′, Santa Rita Mts., Santa Cruz Co., Arizona. [CU]

Adults are similar to the shining species of *Periploca* such as *ceanothiella* and *gleditschiaeella* but are separable by the dark scales with shining yellow and gold reflections being unicolorous rather than being dark bordered. Wing length varies from 3.7 to 5.0 mm.

*Corusca* has been collected in southern Arizona from 10 July to 23 August.

GENUS
## *Synploca* Hodges

*Synploca* Hodges, 1964, *Proc. U.S. Natl. Mus.*, **115**: 295, figs. 8, 14, 16.
Type-species: *Synploca gumia* Hodges, 1964. Original designation.

*Synploca* is a monobasic genus that occurs from western Texas to Arizona. The moth is very similar to many species of *Periploca* and *Ithome* but can be separated from the former by having a series of white spots on the anterior margin of the third segment of the labial palpus and from the latter by the forewing lacking white scales on the anterior and posterior margins at $\frac{2}{3}$ to $\frac{3}{4}$ the length.

Characters of the genus are as follows: labial palpus very smooth scaled, second and third segments nearly equal in length; antenna about $\frac{3}{5}$ length of forewing, pecten absent; ocelli present. Forewing lanceolate, apex acute; $R_4$, $R_5$ and $M_1$ stalked, $M_1$ arising beyond branch of $R_4$ and $R_5$. Hindwing linear, cell open; eight veins present, $R_s$ and $M_1$ stalked. Males with symmetrical genitalia; valva with saccular margin more heavily sclerotized than very lightly sclerotized costal region; aedoeagus very stout, vesica lacking cornuti; uncus bilobed apically. Female with very heavily sclerotized genital plate on anterior margin of eighth abdominal sternum; ductus bursae membranous; corpus bursae with paired signa half-ring shaped; apophyses slender, subequal in length, apophyses anteriores connected by narrow, sclerotized band.

The immature stages are unknown.

*Synploca gumia* Hodges

PL. 5, FIG. 18. TEXT FIGS. 3 *b*; 33 *d*; 47 *d*, *g*.

*Synploca gumia* Hodges, 1964, *Proc. U.S. Natl. Mus.*, **115**: 295, figs. 24, 38, 60.
Type-locality: Madera Canyon, 4880′, Santa Rita Mts., Santa Cruz Co., Arizona. [CU]

*Gumia* can be confused with *Ithome concolorella*, but the forewing of *gumia* is nearly uniformly dark gray, whereas that of *concolorella* has white scales on the anterior and posterior margins at $\frac{2}{3}$ to $\frac{3}{4}$ the length and has black scales followed by white ones at $\frac{2}{3}$ the length of the fold. Wing length varies from 3.3 to 4.1 mm.

*Gumia* has been collected in the Chisos Mountains in western Texas on 6 June, northern Arizona from 25 July to 5 September and the Santa Rita Mountains in southern Arizona from 30 June to 1 November. More than one brood is indicated, but the definite number is unknown.

GENUS
## *Stilbosis* Clemens

*Stilbosis* Clemens, 1860, *Proc. Acad. Nat. Sci. Philadelphia*, **12**: 170.
Type-species: *Stilbosis tesquella* Clemens, 1860. Monotypy.

*Aeaea* Chambers, 1874, *Can. Ent.*, **6**: 73. NEW SYNONYMY.
Type-species: *Aeaea ostryaeella* Chambers, 1874. Monotypy.

*Amaurogramma* Braun, 1919, *Ent. News*, **30**: 261. NEW SYNONYMY.
Type-species: *Amaurogramma extensa* Braun, 1919. Original designation.

*Stilbosis* consists of 34 species of nearly uniform appearing dark-gray or yellow to orange-brown moths. Eight species occur in the East from Massachusetts, southern Ontario and Minnesota south to Florida and eastern Texas. Ten species occur in Arizona, and two occur in coastal California. Most of the other species are in the northern Neotropics, and two species are reported from India. Each species has patches of raised scales on the forewing in the following pattern: one posterad of the fold at $\frac{1}{4}$ the wing length, one on the posterior margin and one costad of the fold slightly beyond $\frac{1}{2}$ the wing length, one costad of the fold just before the end of the fold and four or five small patches along the outer margin.

Superficially, many species of *Stilbosis* and *Chrysopeleia purpuriella* are inseparable; however, the male genitalia of *purpuriella* are highly specialized with very large aedoeagus, reduced valvae and eighth sternum modified as clasping organs. Although the female of *purpuriella* is different from any *Stilbosis* species, the differences may be at the specific level.

Structural characters of *Stilbosis* are as follows: labial palpus recurved, often not reaching vertex; antenna $\frac{1}{2}$ to $\frac{3}{5}$ wing length, pecten a single, slender scale; ocelli present. Forewing narrowly lanceolate, apex acute; 11 or 12 veins present; $R_4$, $R_5$ and $M_1$ stalked; $Cu_2$ absent or developed at margin. Hindwing linear, $R_s$ and $M_1$ stalked, cell open, $M_2$ and $M_3$ very weakly developed or absent. Male genitalia symmetrical or asymmetrical; valva with two or more lobes, sometimes highly asymmetrical; aedoeagus C-shaped in one plane or twisted and coiled, vesica without cornuti; gnathos absent; uncus simple or lobed, rarely absent. Female with ostium bursae on seventh abdominal sternum, seventh abdominal sternum usually modified as pairs of lateral plates, folded flanges or heavily sclerotized bands preceding and succeeding ostium bursae; ductus bursae straight or coiled, unsclerotized to heavily sclerotized; paired signa usually rose-thorn shaped; apophyses relatively slender, subequal in length, apophyses anteriores connected by slender, sclerotized band; ovipositor nonpiercing.

Larvae of the orange-hued species are leaf tiers on Leguminosae. Larvae of the gray species are leaf miners on Betulaceae, Fagaceae and Leguminosae. They place the frass outside of the mine either in a mass or generally over the surface of the mine. Pupation normally occurs in leaf litter outside of the mine.

I am describing four species from southern Arizona that are known from females only, and it is probable that two of them correspond with species known from males only. The similarity of characters other than for genitalia does not allow association of the sexes. When the species have been reared and the adults associated, the proper synonymies can be made.

## KEY TO SPECIES OF *STILBOSIS*

1. Males .................................... 2
— Females .................................... 16

2. Forewing shades of gray .................... 3
— Forewing yellow, orange or orange brown ..... 14

3. Uncus simple .............................. 4
— Uncus bifurcate .......................... 12

4. Valva with apex bifurcate (text figure 34 *b*) .... 5
— Valva divided into lobes (text figure 34 *a*) ...... 6

5. Valva with mesal lobe rounded apically (text figure 34 *b*), Arizona and California .......... *dulcedo* p. 117
— Valva with mesal lobe acute apically (text figure 34 *c*), East Coast west to Arkansas .......... *venifica* p. 117

6. Valva with costa separate, directed dorsally, rest of valva entire (text figure 34 *a*) ...... *juvantis* p. 117
— Valva with costal margin entire, if separate, directed posteriorly, sacculus separate, at least apically (text figure 36 *a*) ............... 7

7. Valvae symmetrical ......................... 8
— Valvae asymmetrical .................. *extensa* p. 119

8. Valva with costa separate, at least apically ..... 9
— Valva with costal margin entire (text figure 34*f*) .................. *venatrix* p. 117

9. Uncus tapering to apex ..................... 10
— Uncus expanded at apex (text figure 35 *b*) .................. *stipator* p. 118

10. Saccus with apex acute (text figure 34 *e*) ...... 11
— Saccus with apex rounded (text figures 35 *a*) .................. *risor* p. 118

11. Valva with lobe extending from posterior part of costa, apex of valva acute (text figure 34 *e*) .................. *ostryaeella* p. 117
— Valva without lobe extending from costa, apex of valva rounded (text figure 34 *d*) .................. *quadricustatella* p. 118

12. Uncus with each arm bifid (text figure 35*f*) .................. *sagana* p. 118
— Uncus with each arm simple (text figure 35 *e*) .................. 13

13. Valvae symmetrical (text figure 35 *c*) ...... *victor* p. 118
— Valvae asymmetrical (text figures 35 *d*, *h*) .................. *rhynchosiae* p. 119

14. Valva with distal lobe cylindrical, apex rounded .................. 15
— Valva with distal lobe twisted, curved, expanded at apex (text figures 36 *c*) ...... .................. *lonchocarpella* p. 122

15. Sacculus narrower than medial part of uncus, ventral margin relatively straight (text figure 36 *b*) .................. *tesquella* p. 121
— Sacculus broader than medial part of uncus, ventral margin sinuate (text figure 36 *e*) ... *nubila* p. 122

16. Forewing yellow, orange or orange brown .... 17
— Forewing shades of gray .................. 19

17. Ostium bursae more than $\frac{2}{5}$ width of seventh sternum (measured at level of greatest width of ostium bursae) .................. 18

— Ostium bursae ⅓ width of seventh sternum (measured at level of greatest width of ostium bursae) (text figure 49 *d*). . . . *lonchocarpella* p. 122

18. Ostium bursae round, ductus bursae becoming very narrow before corpus bursae (text figure 49 *c*). . . . . . . . . . . . . . . . . . . . *tesquella* p. 121

— Ostium bursae somewhat triangular, ductus bursae broad (text figure 49 *b*). . . . . . . . . . . *nubila* p. 122

19. Ostium bursae preceded by narrow, semicircular, sclerotized band (text figure 49 *f*) . . . . . . . . . . . . . . . . . . . . . . . . . . . . . . . . . . *rotunda* p. 121

— Ostium bursae not preceded by such a band. . . 20

20. Ostium bursae near anterior margin of seventh sternum, seventh sternum heavily sclerotized from ostium bursae to anterior margin (text figure 48 *a*). . . . . . . . . . . . . . . . . . . 21

— Ostium bursae medial on seventh sternum, seventh sternum not heavily sclerotized from ostium bursae to anterior margin. . . . . . . 22

21. Sclerotized band preceding ostium bursae with strong longitudinal line, placed to left of center (text figure 48 *a*). . . . . . . . . . . . . . . . *risor* p. 118

— Sclerotized band preceding ostium bursae lacking longitudinal line (text figure 48 *d*) . . . . . . . . . . . . . . . . . . . . . . . . . . . . . . . . . . *sagana* p. 118

22. Ostium bursae preceded and succeeded by broad sclerotized plates (text figure 48 *f*). . . . . . 23

— Ostium bursae not surrounded by broad plates. . . . . . . . . . . . . . . . . . . . . . . . . . . . . . . . . . 25

23. Seventh sternum with series of lateral striae from level of ostium bursae to anterior margin (text figure 48 *f*). . . . . . . . . . . . *rhynchosiae* p. 119

— Seventh sternum without series of lateral striae. . . . . . . . . . . . . . . . . . . . . . . . . . . . . . . . . 24

24. Lightly sclerotized flap extending beyond posterior margin of seventh sternum (text figure 48 *h*). . . . . . . . . . . . . . . . . . . . . . . . . *pagina* p. 119

— Lacking such a flap (text figure 48 *e*). . . . . . *extensa* p. 119

25. Seventh sternum with ostium bursae at apex of conical projection (text figure 49 *a*). . . . *scleroma* p. 120

— Seventh sternum without medial projection (text figure 47 *b*). . . . . . . . . . . . . . . . . . . . . . . . 26

26. Seventh sternum with posterior margin emarginate medially. . . . . . . . . . . . . . . . . . . . . . 27

— Seventh sternum with posterior margin not emarginate. . . . . . . . . . . . . . . . . . . . . . . . . . . . . 30

27. Seventh sternum with pair of sclerotized flaps running from anterior to posterior margin (text figure 48 *c*). . . . . . . . . . . . . . . . . . . 28

— Seventh sternum without such a longitudinal structure, heavily sclerotized region with series of concentric ridges (text figure 47 *b*). . . . . . . . . . . . . . . . . . . . . . . . . . . . . . . . . . 29

28. Seventh sternum with pair of flanges beginning near lateral margin (text figure 48*c*). . *placatrix* p. 118

— Seventh sternum with pair of flanges beginning near middle (text figure 48 *g*). . . . . . . . *victor* p. 118

29. Seventh sternum with posterior margin invaginated medially (text figure 47 *b*). . . . . *dulcedo* p. 117

— Seventh sternum with posterior margin straight (text figure 47 *f*). . . . . . . . . . . . . . *venifica* p. 117

30. Narrow, sclerotized, Y-shaped band extending anteriorly from seventh sternum (text figure 49 *e*). . . . . . . . . . . . . . . . . . . . . . . . *ornatrix* p. 120

— Lacking such a band. . . . . . . . . . . . . . . . . . . . . . 31

31. Seventh sternum with two submedial conical extensions extending from anterior margin almost to posterior margin (text figure 47 *e*). . . . . . . . . . . . . . . . . . . . . . . *ostryaeella* p. 117

— Seventh sternum without such extensions. . . . . 32

32. Seventh sternum with broad scales laterally, heavily sclerotized submedial area running posteriorly from anterior margin to ⅔ length, then angling and extending to lateral margins (text figure 48 *b*). . . . . . . . . . . . . . *stipator* p. 118

— Seventh sternum without broad scales laterally, narrow, submedial sclerotized band running from ostium bursae to anterior margin (text figure 47 *c*). . *quadricustatella* p. 118

*Stilbosis juvantis* (Hodges), NEW COMBINATION

PL. 5, FIG. 19. TEXT FIG. 34 *a*.

*Aeaea juvantis* Hodges, 1964, *Proc. U.S. Natl. Mus.*, **115**: 300, fig. 32.
Type-locality: Madera Canyon, 4880′, Santa Rita Mts., Santa Cruz Co., Arizona. [CU]

Males can be recognized by the combination of symmetrical genitalia and bulbous, free costal lobe on the valva. Females are unknown. Wing length varies from 3.0 to 3.2 mm.

The immature stages are unknown.

*Juvantis* is known from the type-locality, and specimens were collected at light from 1 to 9 July.

*Stilbosis dulcedo* (Hodges), NEW COMBINATION

PL. 5, FIG. 20. TEXT FIGS. 34 *b*; 47 *b*.

*Aeaea dulcedo* Hodges, 1964, *Proc. U.S. Natl. Mus.*, **115**: 301, figs. 29, 48.
Type-locality: Westwood Hills, Los Angeles Co., California. [ANSP]

Males of *dulcedo* are characterized by the symmetrical and simple valvae. Females are most similar to those of *venifica* but have the caudal margin of the seventh sternum emarginate. Wing length varies from 2.5 to 3.5 mm.

Larvae have been reared from California live oak, *Quercus agrifolia* Nee, and canyon live oak, *Q. chrysolepis* Liebmann, and interior live oak, *Q. wislizenii* A. DC. Opler (1974: 41) indicates that the species is univoltine. The blotch mines were found from 1 December to 8 July. Larvae pass the frass through the entry hole and attach it to the leaf. Pupation takes place in a thin, frass-covered cocoon.

*Dulcedo* has been collected in California from Mendocino County south in the Coast Ranges to San Diego County and in Placer County in the Sierra Nevada foothills. Laboratory emergence dates are from 15 April to 15 July according to Opler.

*Stilbosis venifica* (Hodges), NEW COMBINATION

PL. 5, FIGS. 21, 22. TEXT FIGS. 34 *c*; 47 *f*.

*Aeaea venifica* Hodges, 1964, *Proc. U.S. Natl. Mus.*, **115**: 301, fig. 31.
Type-locality: Putnam Co., Illinois. [USNM]

*Venifica* is most similar to *dulcedo* but can be separated as indicated in the key. Wing length varies from 3.0 to 3.5 mm.

The immature stages are unknown, but where *venifica* has been most commonly collected, the dominant tree species are oak and hickory.

*Venifica* has been collected at Barnstable, Massachusetts on 12 and 26 July, Toronto, Ontario on 1 June, District of Columbia area on 20 and 28 July, central Illinois on 30 May to 26 June, Louisiana on 23 April, northwest Arkansas from 23 May to 6 July and the Chisos Mountains in western Texas on 6 June.

*Stilbosis venatrix* (Hodges), NEW COMBINATION

PL. 5, FIG. 23. TEXT FIG. 34 *f*.

*Aeaea venatrix* Hodges, 1964, *Proc. U.S. Natl. Mus.*, **115**: 301, fig. 28.
Type-locality: Madera Canyon, 4880′, Santa Rita Mts., Santa Cruz Co., Arizona. [CU]

*Venatrix* can be recognized as indicated in the key. Only males are known. Wing length varies from 3.4 to 3.7 mm.

The immature stages are unknown.

The type series was collected from 2 to 30 July.

*Stilbosis ostryaeella* (Chambers), NEW COMBINATION

PL. 5, FIG. 24. TEXT FIGS. 34 *e*; 47 *e* (McD 7708).

*Aeaea ostryaeella* Chambers, 1874, *Can. Ent.*, **6**: 74.
Type-locality: Kentucky. [MCZ]

The combination of characters: valva with free costal lobe, uncus with slightly indented apex and symmetrical genitalia will define males of *ostryaeella*. Females of *ostryaeella* can be recognized by the pair of triangular lobes on the seventh sternum. Wing length varies from 2.5 to 3.2 mm.

Larvae have been reared from *Ostrya virginiana* (Mill.) K. Koch. According to Clemens (1872: 27) the larva makes a linear mine between two primary veins and arranges the frass in two lateral rows within the mine. Pupation occurs outside the mine in a cocoon spun in leaf litter on the ground. Mosher (1916: fig. 95) illustrates the pupa.

*Ostryaeella* occurs from eastern Massachusetts and the District of Columbia area west through Ohio and Kentucky to northwest Arkansas. Specimens have been collected from 22 May to 25 July.

*Stilbosis quadricustatella* (Chambers), NEW COMBINATION

PL. 5, FIGS. 25, 26. TEXT FIGS. 34 *d*; 47 *c* (McD. 7708).

*Aeaea quadricustatella* Chambers, 1880, *Jour. Cincinnati Soc. Nat. Hist.*, **2**: 186.
Type-locality: Waco, Texas. [MCZ]

*Quadricustatella* can be separated from other species in the genus as indicated in the key. Wing length varies from 2.5 to 3.3 mm.

The immature stages are unknown.

*Quadricustatella* has been collected in south-central Florida from 1 May to 8 June, in south-central Mississippi on 8, 9 June, northwest Arkansas from 5 June to 22 July and eastern Texas without date. It has been collected commonly at light in Arkansas and Florida.

*Stilbosis stipator* (Hodges), NEW COMBINATION

PL. 5, FIG. 27. TEXT FIGS. 35 *b*; 48 *b*.

*Aeaea stipator* Hodges, 1964, *Proc. U.S. Natl. Mus.*, **115**: 303, figs. 30, 46.
Type-locality: Vail Lake Road, 6500′, 9½ mi SE Flagstaff, Coconino Co., Arizona. [USNM]

Genital characters separating *stipator* from other species of *Stilbosis* are as given in the key. Wing length varies from 3.4 to 3.7 mm.

The immature stages are unknown.

*Stipator* has been collected north and south of the Mogollon Rim in northern Arizona between 5400 feet and 6500 feet. None were collected farther north at elevations of 7200 feet and above. All specimens have been collected at light between 11 July and 1 September.

*Stilbosis risor* (Hodges), NEW COMBINATION

PL. 5, FIG. 28. TEXT FIGS. 35 *a*; 48 *a*.

*Aeaea risor* Hodges, 1964, *Proc. U.S. Natl. Mus.*, **115**: 304, fig. 33.
Type-locality: Madera Canyon, 4880′, Santa Rita Mts., Santa Cruz Co., Arizona. [CU]

*Risor* can be recognized by the characters given in the key. Wing length varies from 3.0 to 3.2 mm. Females have been tentatively associated with males on date range, forewing length and coloration of the head.

The immature stages are unknown.

*Risor* has been collected at elevations of 4880 feet and 5600 feet from 19 to 29 September in Madera Canyon.

*Stilbosis victor* (Hodges), NEW COMBINATION

PL. 5, FIG. 29. TEXT FIGS. 35 *c*, *e*, *i*; 48 *g*, *i*.

*Aeaea victor* Hodges, 1964, *Proc. U.S. Natl. Mus.*, **115**: 304, figs. 27, 27a, 27b, 50.
Type-locality: Cherrydale, Virginia. [USNM]

*Victor* can be recognized by the characters given in the key. Wing length varies from 3.2 to 3.3 mm.

Larvae feed on post oak, *Quercus stellata* Wang. Frass is extruded from the mine and is placed generally over the upper surface of the mine.

*Victor* occurs from the New Jersey pine barrens, south to northern Virginia and west to northwest Arkansas. Field-collected specimens have been taken from 30 May to 11 June. Specimens in the reared type series have a date 22 September 1922 which probably refers to a collection date for the mines.

*Stilbosis placatrix* (Hodges), NEW COMBINATION

PL. 5, FIG. 30. TEXT FIGS. 48 *c*, *j*.

*Aeaea placatrix* Hodges, 1969, *Smithsonian Contrib. Zool.*, **18**: 8, fig. 35.
Type-locality: Devil's Den State Park, Washington Co., Arkansas. [USNM]

*Placatrix* is nearest *victor* on genital characters but can be separated by the seventh sternum having the pair of lateral flanges starting near the lateral margins rather than near the middle of the segment as for *victor*. Wing length is 3.2 mm.

The immature stages are unknown.

The holotype was collected on 20 June.

*Stilbosis sagana* (Hodges), NEW COMBINATION

PL. 5, FIG. 31. TEXT FIGS. 35 *f*; 48 *d*.

*Aeaea sagana* Hodges, 1964, *Proc. U.S. Natl. Mus.*, **115**: 305, fig. 34.
Type-locality: Madera Canyon, 4880′, Santa Rita Mts., Santa Cruz Co., Arizona. [CU]

Males of *sagana* are similar to those of *rhynchosiae* but differ by the bifid apex of each arm of the uncus. Females of *sagana* are nearest those of *risor* but differ in having the basal part of the ductus bursae sclerotized and by having an unlobed, sclerotized plate on the anterior margin of the seventh abdominal sternum. I have associated females with this species on the basis of wing length, head color pattern and date range. Wing length varies from 3.3 to 4.0 mm.

The immature stages are unknown.

*Sagana* has been collected in Madera Canyon from 23 August to 24 October at elevations of 4880 feet and 5600 feet. Most specimens were collected in late September.

*Stilbosis rhynchosiae* (Hodges), NEW COMBINATION
PL. 5, FIG. 32. TEXT FIGS. 35 *d, g, h, j*; 48*f*.

*Aeaea rhynchosiae* Hodges, 1964, *Proc. U.S. Natl. Mus.*, **115**: 305, figs. 35, 35a, 35b, 47.
Type-locality: Chickasaw Forest, near Henderson, Tennessee. [ANSP]

Males of *rhynchosiae* can be separated from those of *extensa* by the bifid apex of the uncus and females from those of *pagina* by the submedial striae on the seventh abdominal sternum. Wing length varies from 3.0 to 3.3 mm.

Larvae are leaf miners on hairy rhynchosia, *Rhynchosia tomentosa* (L.) H. & A., in the Leguminosae. The larva starts a mine at the midrib and ejects the frass which collects at the initial point of entry. More than one mine may be made.

*Rhynchosiae* is known from the type-locality in southwestern Tennessee. Specimens emerged from 28 July to 2 August.

*Stilbosis extensa* (Braun), NEW COMBINATION
PL. 5, FIG. 33. TEXT FIGS. 36 *a*; 48 *e* (McD. 7709).

*Amaurogramma extensa* Braun, 1919, *Ent. News*, **30**: 262.
Type-locality: Loma Linda, California. [ANSP]

Males of *extensa* can be recognized by the combination of asymmetrical genitalia and simple uncus. Diagnostic characters of females are as given in the key. Wing length varies from 3.2 to 3.8 mm.

The immature stages are unknown.

*Extensa* has been collected in California from Contra Costa County south to San Diego County between 14 May and 26 July.

*Stilbosis pagina* Hodges, NEW SPECIES
PL. 6, FIG. 25. TEXT FIG. 48 *h*.

*Stilbosis pagina* Hodges.
Type-locality: Madera Canyon, 4880′, Santa Rita Mts., Santa Cruz Co., Arizona. [CU]

Upper surface as figured. Head: base of tongue, maxillary palpus and frons pale gray with shining yellow and pale purple reflections, apices of individual scales paler gray; vertex and occiput medium gray, scales tipped with pale gray; labial palpus medium to dark gray, individual scales tipped with pale gray to nearly white, base and apex of third segment white, third segment shorter than second segment; antenna with scape and basal two or three segments of shaft dark gray, apices of scales very pale gray, nearly white, rest of shaft with alternate ½ segments dark brown and paler yellowish brown. Foreleg: coxa medium to dark gray, apices of individual scales pale gray, apex of segment pale gray; femur, tibia and tarsus dark gray, nearly black, apices of many scales narrowly pale gray to white, apices of tarsal segments yellowish gray or yellowish brown, slightly paler than bases of segments. Midleg: coxa and trochanter pale yellowish gray; femur, tibia and tarsus medium to dark gray, becoming nearly black on tarsus, most scales with pale apices; tibia with an oblique, pale streak at ⅔ length and apex pale; apices of first, second, fourth and fifth tarsal segments pale gray, nearly white. Hindleg: coxa, trochanter and femur pale yellowish gray with shining yellow and purple reflections; tibia medium to dark gray with apices of scales broadly tipped with pale gray, an oblique white streak running from base of first pair of spurs to dorsal margin and apex white; tarsus becoming darker from first to fifth segments, apex of each segment pale gray to nearly white. Wing length: 4.3 mm. Male genitalia: no specimens available. Female genitalia: as in text figure 48 *h*; seventh abdominal sternum with a pair of lateral, heavily sclerotized plates at ½ length, posterior margin broadly rounded; ductus bursae broad, coiled once; walls of corpus bursae and ductus bursae spiculose; signa paired, each an inwardly directed, rose-thorn shaped spine; apophyses anteriores and posteriores approximately equal in length.

The immature stages are unknown.

TYPE: Holotype: ♀. Madera Canyon, 4880′, Santa Rita Mts., Arizona; 24 July 1959; R. W. Hodges; USNM genital slide 4570; USNM wing slide 4571. CU.

The female genitalia of *pagina* are nearest those of *rhynchosiae* but differ in having paired, sclerotized plates separated by a lightly sclerotized zone on the seventh abdominal sternum. In *rhynchosiae* the plate is solid anteriorly and is narrowly separated posteriorly. Also, the spines of the signa are acute in *pagina*, blunt in *rhynchosiae*.

*Stilbosis ornatrix* Hodges, NEW SPECIES
PL. 6, FIG. 26. TEXT FIG. 49 *e*.

*Stilbosis ornatrix* Hodges.
Type-locality: Madera Canyon, 4880′, Santa Rita Mts., Santa Cruz Co., Arizona. [CU]

Upper surface as figured. Head: tongue pale gray with shining yellow reflections; vertex medium gray with shining yellow and purple reflections, all scales tipped with pale gray; vertex and occiput dark gray with pale-tipped scales; labial palpus mainly dark gray, nearly black, mesal posterior surface yellowish gray, scales tipped with white, second segment shorter than third segment; antenna dark gray, nearly black, scales with pale-gray, nearly white apices on scape, shaft becoming alternating ½ segments of dark gray and pale yellowish gray distally. Foreleg: coxa and trochanter medium gray, scales with pale apices; femur slightly darker gray, scales with pale apices; tibia and tarsus very dark gray, nearly black, most scales with pale apices, apex of tibia and apices of tarsal segments one, two, four and five pale gray to off white. Midleg: coxa and trochanter pale yellow with shining yellow and purple reflections; femur pale gray, becoming slightly darker on distal ½; tibia mainly dark gray, nearly black with an oblique pale-gray to white streak at base, one at middle and at apex; tarsus dark gray, nearly black, base and apex of first segment and apices of second, fourth and fifth segments pale gray, nearly white. Hindleg: coxa, trochanter and femur pale yellowish gray with shining yellow and purple reflections; tibia dark gray, nearly black, scales tipped with pale gray, faint oblique streaks at base, at base of first pair of spurs and at apex; tarsus dark gray, apices of scales narrowly white and scale bases pale, base and apex of first segment and apices of second, third and fourth segments pale gray. Wing length: 3.2–3.4 mm. Male genitalia: no specimens available. Female genitalia: as in text figure 49 *e*; ostium bursae near anterior margin of seventh abdominal sternum, surrounded and followed by two semicircular invaginated areas that extend nearly to posterior margin of segment, surfaces of invaginated areas lined with dark scales arranged in concentric circles; ductus bursae coiled once, moderately heavily sclerotized to ⅝ length; walls of ductus bursae and corpus bursae spiculose; paired signa conical; apophyses anteriores stouter and shorter than apophyses posteriores.

The immature stages are unknown.

TYPES: Holotype: ♀. Madera Canyon, 4880′, Santa Rita Mts., Arizona; 20 July 1959; R. W. Hodges; USNM genital slide 4577. CU. Paratypes: 3 ♀♀. Same locality as for holotype; 10–27 July 1959. Paratypes to CU; USNM.

The shape of the seventh abdominal sternum and the dark scales arranged in concentric, parallel rows will distinguish *ornatrix* from other species of *Stilbosis*.

*Stilbosis scleroma* Hodges, NEW SPECIES
PL. 6, FIG. 27. TEXT FIG. 49 *a*.

*Stilbosis scleroma* Hodges.
Type-locality: Madera Canyon, 4880′, Santa Rita Mts., Santa Cruz Co., Arizona. [CU]

Upper surface as figured. Head: tongue pale yellowish gray with shining yellow and purple reflections; frons pale to medium gray on lower ½, upper ½ very pale gray, all scales with apices paler than bases; scales above eye and on occiput dark gray with pale apices; labial palpus dark gray, nearly black laterally, many scales broadly tipped with white, apex of second segment and base, middle and apex of third segment banded with white; antenna with scape dark gray, nearly black, scales tipped with paler gray at apex, basal ⅗ of shaft alternating ½ segments of pale gray and brown, rest brown with every fourth ½ segment pale gray. Foreleg: coxa and trochanter pale yellowish white with shining yellow and purple reflections; femur medium to dark gray, scales with pale apices; tibia dark gray, nearly black, a white streak just beyond base, some white scales at base of epiphysis and at apex; tarsus with many dark-gray scales, base and apex of first segment and apices of second, third and fifth segments off white. Midleg: coxa, trochanter and femur pale yellowish gray with shining yellow and purple reflections; tibia dark gray with some banding as on first segment; tarsus dark gray, nearly black, scales with pale bases, base and apex of first segment, and apices of second, fourth and fifth segments pale gray to white. Hindleg: coxa, trochanter and femur pale yellowish gray with shining yellow and purple reflections, femur becoming medium gray; tibia and tarsus with color pattern as for midleg and dark areas somewhat paler. Wing length: 3.0 mm. Male genitalia: no specimens available. Female genitalia: as in text figure 49 *a*; ostium bursae at middle of seventh abdominal sternum; basal ½ to ⅔ of ductus bursae moderately heavily sclerotized; walls of ductus bursae and corpus bursae smooth, not spiculose; paired signa rose-thorn shaped; apophyses anteriores slightly more heavily sclerotized and broader than apophyses posteriores.

The immature stages are unknown.

TYPE: Holotype: ♀. Madera Canyon, 4880′, Santa Rita Mts., Arizona; 3 Aug. 1959; R. W. Hodges; USNM genital slide 4591; USNM wing slide 4592. CU.

*Scleroma* can be recognized by the nearly straight, heavily sclerotized part of the ductus bursae, the large ostium bursae slightly beyond ½ the length of the seventh abdominal sternum and the very pale-gray upper part of the frons and vertex.

### *Stilbosis rotunda* Hodges, NEW SPECIES

PL. 6, FIG. 28. TEXT FIG. 49*f*.

*Stilbosis rotunda* Hodges.
Type-locality: Madera Canyon, 4880′, Santa Rita Mts., Santa Cruz Co., Arizona. [CU]

Upper surface as figured. Head: tongue pale gray with yellow and purple reflections; lower part of frons medium gray, scales tipped with pale gray, upper part of frons paler and scales broadly tipped with very pale gray; vertex and occiput medium to dark gray, scales tipped with pale gray; labial palpus mainly dark gray, nearly black, most scales tipped with pale gray to nearly white, apex of second and third segments pale gray to white, dorsal surface of second segment pale, second segment longer than third segment; antenna with scape and base of shaft dark gray, scales tipped with pale gray, rest of shaft alternating ½ segments of yellowish brown and dark gray. Foreleg: mainly dark gray, most scales tipped with very pale gray; apex of tibia and apices of tarsal segments one, two, four and five off white; all surfaces with shining yellow and purple reflections. Midleg: coxa and trochanter pale yellowish gray with shining yellow and purple reflections; femur medium to pale gray, scales tipped with paler gray; tibia dark gray, a faint oblique gray streak at ⅛ length, one at ⅝ length and one at apex; tarsal segments dark gray, apices of first and second segments off white, apices of fourth and fifth segments medium to pale gray. Hindleg: coxa and trochanter gray to yellowish gray with shining yellow and purple reflections; tibia dark gray, an oblique, off-white streak at first pair of spurs and at apex; tarsus dark gray, apices of first, second, third and fourth segments pale gray to off white. Wing length: 3.2–3.8 mm. Male genitalia: no specimens available. Female genitalia: as in text figure 49 *f*; ostium bursae on middle of seventh abdominal sternum; a pair of lateral flanges adjacent to ostium bursae, seventh abdominal sternum concave to margin lateral of the flanges, concavity with heavily pigmented scales; ductus bursae very lightly sclerotized on basal ½, looped once; walls of ductus bursae and corpus bursae spiculose; signa rose-thorn shaped, paired; apophyses heavily sclerotized, apophyses anteriores slightly shorter than apophyses posteriores.

The immature stages are unknown.

TYPES: Holotype: ♀. Madera Canyon, 4880′, Santa Rita Mts., Arizona; 9 July 1959; R. W. Hodges; USNM genital slide 4594. CU. Paratypes: 23 ♀♀. Same locality as for holotype; 2 July–22 Aug. 1959 (22 ♀♀). Same locality except for elevation 5600′; 1 Aug. 1959 (1 ♀). Paratypes to CU; USNM.

The female genitalia ally *rotunda* with *victor*, but the shape of the paired flanges laterad of the ostium bursae will separate the two species.

### *Stilbosis tesquella* Clemens

PL. 5, FIGS. 35, 36. TEXT FIGS. 36 *b*; 49 *c* (McD. 7741).

*Stilbosis tesquella* Clemens, 1860, *Proc. Acad. Nat. Sci. Philadelphia*, **12**: 170.
Type-locality: None given. [Pennsylvania ?] [ANSP]

*Laverna* (?) *quinquicristatella* Chambers, 1881, *Jour. Cincinnati Soc. Nat. Hist.*, **3**: 293.
Type-locality: Amherst, Massachusetts. [lost]

*Tesquella* is very similar to *nubila* but can be separated as indicated in the key and by the foreleg having the third tarsal segment unicolorous, that of *nubila* is dark basally and white apically. Wing length varies from 3.0 to 4.7 mm. Specimens vary in the development of yellowish to yellowish-brown spots at ¾ the wing length.

Busck (1903: 202) stated that the larva ties leaflets of southern hogpeanut, *Amphicarpa bracteata* (L.) Fernald. He described the larva as "Head light yellow; eye spots black; body yellowish white, with thoracic shield, anal plate, thoracic feet, and all tubercles, black brown." Larvae have also been reared from *Lespedeza*.

*Tesquella* occurs from New York, Minnesota and eastern Kansas south to Washington D.C., the southern Appalachians in North Carolina and northwestern Arkansas. Moths have been collected from May through early August.

*Stilbosis nubila* Hodges

PL. 1, FIG. 64; PL. 5, FIGS. 34, 37. TEXT FIGS. 3 *c*; 36 *e*; 49 *b*.

*Stilbosis nubila* Hodges, 1964, *Proc. U.S. Natl. Mus.*, **115**: 311, figs. 23, 40.
Type-locality: Madera Canyon, 4880′, Santa Rita Mts., Santa Cruz Co., Arizona. [CU]

In addition to characters cited previously *nubila* is somewhat longer in wing length (4.0–5.2 mm) than is *tesquella* (3.0–4.7 mm). The pair of spots at $\frac{3}{4}$ to $\frac{4}{5}$ the wing length vary in development from faint and discrete to strong and coalescing.

Nothing is known of the immature stages.

*Nubila* occurs from Utah south through Arizona and perhaps into Mexico. Most specimens have been collected in the west fork of Oak Creek, Arizona from 4 to 19 July. In southern Arizona moths have been collected from 1 August to 10 October.

*Stilbosis lonchocarpella* Busck

PL. 5, FIG. 38. TEXT FIGS. 36 *c*; 49 *d*.

*Stilbosis lonchocarpella* Busck, 1934, *Ent. Americana*, **13** (n.s.): 157.
Type-locality: Santiago de las Vegas, Cuba. [USNM]

*Lonchocarpella* can be separated from the other species of *Stilbosis* by the nonmetallic, yellowish-brown forewings. Wing length varies from 3.3 to 3.8 mm.

Busck (1934: 157) reported the larva feeding on *Lonchocarpus sericeus* H. B. & K., an herbaceous legume which does not occur in Florida.

*Lonchocarpella* has been collected on Key Largo, Florida, 16 December–20 April, and in Cuba.

GENUS
## *Chrysopeleia* Chambers

*Chrysopeleia* Chambers, 1874, *Can. Ent.*, **6**: 72.
Type-species: *Chrysopeleia purpuriella* Chambers, 1874. Monotypy.

The single species of *Chrysopeleia* occurs in eastern North America. Superficially, the adults are inseparable from those of the gray species of *Stilbosis* but can be separated on genital characters as indicated under *Stilbosis*.

Characters of the genus are as follows: head smooth scaled; labial palpus recurved, second segment longer than third; antenna $\frac{2}{3}$ length of forewing, pecten present; ocelli present. Forewing narrowly lanceolate, apex acute; 11 or 12 veins present; $R_4$, $R_5$ and $M_1$ stalked; $Cu_2$ present at margin. Hindwing linear; cell open; $M_1$ stalked with $R_s$. Male with seventh abdominal sternum heavily sclerotized; eighth abdominal sternum heavily sclerotized, asymmetrical as two separate, lateral plates each with long, stout setae on the posterior margin; genitalia with valvae reduced, asymmetrical; aedoeagus massive relative to rest of genitalia; uncus relatively slender, slightly curved. Female with ostium bursae in middle of seventh abdominal sternum, a pair of asymmetrical lateral folds slightly anterad and laterad of ostium bursae; basal $\frac{1}{2}$ to $\frac{3}{5}$ of ductus bursae curved, heavily sclerotized, rest directed anterad and nearly membranous; corpus bursae nearly membranous, walls spiculose; signa paired; apophyses slender, apophyses anteriores shorter than apophyses posteriores and joined by sclerotized band dorsally.

The larvae may be leaf miners on *Robinia*.

*Chrysopeleia purpuriella* Chambers

PL. 5, FIG. 39. TEXT FIGS. 3 *d*; 36 *d, f*; 50 *a* (McD. 7707).

*Chrysopeleia purpuriella* Chambers, 1874, *Can. Ent.*, **6**: 73.
Type-locality: Kentucky. [MCZ]

*Purpuriella* can be confused with any of the eastern, gray species of *Stilbosis* but can be separated on genital characters as indicated in the key to genera. Wing length varies from 3.0 to 3.8 mm.

Chambers (1880: 64) reared *purpuriella* from *Robinia pseudoacacia* L. Larval mines, tentatively determined as those of *purpuriella*, were small, thin silk webs at the juncture of a vein with the midrib; however, he did not segregate these particular mines and questioned whether they might be the early stage mines of *Filatima pseudacaciella* (Chambers). Chapman (*in litt.*) reared adults from cocoons that were attached to small branches of apple. Even though the cocoons were abundant, no feeding larvae were found.

*Purpuriella* occurs from New York south to Washington, D.C. and west to northwestern Arkansas. Field-collected specimens have been taken from 3 to 25 July.

GENUS
## *Walshia* Clemens

*Walshia* Clemens, 1864, *Proc. Ent. Soc. Philadelphia*, **2**: 418.
Type-species: *Walshia amorphella* Clemens, 1864. Monotypy.

*Walshia* is a genus of 11 small moths that occur in North America. Several undescribed species are known from South America. Most species have the color pattern on the forewing consisting of a broad dark-gray, nearly black basal band followed by a pale-yellow or yellowish-brown area and becoming darker to the apex. *Walshia* is allied with *Stilbosis* and *Nepotula*. It differs from *Stilbosis* by the forewing having $M_1$ separate and the hindwing having the cell closed and $M_1$ separate from $R_s$. *Walshia* differs from *Nepotula* by the forewing having $Cu_2$ well defined and the hindwing having the cell closed. In males the tegumen is narrow and the uncus is poorly defined in *Walshia*, whereas the tegumen is broad and the uncus is heavily sclerotized in *Nepotula*.

Characters of *Walshia* are as follows: labial palpus with second and third segments subequal in length; ocelli present; antenna with pecten consisting of a single scale at the base of the first segment. Forewing with 12 veins; patches of raised scales present; $M_1$ separate from $R_{4+5}$; $Cu_1$ and $Cu_2$ separate basally, $Cu_2$ well defined. Hindwing with eight veins, cell closed, $R_s$ and $M_1$ separate. Male genitalia symmetrical or asymmetrical; valva with well-defined fracture zones; aedoeagus C-shaped; tegumen and vinculum narrow, heavily sclerotized; uncus triangular, lightly sclerotized. Female with apophyses slender, ovipositer non-piercing, apophyses anteriores joined by a sclerotized band; ostium bursae well defined, usually at anterior margin of seventh abdominal sternum; bursa copulatrix spiculose, signa absent or present and paired; sixth sternum sometimes with patches of modified scales laterally.

Larvae are stem gall makers, usually on Leguminosae, rarely on members of other plant families.

## KEY TO SPECIES OF *WALSHIA*

1. Males . . . . . . . . . . 2
— Females . . . . . . . . . . 9

2. Valvae symmetrical . . . . . . . . . . 3
— Valvae asymmetrical . . . . . . . . . . 8

3. Valva sublinear or quadrate, costal area not expanded (text figures 37 *a, c*) . . . . . . . . . . 4
— Valva otherwise, with well-developed costal area . . . . . . . . . . 5

4. Valva quadrate (text figure 37 *a*) . . . . . . . . . *elegans* p. 124
— Valva sublinear (text figure 37 *c*) . . . . *miscecolorella* p. 124

5. Valva with distal lobe (text figure 37 *d*) . . . . . . . . 6
— Valva without distal lobe (text figure 36 *g*) . . . . . . . . . . *particornella* p. 124

6. Valva with distal lobe ⅔ basal width (text figure 37 *d*) . . . . . . . . . . *amorphella* p. 125
— Valva with distal lobe ⅓ basal width (text figure 37 *b*) . . . . . . . . . . 7

7. Valva with a heavily sclerotized band on basal ½ (text figure 38 *a*) . . . . . . . . . . *exemplata* p. 126
— Valva without heavily sclerotized band on basal ½ (text figure 37 *b*) . . . . . . . . . . *floridensis* p. 125

8. Right valva with distal lobe slightly smaller than that on left valva (text figure 38 *b*) . . *similis* p. 126
— Right valva with distal lobe much smaller than that on left valva (text figure 37 *e*) . . . . *dispar* p. 126

9. Signa present and paired . . . . . . . . . . 10
— Signa absent . . . . . . . . . . 11

10. Ostium bursae near posterior margin of seventh abdominal sternum (text figure 50 *e*) . . . . . . . . . . *particornella* p. 124
— Ostium bursae near middle of seventh abdominal sternum (text figure 50 *g*) . . . . . *elegans* p. 124

11. Ductus bursae broadening gradually to tear drop shaped corpus bursae (text figure 50 *f*) . . . . . . . . . . *amorphella* p. 125
— Ductus bursae not broadening gradually to corpus bursae, corpus bursae ellipsoidal or spheroidal, well set off from ductus bursae . . . . 12

12. Ductus bursae directed to left at ostium bursae, a heavily sclerotized band on left margin at base (text figure 50 *c*) . . . . . *miscecolorella* p. 124
— Ductus bursae directed anteriorly at ostium bursae, without heavily sclerotized band on left margin . . . . . . . . . . 13

13. Ductus bursae with narrow, heavily sclerotized, ventral band at base; corpus bursae spheroidal (text figure 50 *b*) .......... *floridensis* p. 125
— Ductus bursae without sclerotized band, corpus bursae spheroidal or ellipsoidal ....... 14

14. Corpus bursae spheroidal (text figure 50 *d*) .................................. *exemplata* p. 126
— Corpus bursae ellipsoidal (text figures 51 *a*, *b*) ...................................... 15

15. Eighth abdominal segment with caudal row of setae having midventral gap (text figure 51 *a*) ................................ *similis* p. 126
— Eighth abdominal segment with row of setae continuous or with very small midventral gap (text figure 51 *b*) ............ *dispar* p. 126

### *Walshia particornella* (Busck)

PL. 4, FIG. 50. TEXT FIGS. 36 *g*; 50 *e* (McD. 7721).

*Perimede particornella* Busck, 1919, *Proc. Ent. Soc. Washington*, **21**: 96.
Type-locality: Plummers Island, Maryland. [USNM]

*Particornella* is very similar to *elegans* but can be separated as indicated under *elegans*. It is also similar to many species of *Perimede* from which it differs by the closed cell in the hindwing and genital differences as given in the key to genera. Wing length varies from 5.0 to 6.8 mm. Fresh specimens have extremely dark forewings with raised scales; worn specimens are much paler and lack the patches of raised scales.

The immature stages are unknown.

*Particornella* occurs from Washington, D.C., central Illinois and northwestern Arkansas south to coastal Mississippi and Texas. Adults have been collected at light from 14 April in Mississippi to mid-July in Illinois and Washington, D.C.

### *Walshia elegans* Hodges, NEW SPECIES

PL. 6, FIG. 29. TEXT FIGS. 37 *a*; 50 *g*.

*Walshia elegans* Hodges.
Type-locality: Fluker, Tangipahoa Parish, Louisiana. [USNM]

Upper surface as figured. Head: tongue very dark gray basally, becoming paler distally; labial palpus nearly unicolorous dark gray, scale bases pale; frons dark gray with shining yellow and purple reflections; vertex and occiput darker gray with less shining reflections than frons; antenna nearly black, distal nine segments white or yellowish white. Foreleg very dark gray, apex of fourth tarsal segment and entire fifth segment medium gray. Midleg much as for foreleg but with apices of tarsal segments pale gray. Hindleg much as for midleg but slightly paler, apex of tibia, base and apex of first tarsal segment and apices of remaining segments pale gray. Wing length: 5.2 mm. Male: as in text figure 37 *a*; eighth sternum with heavily sclerotized band on anterior margin; eighth tergum a narrow, heavily sclerotized band on anterior margin; genitalia symmetrical; valva with nearly parallel anterior and posterior margins, saccular margin slightly sinuous, a slightly downcurved point on ventrodistal margin; aedoeagus relatively stout, apex dilated. Female: as in text figure 50 *g*; ostium bursae at middle of seventh abdominal sternum; signa paired, each with a strong, inwardly directed, curved spine.

The immature stages are unknown.

TYPES: Holotype: ♂. Louisiana, Tangipahoa Parish, Fluker; 13 Sept. 1971; G. Strickland; SBA slide 224. USNM. Paratype: 1 ♀. Arkansas, Devil's Den State Park, Washington Co.; 13 July 1966; R. W. Hodges. Paratype to USNM.

*Elegans* is very similar to *particornella*. It can be separated from *particornella* by the pale apices of the tarsal segments and quadrate valvae. Also, the scale tufts of the forewings seem to be better developed than in *particornella*.

### *Walshia miscecolorella* (Chambers) (Sweetclover Root Borer*)

PL. 1, FIG. 63; PL. 4, FIGS. 53–59. TEXT FIGS. 37 *c*; 50 *c* (McD. 7743).

*Laverna miscecolorella* Chambers, 1875, *Can. Ent.*, **7**: 51.
Type-locality: Bosque Co., Texas. [MCZ]

*Laverna miscecalonella*, misspelling.

In the East *miscecolorella* can be confused with most other species in the genus; therefore, genitalia should be examined for recognition. It is the only species of *Walshia* that occurs in the West. Forewing color pattern varies greatly among populations but tends to be consistent within populations. Specimens from Nebraska and northwestern

Arkansas are the smallest known for the species. Wing length varies from 4.5 to 9.0 mm. Females from Alexandria, Virginia have the heavily sclerotized part of the ductus bursae longer than observed elsewhere.

Larvae are stem and root borers in several legumes, including *Lupinus arboreus* Sims, *L. chamissionis* Eschschoff, *Astragalus* species, *Arachis* species, six species of *Melilotus* and *Cirsium vulgare* (Savi) Tenore in the Compositae. Manglitz, Gorz and Stevens (1971: 1154) described the life history in southeastern Nebraska. Two generations occur per year. Overwintering occurs in the larval stage. Pupation normally is in a small chamber in the soil outside the root or lower stem but rarely takes place in the stem. According to Manglitz *et al.* *miscecolorella* is a potential pest on sweetclover in the north-central United States.

*Miscecolorella* occurs from Washington, D.C., the Prairie Provinces and southern British Columbia south to Florida and southern California. Moths have been collected from spring through fall over much of the range. Extremes in dates are 13 January in Florida and 23 November in California.

### *Walshia amorphella* Clemens

PL. 4, FIGS. 51, 52. TEXT FIGS. 3 *e*; 37 *d*; 50 *f* (McD. 7743).

*Walshia amorphella* Clemens, 1864, *Proc. Ent. Soc. Philadelphia*, **2**: 419.
Type-locality: Illinois. [ANSP]

Most specimens of *amorphella* are somberly colored. Major differences in forewing coloration apparently result from wear. The key based on genital characters should be used for identification. Males can be determined by brushing the tip of the abdomen to see the shape of the valvae. Wing length varies from 6 to 7 mm.

The larva is a stem gall maker in *Amorpha fruticosa* L. Riley (1870: 132) and Brandhorst (1962: 477) described the life history. The moth is univoltine with adults emerging in April and May. Eggs are deposited on the stem, and the larva burrows into the stem. Overwintering occurs in the larval stage, and the gall is most conspicuous at this time. The plant normally dies beyond the gall, and Brandhorst supposed that the dead, dried plant tissue was necessary for larval survival. Galls from living stems yielded dead larvae. Pupation occurs in the gall about three weeks before adults emerge.

*Amorphella* has a limited range from northern Illinois, southeastern Minnesota and eastern Kansas south to northeastern Texas. Adults are found in late April and early May.

### *Walshia floridensis* Hodges, NEW SPECIES

PL. 6, FIG. 30. TEXT FIGS. 37 *b*; 50 *b*.

*Walshia floridensis* Hodges.
Type-locality: L. Alfred, Florida. [USNM]

Upper surface as figured. Head: tongue pale yellow to yellowish gray; maxillary palpus pale yellowish gray; labial palpus medium gray on outer surface of second and third segments, apex of second segment with pale-orange scales, apex of third segment with dark-gray scales, medial surface paler than lateral surface, all scales tipped with pale gray; antenna dark gray, several ½ segments medium yellowish gray dorsally, pecten a single seta; frons and lower part of vertex pale gray with faint, shining purple reflections, rest of vertex and occiput brown with some shining purple reflections. Foreleg: mainly dark gray, apex of coxa and trochanter pale gray, many scales on distal ½ of epiphysis pale gray, apices of first, second, fourth and fifth tarsal segments pale gray. Midleg: much as for foreleg: a few pale scales beyond midlength of tibia, apex pale yellowish gray, spur slightly paler than dark part of tibia; apices of first, second, fourth and fifth tarsal segments pale gray. Hindleg: coxa mainly pale yellow; trochanter and femur medium to dark gray, many pale-yellow scales on anterior surface; tibia very dark gray, a white streak at base of each pair of spurs, spurs dark, individual scales pale-gray tipped; apex of each tarsal segment pale gray to off white. Abdomen: yellowish orange or orange brown dorsally, posterior margins of segments shining pale gray to off white, genital tuft yellowish white to off white; ventral surface mainly off white, second, third, fourth and fifth segments with some brown scales laterally, fifth and sixth segments with numerous gray scales. Wing length: 4.2–6.0 mm. Male genitalia: as in text figure 37 *b*; valvae symmetrical, saccular margin lacking heavily sclerotized band in first fracture zone area, distal lobe with acute projection on ventral margin; aedoeagus slightly sinuous in posterior view. Female: as in text figure 50 *b*; ostium bursae on anterior margin of seventh abdominal sternum; ductus bursae with heavily sclerotized ring at ⅓ length followed by heavily spiculose area to corpus bursae; corpus bursae spherical, walls spiculose; apophyses anteriores and posteriores nearly equal in length;

sixth abdominal sternum with paired lateral zones bearing broad, dark scales.

Larvae were reared from *Petalostemon pinnatum* (Walter) Blake in the Leguminosae. Label data do not indicate whether the larva makes a gall.

TYPES: Holotype: ♂. #143; L. Alfred, Florida; 31 Dec. 1929; L. J. Bottimer; on *Kuhnistera pinnata*. USNM. Paratypes: 24 ♂♂, 9 ♀♀. Same data as for holotype; Nov. 1929–Jan. 1930. Paratypes to USNM.

The type series is relatively uniform in color, pattern and shade. Adults can be confused with several species of *Walshia*, and the genitalia should be examined for determination.

*Walshia exemplata* Hodges
PL. 4, FIG. 62. TEXT FIGS. 38 *a*; 50 *d*.

*Walshia exemplata* Hodges, 1961, *Bull. Brooklyn Ent. Soc.*, **56**: 73, fig. 11.
Type-locality: Plummers Island, Maryland. [LACM]

*Exemplata* is known from four males and one female that vary in color from orange brown to nearly black. Genital characters given in the key should be examined for identification. The male genitalia are nearest those of *floridensis*. Wing length varies from 6.5 to 7.0 mm.

The immature stages are unknown.

*Exemplata* has been collected in Maryland, northwestern Arkansas in July and McClellanville, South Carolina, Jackson, Mississippi on 24, 25 September and Fluker, Louisiana on 18 September.

*Walshia similis* Hodges
PL. 4, FIG. 61. TEXT FIGS. 38 *b*; 51 *a*.

*Walshia similis* Hodges, 1961, *Bull. Brooklyn Ent. Soc.*, **56**: 71, figs. 4, 10, 13.
Type-locality: Cherry Hill Recreation Area, 2000′, Route 107, Oconee Co., South Carolina. [CU]

*Similis* is nearest *dispar* on genital characters but can be separated as indicated in the key and by reference to the text figures. Moths vary in the amount of dark scales on the forewings; some specimens are very dark. Wing length varies from 5.3 to 6.5 mm.

The immature stages are unknown.

*Similis* is known from Martha's Vineyard, Massachusetts, southern Michigan and north-central Illinois south to South Carolina. Moths have been collected from 1 June to 30 September with most records from July and August.

*Walshia dispar* Hodges
PL. 4, FIG. 60. TEXT FIGS. 37 *e*; 51 *b*.

*Walshia dispar* Hodges, 1961, *Bull. Brooklyn Ent. Soc.*, **56**: 72, figs. 1, 8, 16.
Type-locality: Six Mile Creek, Ithaca, New York. [CU]

*Dispar* is nearest *similis* in genital characters but can be separated as indicated in the key. Males have the most asymmetrical valvae of any *Walshia* species. Specimens vary in the amount of dark scales on the head, thorax and forewings. Wing length varies from 6 to 7 mm.

The immature stages are unknown.

*Dispar* occurs from Ottawa, Ontario nearly due south to Washington, D.C. and Hot Springs, Virginia. Specimens have been collected from 20 June to 19 August.

GENUS
*Nepotula* Hodges

*Nepotula* Hodges, 1964, *Proc. U.S. Natl. Mus.*, **115**: 296, figs. 2, 16.
Type-species: *Nepotula secura* Hodges, 1964. Original designation.

*Nepotula* is a monobasic genus whose single species occurs in northwestern Florida. It is closely allied with *Walshia*, and differences are given under *Walshia*.

Characters of *Nepotula* are as follows: labial palpus with second and third segments subequal in length; ocelli present; antenna $\frac{3}{5}$ to $\frac{2}{3}$ length of forewing, pecten apparently absent. Forewing: with patches of raised scales; lanceolate; 12 veins present; cell closed; $M_1$ separate from $R_{4+5}$; $Cu_2$ weakly developed at margin. Hindwing: lanceolate; cell open but crossvein between $M_1$ and $M_2$ weakly developed; $R_s$ and $M_1$ separate. Male with symmetrical genitalia; valva with a costal fold extending apically, a broad apical lobe; aedoeagus C-shaped, heavily sclerotized; uncus short, heavily sclerotized, triangular. Female unknown.

The immature stages are unknown.

*Nepotula secura* Hodges
PL. 4, FIG. 63. TEXT FIGS. 3 *f*; 38 *c*.

*Nepotula secura* Hodges, 1964, *Proc. U.S. Natl. Mus.*, **115**: 296, figs. 17, 61.
Type-locality: Pensacola, Florida. [USNM]

*Secura* can be recognized by the colored figure and by the illustration of the male genitalia. The forewing has three patches of raised scales in a slightly oblique line at $\frac{1}{2}$ the wing length; two are anterad of the fold, and one is posterad of the fold. The hindtibia is cream white on the anterior $\frac{1}{2}$ and dark gray on the posterior $\frac{1}{2}$. Wing length: 7.5 mm.

The single specimen of *secura* was collected on 28 October.

GENUS

## *Afeda* Hodges, NEW GENUS

Gender: Feminine.
Type-species: *Afeda biloba* Hodges. New species.

*Afeda* is a monobasic genus with the single species known from Key Largo, Florida. Wing shape and venation ally *Afeda* with *Walshia* and *Nepotula*, particularly *Walshia*. *Afeda* differs by lacking raised scales on the forewing, the male genitalia lacking an uncus and the female genitalia having relatively stout apophyses and well-developed signa.

Characters of the genus are: labial palpus with second segment slightly longer than third and thickened with scales on distal $\frac{1}{2}$; ocelli present; antenna $\frac{3}{4}$ length of forewing, pecten absent. Mid- and hindtibiae and tarsi dorsoventrally flattened with scales. Forewing lanceolate, 12 veins present, raised scales absent, $M_1$ separate from $R_{4+5}$; $Cu_2$ well developed. Hindwing narrowly lanceolate, cell closed, eight veins present, $R_s$ and $M_1$ separate. Male with symmetrical genitalia; valva bilobed on distal $\frac{1}{2}$, each lobe with numerous, stout setae apically; aedoeagus massive relative to rest of genitalia, somewhat C-shaped; uncus absent. Female with ostium bursae dorsad of sclerotized, somewhat triangular plate on anterior margin of seventh abdominal sternum; wall of bursa copulatrix spiculose; ductus bursae coiled once; signa paired, somewhat saddle shaped; apophyses anteriores shorter than apophyses posteriores, each stout, no sclerotized band connecting apophyses anteriores.

The immature stages are unknown.

### *Afeda biloba* Hodges, NEW SPECIES

PL. 6, FIG. 31. TEXT FIGS. 3 *g*; 38 *e*; 51 *c*.

*Afeda biloba* Hodges.
Type-locality: Tavernier, Monroe Co., Florida. [USNM]

Upper surface as figured. Head: base of tongue and maxillary palpus pale gray or yellowish gray; labial palpus with lateral surface orange brown on second and third segments, a few pale-tipped scales at apex of second segment, medial surface somewhat paler than lateral surface, base of third segment pale gray; frons, vertex and occiput pale gray medially, scales bordering eye brown, a row of dark-brown scales behind eye; basal segment of antenna pale gray to yellowish gray, shaft alternating $\frac{1}{2}$ segments of yellowish gray and darker yellowish brown. Foreleg: coxa and trochanter mottled medium and dark brown with some pale-tipped scales on outer margin; femur pale gray on basal $\frac{1}{2}$ anteriorly, dark brown posteriorly, distal $\frac{1}{2}$ medium gray, scales tipped with pale gray; tibia brown on basal $\frac{2}{5}$, followed by a pale-gray band at middle, brown beyond it, pale gray at apex; first segment of tarsus gray brown, pale at base and apex, last four segments dark brown, apices slightly paler. Midleg: coxa, trochanter and femur mainly pale gray, many scales darker before apices, femur with many brown scales at apex; tibia mainly grayish brown, a pale-gray scale tuft at $\frac{3}{8}$ length and one at apex, spurs brown; tarsus brown, slightly paler at apices. Hindleg: much as for midleg; tibia medium brown at base, followed by broad, pale-gray streak to base of first pair of spurs, then brown to slightly darker brownish-gray band at apex, first pair of spurs brown, shorter one the darker, distal spurs medium gray; tarsus with first segment medium brown at base becoming gray brown apically and on ventral margin, remaining segments similar but with pale scales predominating. Wing length: 4.3–5.0 mm. Male genitalia: as in text figure 38 *e*; valva with two parallel lobes distally, extending $\frac{1}{2}$ length of valva, each with numerous, stout setae apically; vinculum heavily sclerotized; tegumen less heavily sclerotized than vinculum, narrow; uncus absent; aedoeagus massive, curved, cornuti absent. Female: as in text figure 51 *c*; seventh abdominal sternum with somewhat triangular, sclerotized plate on anterior margin, margin with two lobes, each with short setae; ostium bursae at posterior margin of triangular plate; ductus bursae coiled once, walls spiculose; corpus bursae with spiculose walls; signa paired, somewhat saddle shaped; apophyses heavily sclerotized; ovipositor somewhat modified, perhaps for piercing.

TYPES: Holotype: ♂. Tavernier, Monroe Co., Florida; 6 Oct. 1955; J. M. Todd; USNM genital slide 4877. USNM. Paratypes: 1 ♂, 2 ♀♀. Same data as for holotype; 2, 6, 9 Oct. Paratypes to CPK; USNM.

In general appearance *biloba* might be confused with *Obithome punctiferella*, but *biloba* lacks raised scales on the forewing, and the cell in the hindwing is closed.

GENUS
## *Perimede* Chambers

*Perimede* Chambers, 1874, *Can. Ent.*, **6**: 51. Type-species: *Perimede erransella* Chambers, 1874. Monotypy.

*Perimede* is a nearctic genus with 11 species that occur from Nova Scotia and southern Quebec, south to Florida and west to Arizona and south to Guerrero, Mexico. Male genital characters ally *Perimede* somewhat with *Sorhagenia*, but numerous differences are present such as the divided uncus, the well-developed aedoeagus, the glands associated with the valvae and the Y-shaped sclerites on the eighth abdominal sternum of *Perimede.* All species are similar in forewing maculation with dark-gray ground color and dark spots, one at $\frac{1}{2}$ the length of the fold, one at the end of the cell and one at the apex, each preceded or followed by white scales. Additional dark spots may be present.

Characters of the genus are as follows: labial palpus with second segment longer than third segment; ocelli present; antenna $\frac{1}{2}$ to $\frac{2}{3}$ length of forewing; pecten present or absent. Forewing lanceolate; raised scales absent; 12 veins present; $R_{4+5}$ and $M_1$ separate; $Cu_2$ well developed, arising before $\frac{3}{4}$ length of cell. Hindwing narrowly lanceolate; cell open; eight veins present; $R_s$ and $M_1$ arising separately, $M_1$ curved toward $R_s$ at base. Male with eighth abdominal sternum having three sclerites in shape of Y; a pair of glands arising between eighth and ninth abdominal sterna very closely associated with genitalia, each gland ending in a spine; genitalia symmetrical; valva simple, generally cylindrical; vinculum short, heavily sclerotized; tegumen moderately long medially, less heavily sclerotized than vinculum; aedoeagus short, slightly curved, ankylosed with juxta; uncus bilobed. Female with slender apophyses, apophyses anteriores not joined by sclerotized band; ovipositor nonpiercing; bursa copulatrix very lightly sclerotized, walls spiculose; ostium bursae well defined, usually on posterior part of seventh abdominal sternum; signa paired, each a single spine arising from a sclerotized base; seventh sternum symmetrical or asymmetrical.

Larvae have been reared from leaves of *Taxodium* Rich., *Ulmus* L., *Quercus* L. and *Carya* Nutt.

### KEY TO SPECIES OF *PERIMEDE*

1. Males . . . . . . . . . . 2
— Females . . . . . . . . . . 9

2. Accessory glands associated with genitalia with symmetrical spines . . . . . . . . . . 3
— Accessory glands associated with genitalia with asymmetrical spines . . . . . . . . . . 7

3. Spines on accessory glands shorter than valvae . . . . . . . . . . 4
— Spines on accessory glands as long or longer than valvae . . . . . . . . . . 6

4. Apex of valva drawn to point ventrally, spine on accessory gland curved at $\frac{1}{2}$ to $\frac{2}{3}$ length (text figure 40 *a*) . . . . . . . . . . *circitor* p. 130
— Apex of valva rounded; spine on accessory gland nearly straight, slightly rounded at $\frac{1}{3}$ length . . . . . . . . . . 5

5. Accessory gland with posterior margin nearly straight ventrad of spine (text figure 39 *g*) . . . . . . . . . . *parilis* p. 130
— Accessory gland with posterior margin curved ventrad of spine (text figure 39 *c*) . . *grandis* p. 130

6. Accessory glands with sinuate spines, apex of valva rounded (text figure 39 *f*) . . . . . . . . . . *latris* p. 130
— Accessory glands with straight spines, apex of valva drawn to point (text figure 38 *d*) . . . . . . . . . . *erransella* p. 129

7. Spine on right accessory gland coiled (text figure 39 *b*) . . . . . . . . . . *battis* p. 129
— Spine on right accessory gland not coiled (text figure 40 *b*) . . . . . . . . . . 8

8. Uncus with apices blunt; valva narrow, saccular margin with an expanded shoulder near base (text figure 40 *b*) . . . . . . . *ricina* p. 131

— Uncus with apices acute; valva broader, saccular margin without a distinct shoulder near base (text figure 40 *h*) . . . . . . . . . . . . . *falcata* p. 131

9. Seventh sternum symmetrical . . . . . . . . . . . . . . 11
— Seventh sternum asymmetrical . . . . . . . . . . . . . 10

10. Ostium bursae at middle of seventh sternum (text figure 52 *c*) . . . . . . . . . . . . . . . . . . . . . . . *ricina* p. 131
— Ostium bursae at apex of seventh sternum (text figure 51 *e*) . . . . . . . . . . . . . . . . . . . . . . . *latris* p. 130

11. Ostium bursae extending beyond posterior margin of seventh tergum, on raised prominence (text figure 52 *d*) . . . . . . . . . . . . . . . . *falcata* p. 131
— Ostium bursae not extending beyond posterior margin of seventh tergum, on raised prominence or not . . . . . . . . . . . . . . . . . . . . . . . . 12

12. Lamella antevaginalis narrowly triangular, extending beyond anterior margin of seventh sternum (text figure 51 *h*) . . . . . . . . . . . . . . . *parilis* p. 130
— Lamella antevaginalis variously shaped, not extending beyond anterior margin of seventh sternum . . . . . . . . . . . . . . . . . . . . . . . . . . 13

13. Ostium bursae open to ventral surface, lamella antevaginalis a sclerotized rim (text figure 51 *g*) . . . . . . . . . . . . . . . . . . . . . . . . . . . . . 14
— Ostium bursae not immediately open to ventral surface, lamella antevaginalis extending beyond ostium bursae ventrally making a flange (text figure 51 *f*) . . . . . . . . . . . . 16

14. Ostium bursae laterally compressed; lamella antevaginalis with a narrow, medial slit, extending to posterior margin of seventh sternum (text figure 51 *g*) . . . . . . . . *circitor* p. 130
— Ostium bursae broad . . . . . . . . . . . . . . . . . . . . . . 15

15. Lamella antevaginalis broadest at center, ostium bursae nearly at posterior margin of seventh sternum, no sclerotized band posterad of opening (text figure 52 *b*) . . . . . . . *maniola* p. 131
— Lamella antevaginalis evenly narrow, a sclerotized band between ostium bursae and posterior margin of seventh sternum (text figure 52 *a*) . . . . . . . . . . . . . . . . . . . . . . . . . . . *erema* p. 131

16. A pair of invaginations containing scales anterior to ostium bursae (text figure 51 *f*) . . *battis* this page
— Lacking such invaginations (text figure 51 *d*) . . . . . . . . . . . . . . . . . . . . . . . . . . . . *erransella* this page

### *Perimede erransella* Chambers

PL. 5, FIG. 45. TEXT FIGS. 3 *h*; 38 *d*; 51 *d* (McD. 7720).

*Perimede erransella* Chambers, 1874, *Can. Ent.*, **6**: 52.
Type-locality: Covington, Kentucky. [MCZ]

*Erransella* is the commonest species in the genus, and within its geographic range it can be confused with *parilis*, *circitor*, *erema* and *maniola*. Male genital characters ally it most closely with *circitor* from which it is distinguished by the long, terminal spines on the accessory glands. Females of *erransella* are distinct by the shape of the genital plate which has a posterior, conical, medial extension ventrad of the ostium bursae. Most variation in maculation is a result of wear: fresh specimens are very dark gray; worn specimens can be yellowish gray. Wing length varies from 4 to 6 mm.

Larvae have been collected on *Liriodendron tulipifera* L., *Carya tomentosa* Nutt., *Quercus velutina* Lamarck, *Q. coccinea* Muenchh., *Taxodium distichum* (L.) Rich. and *Ulmus* species.

*Erransella* occurs from Nova Scotia and southern Quebec south to Florida and Louisiana. In the northern part of its range moths have been collected from May to September. In Florida they may be present throughout the year.

### *Perimede battis* Hodges

PL. 5, FIG. 52. TEXT FIGS. 39 *b*, *h*; 51 *f*.

*Perimede battis* Hodges, 1962, *Proc. Ent. Soc. Washington*, **64**: 153, figs. 5, 10, 13.
Type-locality: Madera Canyon, 4880′, Santa Rita Mts., Santa Cruz Co., Arizona. [CU]

*Battis* is most nearly related to *erransella*, *latris* and *grandis*. Genital differences are given in the key. Within its geographic range *battis* can be separated from *latris* by the third segment of the labial palpus being gray laterally; that of *latris* is white laterally. Color variation is a result of wear. Fresh specimens are very dark gray. Wing length varies from 3.4 to 5.2 mm.

The immature stages are unknown.

*Battis* is known from three localities in Arizona:

Pine, the Santa Rita Mountains and Peña Blanca Canyon, and it has been collected from 20 July to 17 September.

### *Perimede latris* Hodges

PL. 5, FIG. 51. TEXT FIGS. 39 *e,f*; 51 *e*.

*Perimede latris* Hodges, 1962, *Proc. Ent. Soc. Washington*, **64**: 152, figs. 2, 8, 14.
Type-locality: Madera Canyon, 4880′, Santa Rita Mts., Santa Cruz Co., Arizona. [CU]

Superficially, *latris* is very similar to *battis*, but it can be separated from *battis* as indicated in the key and under *battis*. It is a far less common species in Madera Canyon, and adults fly in a more restricted time period. Wing length varies from 3.5 to 5.1 mm.

The immature stages are unknown.

*Latris* is known from the type-locality, and specimens have been collected from 24 July to 22 August.

### *Perimede grandis* Hodges, NEW SPECIES

PL. 6, FIG. 32. TEXT FIGS. 39 *c, d*.

*Perimede grandis* Hodges.
Type-locality: McKittrick Canyon, 5000′, Guadalupe Mts., Culberson Co., Texas. [USNM]

Upper surface as figured. Head: base of tongue dark gray, scales becoming yellowish gray by ¼ length of tongue; maxillary palpus gray with shining yellow reflections; labial palpus mainly dark gray, scales tipped with very pale gray, dorsomedial surface of second segment yellowish white; frons dark gray with pale-tipped scales laterally, pale gray with shining yellow and purple reflections medially; vertex and occiput dark gray, scales tipped with pale gray, middle of vertex with continuation of pale area of frons; antenna dark gray, most scales unicolorous. Foreleg nearly uniformly dark gray, scale bases pale, inner surface of epiphysis yellowish gray, apex of first tarsal segment paler than rest of segment. Midleg much as for foreleg, coxa and trochanter with more pale scales. Hindleg mainly dark gray, many scale bases pale; tibia with gray streak at base of first pair of spurs, base of each of the shorter spurs yellowish white to white; base and apex of first tarsal segment pale gray to off white, apices of second and third segments pale gray to off white. Wing length: 8.0 mm. Male: as in text figures 39 *c, d*; eighth sternum a tripartite sclerotized unit, generally Y-shaped, distal sclerites nearly semicircular; accessory glands with heavily sclerotized parts symmetrical, posterior margin strongly sinuous, spine nearly straight and from ½ to ⅔ length of valva; valva slender with rounded apex; juxta very heavily sclerotized, a pair of lobes laterad of base of aedoeagus; aedoeagus slightly curved, ventral part heavily sclerotized; uncus with lobes separate for ½ length. Female: no specimens available.

The immature stages are unknown.

TYPE: Holotype: ♂. Texas, Culberson Co., Guadalupe Mts., McKittrick Canyon, 5000′; 23 May 1973; R. W. Hodges; USNM genital slide 4723. USNM.

The large size and male genitalia immediately separate *grandis* from other species of *Perimede*. Whether *grandis* has a northern or southern distribution is not known.

### *Perimede parilis* Hodges

PL. 5, FIG. 48. TEXT FIGS. 39 *a, g*; 51 *h*.

*Perimede parilis* Hodges, 1969, *Smithsonian Contrib. Zool.*, **18**: 12, figs. 6, 21–23, 25, 41.
Type-locality: Devil's Den State Park, Washington Co., Arkansas. [USNM]

*Parilis* can be confused with most of the eastern species of *Perimede* and can be separated from them with certainty by genital characters as given in the key. Observed variation is slight with some differences in the amount of light and dark scales on the head. Fresh specimens are very dark gray; worn ones are very pale, almost shining yellowish gray. Wing length varies from 4.0 to 5.2 mm.

The immature stages are unknown.

*Parilis* has been collected in central Illinois and northwestern Arkansas from 7 June to 14 August.

### *Perimede circitor* Hodges

PL. 5, FIG. 49. TEXT FIGS. 40 *a,f*; 51 *g*.

*Perimede circitor* Hodges, 1969, *Smithsonian Contrib. Zool.*, **18**: 13, figs. 24, 26, 43, 45.
Type-locality: Devil's Den State Park, Washington Co., Arkansas. [USNM]

*Circitor* is similar to many other species in the genus. The male genitalia are most nearly like those of *erransella* but differ as indicated in the key. Additionally, the apices of the uncus are rounded in *circitor*, acute in *erransella*. The cleft posterior margin of the eighth abdominal sternum with the ostium bursae at the base of the cleft is distinctive for females of *circitor*. Specimens vary very little in maculation. Wing length varies from 3.5 to 4.5 mm.

The immature stages are unknown.

*Circitor* has been collected in central Illinois and northwestern Arkansas from 3 June to 2 August.

### *Perimede erema* Hodges

PL. 5, FIG. 47. TEXT FIG. 52 *a*.

*Perimede eremos* Hodges, 1969, *Smithsonian Contrib. Zool.*, **18**: 12, figs. 40, 42.
Type-locality: Devil's Den State Park, Washington Co., Arkansas. [USNM]

In addition to genital characters given in the key, *erema* usually can be recognized by the upper part of the frons and all the vertex and occiput being pale yellowish gray. Most other species have these areas dark gray. Wing length of the two specimens is 3.5 and 3.9 mm.

Nothing is known of the immature stages.

*Erema* has been collected in northwestern Arkansas on 17 and 27 June.

### *Perimede maniola* Hodges

PL. 5, FIG. 50. TEXT FIG. 52 *b*.

*Perimede maniola* Hodges, 1969, *Smithsonian Contrib. Zool.*, **18**: 14, figs. 44, 46.
Type-locality: Putnam Co., Illinois. [USNM]

*Maniola* is similar to many of the eastern species of *Perimede* and can be separated on female characters given in the key. Males are unknown. Wing length varies from 4.2 to 4.8 mm.

The immature stages are unknown.

The three known specimens were collected at the type-locality on 11 July and 17, 19 August.

### *Perimede ricina* Hodges

PL. 4, FIG. 65. TEXT FIGS. 40 *b–d*; 52 *c*.

*Perimede ricina* Hodges, 1962, *Proc. Ent. Soc. Washington*, **64**: 150, figs. 1, 6.
Type-locality: Ithaca, New York. [CU]

*Ricina* and *falcata* are very closely allied and appear much alike. The forewings are slightly broader than those of the other species, and there is a row of yellowish-gray scales on the fringe just below the apex of the forewings. *Ricina* differs from *falcata* by the hindleg having unicolorous third and fourth tarsal segments, those of *falcata* are yellowish gray to white apically and contrast with the dark-gray bases. Variation appears to be limited and is expressed mainly in size: wing length, 5–6 mm. Worn specimens are pale yellowish brown.

The immature stages are unknown.

*Ricina* is known from Martha's Vineyard and Barnstable, Massachusetts, Ithaca, New York and central Illinois south to the Washington, D.C. area and eastern Kentucky. Specimens have been collected from 4 July to 7 September.

### *Perimede falcata* Braun

PL. 5, FIG. 46. TEXT FIGS. 40 *h*; 41 *f*; 52 *d* (McD. 7722).

*Perimede falcata* Braun, 1919, *Ent. News*, **30**: 263.
Type-locality: Cincinnati, Ohio. [ANSP]

*Falcata* and *ricina* have a small number of yellowish-gray scales in the fringe of the outer margin of the forewing just below the apex. Characters separating *falcata* from other species are given in the key and in the discussion of *ricina*. Fresh specimens are dark gray brown with a pair of oblique streaks, each running from the anterior or posterior margin toward the apex, at $\frac{3}{4}$ the wing length. Worn specimens are pale yellowish gray. Wing length varies from 3.8 to 6.2 mm.

The immature stages are unknown.

*Falcata* occurs from eastern Massachusetts, Ithaca, New York and central Illinois south to south-central Florida and south-central Louisiana. Northern specimens have been collected from 12 June to 3 August. Floridian specimens have been collected from 12 February to 30 December. A small series from Louisiana was collected from 14 to 28 April.

## GENUS *Sorhagenia* Spuler

*Sorhagenia* Spuler, 1910, *Die Schmetterlinge Europas*, **2**: 384, fig. 141.
Type-species: *Elachista rhamniella* Zeller, 1839. Monotypy.

*Cystioecetes* Braun, 1915, *Can. Ent.*, **47**: 194, fig. 21.
Type-species: *Cystioecetes nimbosus* Braun, 1915. Original designation.

*Sorhagenia* is a holarctic genus with eight included species. Three species occur in the western Palearctic Region, two in central North America and three in the American West. *Sorhagenia* species are very similar to the gray species of *Stilbosis* and *Chrysopeleia*, but *Sorhagenia* differs by the forewing having $M_1$ arising before the fork of $R_4$ and $R_5$ and the hindwing with the cell closed and $M_1$ free or short stalked with $R_5$. The male genitalia of

*Sorhagenia* species, with three-lobed valvae, reduced aedoeagus, modified vinculum and reduced uncus, could have been derived from *Stilbosis*.

Characters of *Sorhagenia* are as follows: labial palpus with second and third segments subequal in length, second segment slightly tufted at apex; antenna $\frac{3}{5}$ to $\frac{3}{4}$ length of forewing, pecten of a single scale present; ocelli present. Forewing lanceolate, apex acute, raised scales present; 12 veins usually present; $M_1$ stalked with $R_{4+5}$, separating about $\frac{1}{2}$ distance from end of cell to forking of $R_{4+5}$; $Cu_2$ present or absent at wing margin. Hindwing narrowly lanceolate, eight veins present, $R_s$ and $M_1$ stalked or separate, cell closed. Male with eighth abdominal sternum heavily sclerotized, often with a row of stout setae on posterior margin; genitalia symmetrical; valva with three lobes, costal, medial, and saccular, medial lobe usually smallest and most heavily sclerotized, saccular lobe widely varying in size; aedoeagus short, basal part larger in diameter than distal part, weakly ankylosed in juxtal area; vinculum with sacculus developed as a quadrate or triangular extension, a medial flange variously modified; tegumen slender, heavily sclerotized; uncus short, almost free basally. Female with ostium bursae well defined, usually preceded by triangular or conical genital plate; ductus bursae membranous with spiculose walls or sclerotized, often pleated on distal part; corpus bursae with spiculose walls; signa paired; apophyses slender, equal in length, apophyses anteriores joined by sclerotized band; ovipositor nonpiercing.

According to Malicky and Sobhian (1971: 189) larvae of the three European species feed on species of *Rhamnus* Tourn., Rhamnaceae; the larva of *nimbosa* has been reared from *Rhamnus* species. The larvae are leaf miners, leaf tiers or flower feeders. Plants in the Rhamnaceae are likely to be hosts for the other species in our area.

## KEY TO SPECIES OF *SORHAGENIA*

1. Males .................................... 2
— Females .................................... 6

2. Valva with costal lobe longer than saccular lobe .................................... 3
— Valva with costal lobe shorter than saccular lobe .................................... 5

3. Valva with costal lobe expanded distally, broadly rounded at apex (text figure 41 *h*); Midwest .................................... *baucidis* p. 133
— Valva with costal lobe not expanded distally, apex narrowly rounded; West .......... 4

4. Vinculum with medial lobe wider than long, lateral margin with ridges (text figure 40 *g*); eighth sternum longer than wide. . . *nimbosa* this page
— Vinculum with medial lobe longer than wide, lateral margin smooth (text figure 41 *d*); eighth sternum wider than long. . . . . *cracens* p. 133

5. Medial lobe of valva relatively constant in width, slightly expanded at apex; costal lobe slightly expanded before apex, then tapering to sharply pointed apex (text figure 41 *a*) .................................... *pexa* p. 134
— Medial lobe of valva expanded distally, more than twice medial width; costal lobe tapering evenly to narrowly rounded apex (text figure 41 *c*) .................................... *daedala* p. 133

6. Signa absent .................................... *pexa* p. 134
— Signa present .................................... 7

7. Pregenital plate rectangular (text figure 52 *e*) .................................... *nimbosa* this page
— Pregenital plate triangular .................................... 8

8. Ductus bursae slender, nearly uniformly wide; sclerotized, anterior arms of pregenital plate narrow, posterior margin excavated (text figure 52 *f*) ............ *daedala* p. 133
— Ductus bursae moderately wide, distal part nearly twice as wide as basal, sclerotized part; sclerotized, anterior part of pregenital plate broad, posterior margin truncated (text figure 52 *h*) .................... *baucidis* p. 133

### *Sorhagenia nimbosa* (Braun)

PL. 5, FIG. 40. TEXT FIGS. 3 *i*; 40 *e*, *g*; 52 *e* (McD. 7710).

*Cystioecetes nimbosus* Braun, 1915, *Can. Ent.*, **47**: 195, fig. 22.
Type-locality: Mills College, Alameda Co., California. [ANSP]

In western North America *nimbosa* can be confused with *daedala* and *cracens* but can be separated from each as indicated in the key. Wing length varies from 3.9 to 5.6 mm, and the forewing varies in intensity of dark-gray scales. In some specimens the basal part of the forewing is contrastingly pale gray; in others the forewing is uniformly dark gray.

Larvae have been reared from *Rhamnus purshiana* D.C. and *R. californica* Esch. Braun (1915: 195) illustrated the larval chamber. The caterpillar ties a leaf longitudinally, and the leaf bulges between the major veins giving the appearance of a gall. Pupation occurs in a cocoon often spun outside the larval chamber. Overwintering occurs as an egg or larva.

*Nimbosa* occurs from northwestern Washington and northern Idaho south to San Bernardino County, California. Most specimens have been reared, and emergence dates are 18 June to 8 July. Field-collected specimens have collection dates of 7 June to 7 September.

### *Sorhagenia cracens* Hodges, NEW SPECIES

PL. 6, FIG. 33. TEXT FIGS. 41 *b, d.*

*Sorhagenia cracens* Hodges.
Type-locality: Spirit L., Idaho. [USNM]

Upper surface as figured. Head: base of tongue medium to pale gray with shining yellow reflections; frons pale gray medially, dark gray bordering eye, with some shining yellow and purple reflections; vertex and occiput darker gray, some scales tipped with pale gray; labial palpus with second segment medium to dark gray, most scales tipped with paler gray, apex off white, third segment absent on type; antenna mottled dark and medium gray, scape dark gray. Foreleg: mainly dark gray, scales with shining yellow and purple reflections, tibia with pale-gray scales at base of epiphysis and at apex of segment, apices of tarsal segments pale gray to off white. Midleg: coxa and trochanter yellowish gray with shining yellow and purple reflections; femur, tibia and tarsus much as for foreleg. Hindleg: coxa, trochanter and femur pale yellowish gray with some darker gray scales and very pale off-white scales and shining yellow and purple reflections; tibia with white-tipped scales at base, middle and apex, spurs off white; tarsal segments dark gray basally, off white to white apically, base of first segment off white to white. Wing length: 4.7 mm. Male: as in text figures 41 *b, d*; eighth abdominal sternum heavily sclerotized, $1\frac{1}{2}$ times wider than long, stout scales on posterior margin; valva with saccular lobe much shorter than costal lobe, relatively slender, costal lobe very slender, apex rounded; vinculum with saccal region quadrate, medial lobe parallel margined and smooth laterally. Female: no specimens available.

The immature stages are unknown.

TYPE: Holotype: ♂. Spirit L., Idaho; 16–23 July; USNM genital slide 1368. USNM.

Characters of the male genitalia as illustrated and indicated in the key will separate *cracens* from *nimbosa*, its closest ally. The larval host is likely to be a species of *Rhamnus*.

### *Sorhagenia daedala* Hodges

PL. 5, FIG. 41. TEXT FIGS. 41 *c*; 52 *f*.

*Sorhagenia daedala* Hodges, 1964, *Proc. U.S. Natl. Mus.*, **115**: 314, figs. 21, 44.
Type-locality: Mt. Shasta City, Siskiyou Co., California. [USNM]

*Daedala* is closely allied with *nimbosa* and *cracens*. Differences are as given in the key. The middle part of the frons is pale and more uniformly gray in *daedala* than in the other two species. Wing length varies from 4.8 to 5.0 mm.

The immature stages are unknown.

*Daedala* has been collected at Kettle Falls, Stevens County, Washington on 21 July and at the type-locality on 4 July.

### *Sorhagenia baucidis* Hodges

PL. 5, FIG. 42. TEXT FIGS. 41 *g, h*; 52 *h*.

*Sorhagenia baucidis* Hodges, 1969, *Smithsonian Contrib. Zool.*, **18**: 9, figs. 3, 16, 17, 37, 38.
Type-locality: Putnam Co., Illinois. [USNM]

*Baucidis* is closely allied with *nimbosa*, *daedala* and *cracens* on genital characters and color. Genital differences separating *baucidis* from these species are given in the key. From the sympatric species *pexa*, *baucidis* differs by the gray color of the forewings and smaller wing length. Because these small moths have been so poorly collected, the genitalia of gray specimens of *Sorhagenia* from the Midwest should be examined on the chance that other species are present. Wing length varies from 4.8 to 5.0 mm.

Nothing is known of the immature stages.

*Baucidis* has been collected at the type-locality from 17 June to 2 July.

*Sorhagenia pexa* Hodges
PL. 4, FIG. 64. TEXT FIGS. 41 *a, e*; 52 *g*.

*Sorhagenia pexa* Hodges, 1969, *Smithsonian Contrib. Zool.*, **18**: 9, figs. 4, 18, 20, 39.
Type-locality: Devil's Den State Park, Washington Co., Arkansas. [USNM]

*Pexa* differs from other species of *Sorhagenia* by the yellowish-brown ground color of the head, thorax and forewings. The few specimens vary little in color pattern. Wing length varies from 5.8 to 6.8 mm.

The immature stages are unknown.

*Pexa* has been collected at the type-locality from 20 June to 13 July and in central Illinois on 6 and 10 July.

GENUS
*Ithome* Chambers

*Eriphia* Chambers, 1875, *Can. Ent.*, **7**: 55.
Type-species: *Eriphia concolorella* Chambers, 1875. Monotypy.

NOTE—*Eriphia* Chambers, 1875, is preoccupied by *Eriphia* Latreille, 1817, in the Crustacea.

*Ithome* Chambers, 1875, *Can. Ent.*, **7**: 93.
Type-species: *Ithome unimaculella* Chambers, 1875, considered to be a synonym of *Eriphia concolorella* Chambers, 1875. Monotypy.

*Ithome* is a New World genus. Its 17 species occur from Tennessee, North Carolina and the Southwest south through the Antilles and Central America to Peru and the Amazon Basin of Brasil. An undescribed species occurs in central Chile, and *concolorella* has been introduced into the Hawaiian Islands. *Ithome* is related to *Perimede* and *Obithome* but adults can be separated by the forewing having vein $M_1$ long stalked with $R_{4+5}$. In *Perimede* and *Obithome* $M_1$ is a separate vein. Males of *Ithome* differ by the eighth abdominal sternum bearing a pair of valvalike structures distolaterally; females differ by having massive apophyses, and the apophyses anteriores not connected by a dorsal, sclerotized band.

Characters of *Ithome* are as follows: labial palpus very smooth scaled, second segment slightly longer than third; antenna $\frac{3}{5}$ to $\frac{3}{4}$ length of forewing, pecten of a single scale, present; ocelli present. Forewing lanceolate, apex acute, raised scales absent, 12 veins present, $M_1$ long stalked with $R_{4+5}$, $Cu_1$ and $Cu_2$ slightly downcurved and parallel from end of cell. Hindwing narrowly lanceolate, nearly linear, cell open, eight veins present, $R_s$ and $M_1$ long stalked, a tuft of scales on costal margin at $\frac{1}{4}$ wing length. Male: seventh abdominal segment sometimes with posterior margins heavily sclerotized and modified; eighth sternum with three separate sclerites, basal one heavily sclerotized, often bearing a pair of lateral, anteriorly directed rods, distal sclerites symmetrical or asymmetrical, valvalike; eighth tergum hood shaped, apex sometimes deeply excavated, usually supported by sclerotized rod in intersegmental membrane between eighth and ninth segments; genitalia strongly asymmetrical, right valva usually a single, cylindrical lobe; left valva often with three, varyingly developed lobes; aedoeagus large, usually longer than valvae; cornuti present or absent; vinculum usually short, narrow and heavily sclerotized; tegumen very short, heavily sclerotized, sometimes with a transtillar band; subscaphium sometimes present; uncus a lightly sclerotized, triangular flap. Female: ostium bursae on seventh abdominal sternum, sometimes difficult to discern; ductus bursae slender or broad; corpus bursae lightly sclerotized; bursa copulatrix small, shorter than apophyses anteriores; apophyses massive, heavily sclerotized, apophyses anteriores usually not connected by sclerotized band; ovipositor piercing.

Larvae of three species have been reared from plants in the Leguminosae, and one has been reared from a plant in the Polygonaceae. According to Namba (1956: 97) the first and second instar larvae of *concolorella* feed within flowers. Pupation occurs in a cocoon outside the flowers, probably on the ground.

## KEY TO SPECIES OF *ITHOME*

1. Males ........................................ 2
— Females ........................................ 8

2. Eighth abdominal sternum with symmetrical, posterior lobes (text figure 42 *h*) ........... 4
— Eighth abdominal sternum with asymmetrical, posterior lobes (text figure 42 *a*) ...... 3

3. Eighth sternum with right, posterior plate bilobed (text figure 42 *c*) ........... *curvipunctella* p. 135
— Eighth sternum with right, posterior plate one lobed (text figure 42 *a*) .......... *concolorella* p. 135

4. Posterior margin of eighth abdominal tergum rounded or slightly indented medially . . . . 5
— Posterior margin of eighth abdominal tergum deeply excavated medially (text figure 42 *h*) . . . . *ferax* p. 136

5. Seventh sternum with posterior margin sclerotized, bilobate, each lobe with acute apex (text figure 42 *b*) . . . . 6
— Seventh sternum not so modified . . . . 7

6. Apex of eighth tergum indented medially, sclerotized rod between eighth tergum and genitalia with heavily sclerotized basal and distal forks (text figure 42 *b*) . . . . *lassula* p. 137
— Apex of eighth tergum rounded with very small medial indentation, sclerotized rod between eighth tergum and genitalia with heavily sclerotized basal broadening, no distal expansion (text figure 43 *a*) . . . . *simulatrix* p. 137

7. Aedoeagus with row of curved teeth on dorsal margin from $\frac{1}{2}$ to $\frac{3}{4}$ length (text figure 42 *e*) . . . . *aquila* p. 136
— Aedoeagus without teeth on dorsal margin (text figure 42 *d*) . . . . *edax* p. 136

8. Lamella antevaginalis with rounded posterior margin, lamella postvaginalis triangular (text figure 53 *e*) . . . . *ferax* p. 136
— Seventh sternum otherwise . . . . 9

9. Ostium bursae circular, apophyses anteriores joined by sclerotized band (text figure 53 *d*) . . . . 10
— Ostium bursae otherwise, apophyses anteriores not joined by sclerotized band . . . . 11

10. Base of ductus bursae sclerotized, ostium bursae at $\frac{2}{3}$ length of seventh sternum (text figure 53 *a*) . . . . *curvipunctella* this page
— Base of ductus bursae not sclerotized, ostium bursae at $\frac{1}{3}$ length of seventh sternum (text figure 53 *d*) . . . . *concolorella* this page

11. Ductus bursae slender, signa paired (text figure 53 *g*) . . . . *lassula* p. 137
— Ductus bursae broad, scarcely set off from corpus bursae, signa absent (text figure 53 *b*) . . . 12

12. Ostium bursae as in text figure 53 *b*, apophyses posteriores uniformly broad anteriorly . . . . *aquila* p. 136
— Ostium bursae as in text figure 53 *c*, apophyses posteriores becoming broader anteriorly . . . . *edax* p. 136

### *Ithome concolorella* (Chambers)

PL. 2, FIG. 19; PL. 5, FIG. 53. TEXT FIGS. 3 *j*; 42 *a*, *i*; 53 *d* (McD. 7711, 7712).

*Eriphia concolorella* Chambers, 1875, *Can. Ent.*, **7**: 55.
Type-locality: Texas. [MCZ]

*Ithome unimaculella* Chambers, 1875, *Can. Ent.*, **7**: 94.
Type-locality: Texas. [MCZ]

*Concolorella* and *curvipunctella* are closely allied. In males the left lobe of the eighth abdominal sternum and the left valva differ as shown in the text figures. A large series of specimens from northwest Arkansas shows little variation in maculation. Worn moths are much paler than fresh ones. Wing length varies from 2.5 to 4.3 mm.

Namba (1956: 95–100) gave a full report on the life history and illustrated larvae and pupae of Hawaiian specimens. *Concolorella* is a pest species in Hawaii on *Acacia farnesiana* (L.) Willd. and *Prosopis chilensis* (Mol.) Stuntz by reducing the production of flowers. A smaller nectar source is available to honey bees, and the yield of honey is lowered. Some moths were reared from mesquite, *Prosopis* species in Arizona. In Arkansas I observed many adults at the flowers of *Schrankia nuttalli* (D.C.) Standley in the late afternoon. The female pierces the flower and deposits an egg inside. First and second instar larvae feed in the first flower; subsequent instars feed on additional flowers. Pupation occurs in a cocoon outside the flower.

*Concolorella* occurs from southern Ohio and central Illinois south to Florida, Texas and southern Arizona. Adults have been collected from June to August in the North and from 4 August to 12 October in southern Arizona.

### *Ithome curvipunctella* (Walsingham)

PL. 5, FIG. 54. TEXT FIGS. 42 *c*, *g*; 53 *a*.

*Anybia curvipunctella* Walsingham, (1892), *Proc. Zool. Soc. London*, 1891: 538.
Type-locality: St. Vincent (windward side), West Indies. [BMNH]

*Eriphia quinquepunctata* Forbes, 1931, *Jour. Dept. Agric. Porto Rico*, **4**: 361, pl. 46, fig. 35.
Type-locality: Puerto Real, Vieques Island, Puerto Rico. [CU]

*Curvipunctella* is closely allied with *concolorella* but can be separated on genital characters as indicated in the key. The antennae are sexually dimorphic: males have the scape dark gray and the shaft pale yellowish gray; females have the antenna dark gray except for the pale yellowish-gray apices of the scape and shaft. Males have the hindwings with dark-gray sex scales on dorsal and ventral surfaces. Wing length varies from 2.8 to 3.7 mm.

Larvae have been reared from *Coccoloba uvifera* (L.) Jacq. in the Polygonaceae on Ramrod Key, Florida. The specimen labels do not indicate what the larvae were doing.

*Curvipunctella* occurs from central Florida south to St. Vincent in the West Indies. Near Sarasota adults were collected in April and May.

### *Ithome ferax* Hodges

PL. 5, FIG. 57. TEXT FIGS. 42 *h*; 53 *e*.

*Ithome ferax* Hodges, 1962, *Jour. Lep. Soc.*, **15**: 89, figs. 4, 9.
Type-locality: Siesta Key, Sarasota Co., Florida. [CU]

Male genital characters ally *ferax* with *lassula*, *edax*, *simulatrix* and *aquila*, but the deeply excavated posterior margin of the eighth abdominal tergum is highly distinctive. The female genitalia, as illustrated, are also distinctive. Variation in the short type series is limited. Wing length varies from 3.9 to 5.0 mm. This species is not sexually dimorphic in antennal coloration or hindwing surface scaling.

The immature stages are unknown.

*Ferax* has been collected on Siesta Key and at St. Petersburg, Florida from 4 January to 12 April.

### *Ithome edax* Hodges

PL. 5, FIG. 55. TEXT FIGS. 42 *d,f*; 53 *c*.

*Ithome edax* Hodges, 1962, *Jour. Lep. Soc.*, **15**: 88, figs. 5, 11.
Type-locality: Brownsville, Texas. [USNM]

*Edax* is most closely allied with *aquila* from which it can be separated by characters given in the key and comparison with the line drawings. Fresh specimens of *edax* have very dark-gray forewings; worn specimens have the forewings blotchy pale yellowish gray and dark gray. The vertex and occiput are medium to dark gray or pale gray. Wing length varies from 3.0 to 3.6 mm. The antennae and hindwings are not sexually dimorphic in color.

Larvae have been reared from blossoms of 'ebony', *Pithecellobium flexicaule* (Bentham) Coulter.

*Edax* has been collected at Brownsville and Corpus Christi, Texas from 8 May to 7 June.

### *Ithome aquila* Hodges, NEW SPECIES

PL. 5, FIG. 58. TEXT FIGS. 42 *e*; 53 *b*.

*Ithome aquila* Hodges.
Type-locality: Tavernier, Plantation Key, Florida. [UCB]

Upper surface as figured. Head: base of tongue and maxillary palpus medium gray with shining yellow and purple reflections; frons mainly medium gray with shining yellow reflections, some scales in front of eye darker gray; vertex and occiput very dark gray with shining yellow reflections, some scales on vertex tipped with very pale gray; labial palpus dark gray on lateral surface of second and third segments, base and apex of third segment pale yellow to yellowish gray, third segment with row of pale-yellow spots on anterior margin, dorsal and inner surface of second segment pale yellowish gray; antenna dark gray basally, becoming pale yellowish gray on distal $\frac{1}{5}$. Foreleg mainly dark gray, epiphysis and apex of tibia pale gray, apices of first, second and fourth tarsal segments and all of fifth tarsal segment pale yellowish gray. Midleg: coxa and trochanter mainly pale yellow with some medium-gray scales; femur, tibia and tarsus dark gray, scale bases often pale gray, tibia with a few pale-yellow scales just beyond $\frac{1}{2}$ length and at apex; apices of first, second, fourth and all of fifth tarsal segments pale yellow. Hindleg: coxa, trochanter mainly pale yellow with some gray scales; femur yellow with more gray scales; tibia mainly dark gray on lateral surface, scale bases pale gray, a pale-yellow streak at base of first pair of spurs and at apex, spurs pale gray with darker scales on posterior surface, long scales on dorsal surface pale yellow; base and apex of first tarsal segment, apices of second, third and fourth segments and all of fifth segment pale yellowish gray. Abdomen mainly yellowish gray with some darker gray scales on dorsal surface, all surfaces with shining yellow and purple reflections. Wing length: 3.1–3.5 mm. Male: as in text figure 42 *e*; posterior margin of seventh abdominal sternum rounded and slightly indented medially; posterior margin of seventh abdominal tergum convex; eighth abdominal sternum as illustrated; eighth abdominal tergum elongated; sclerotized rod between eighth tergum and genital capsule with broadened and flattened basal (as seen in illustration) part, distal part slightly flared at $\frac{4}{5}$ length and tapering to acute apex; right valva with main lobe broader and more heavily sclerotized than individual lobes of left valva; left valva trilobate, saccular lobe slender throughout and with few setae on dorsal margin, medial lobe broad basally, slender medially and becoming broad before apex, costal lobe short with series of setae at apex; aedoeagus nearly twice length of right valva, series of curved, toothlike structures on dorsal margin; tegumen very heavily sclerotized, heavily sclerotized band in transtillar region; subscaphium present; uncus much less heavily sclerotized than tegumen, tapering to slender point. Female: as in text figure 53 *b*; ostium bursae on posterior $\frac{1}{3}$ of seventh abdominal sternum, ostial margin curved as illustrated; ductus bursae broad to corpus bursae, wall spiculose; signa paired; lateral margin of papillae anales serrate; apophyses posteriores much longer than apophyses anteriores and slightly broadened anteriorly; apophyses anteriores not connected by sclerotized band.

Larvae feed on flowers of *Pithecellobium* (=*Pithecolobium, auct.*) *unguis-cati* (L.) Bentham.

TYPES: Holotype: ♂. Florida: Tavernier, Plantation Key; 30 XI 1970; J. Powell 7016; em. by 5 I 1971; reared from flowers *Pithecellobium unguis-cati*; J. A. Powell slide 3054. UCB. Paratypes: 5 ♂♂, 3 ♀♀. Same data as for holotype (3 ♀♀). Everglade, Florida; 8–15 April (1 ♂). Key Largo Key, Monroe Co., Florida; Mrs. Spencer Kemp (3 ♂♂). Siesta Key, Sarasota Co., Florida; 12 March 1953; C. P. Kimball (1 ♂). Paratypes to CPK; UCB; USNM.

Male abdominal and genital characters ally *aquila* most closely with *edax* from which it is separated by the differently shaped distal lobes of the eighth abdominal sternum and the presence of toothlike structures on the dorsal surface of the aedoeagus. Females of *aquila* are separable from those of *edax* by the differently shaped ostium bursae and surrounding sclerotized plates and by having signa. *Aquila* varies in the proportion of light and dark scales on the frons, vertex and occiput. Some specimens, as described for the type, are very dark; others have the frons and lower part of the vertex pale yellowish gray and the upper part of the vertex and the occiput medium gray. As specimens become worn, the forewings, head and thorax become pale yellowish gray. *Aquila* is not sexually dimorphic in antennal color and hindwing scale pattern.

### *Ithome lassula* Hodges

PL. 5, FIG. 56. TEXT FIGS. 42 *b*; 53 *g*.

*Ithome lassula* Hodges, 1962, *Jour. Lep. Soc.*, **15**: 89, figs. 3, 10.
Type-locality: Key West, Florida. [USNM]

The shape of the seventh sternum in the male and general configuration of the genitalia associate *lassula* with *simulatrix*. Differences are given in the key and as illustrated. Fresh specimens have dark-gray head, thorax and forewings; worn specimens are yellowish gray to yellowish brown. Wing length varies from 3.6 to 4.3 mm. The male antenna usually is medium gray basally, becoming pale yellowish gray distally. In females the antenna is dark gray basally, becoming pale yellowish gray distally. Males do not have sex scaling on the hindwings.

Larvae have been reared on *Leucaena glauca* (L.) Bentham on Key West.

Reared specimens emerged from 16 March to 19 April at Key West. An adult was collected in the Sierra Diablo northwest of Van Horn, Texas on 26 May.

### *Ithome simulatrix* Hodges, NEW SPECIES

PL. 6, FIG. 34. TEXT FIG. 43 *a*.

*Ithome simulatrix* Hodges.
Type-locality: Everglade, Florida. [USNM]

Upper surface as figured. Head: base of tongue, maxillary palpus and frons pale yellowish gray with shining yellow and purple reflections; labial palpus with lateral surface of second and third segments dark gray, apex of second segment and base and apex of third segment pale yellow, an irregular row of pale-yellow dots on anterior surface of third segment, mesial surface of second segment and contiguous posterior surface of third segment pale yellow; antenna with scape mainly dark gray, apex pale yellow, shaft medium gray becoming pale yellowish gray distally in male (female unknown). Foreleg: coxa and trochanter medium to pale yellowish gray; femur, tibia and tarsus dark gray, a few yellowish-gray scales on epiphysis, apex of tibia, base of first tarsal segment and apex of fifth tarsal segment. Midleg: coxa, trochanter and femur pale yellowish gray with shining yellow and purple reflections; tibia dark gray, an off-white band at $\frac{1}{5}$ length, one at middle, apex off white, apical spurs yellowish white to white; tarsus with first three segments mainly dark gray, apex of each pale yellowish gray to white, fourth and fifth segments yellowish gray to white. Hindleg: absent. Wing length: 3.2 mm. Male: as in text figure 43 *a*; seventh abdominal sternum with posterior margin having two, acute lobes; seventh abdominal tergum with posterior margin strongly produced medially; eighth abdominal sternum and tergum as in text figure 43 *a*; sclerotized rod in membrane between eighth tergum and tegumen with expanded base (as seen in figure) and tapering to acute apex; right valva bilobate, saccular lobe as long as aedoeagus, costal short with series of short, stout setae at apex; left valva trilobate, saccular and medial lobes nearly equal in length, costal lobe short with series of setae at apex; aedoeagus angled near base; vesica with numerous, small cornuti; tegumen relatively broad distally, a heavily sclerotized band in transtillar region; subscaphium lightly sclerotized; uncus triangular. Female unknown.

The immature stages are unknown.

TYPE: Holotype: ♂. Everglade, Florida; 8–15 April; USNM genital slide 1372. USNM.

*Simulatrix* is very near *Ithome pernigrella* (Forbes), NEW COMBINATION, [described as *Eriphia pernigrella* Forbes, 1931, from Puerto Rico] on genital characters. The male of *simulatrix* lacks sex scaling on the hindwings, and the medial lobe of the left valva is right angled apically. The male of *pernigrella* has dark-gray, nearly black sex scales on the hindwings, and the medial lobe of the left valva is slightly curved apically.

GENUS
## *Obithome* Hodges

*Obithome* Hodges, 1964, *Proc. U.S. Natl. Mus.*, **115**: 316, figs. 1, 15.
Type-species: *Mompha punctiferella* Busck, 1906. Original designation.

*Obithome* is a monobasic genus whose single species occurs in coastal Texas. It is allied with *Ithome* and *Eritarbes* Walsingham. *Obithome* differs from them by having the forewing with $M_1$ separate; *Ithome* has the forewing with $M_1$ long stalked with $R_{4+5}$; and *Eritarbes* has the forewing with $M_1$ and $M_2$ stalked with $R_{4+5}$.

Characters of *Obithome* are as follows: labial palpus smooth scaled, second segment with slight anteroventral tuft, third segment shorter than second; ocelli present; antenna with pecten a basal scale. Forewing lanceolate, apex acute, with black raised scales, cell closed, $M_1$ separate. Hindwing narrowly lanceolate, apex acute, cell open, eight veins present, $R_s$ and $M_1$ stalked. Male: eighth abdominal segment with sternum a broadened lobe, pleural region with a pair of lateral lobes, tergum short; genitalia highly asymmetrical; right valva bilobate, costal lobe curved, apex acute, main lobe much broader than costal lobe; left valva a single lobe with acute apex; aedoeagus straight on basal $\frac{2}{3}$, then offset, distal part parallel with basal part; vinculum very heavily sclerotized, asymmetrical; tegumen short, expanded and slightly lobate posterodistally; uncus absent. Female: ostium bursae near anterior margin of seventh abdominal sternum, preceded by heavily sclerotized plate; ductus bursae with expanded lobe on left side at base, nearly membranous, a sclerotized zone at juncture with corpus bursae; wall of corpus bursae nearly membranous; paired signa, each a series of inwardly projecting spines arranged in curved rows; apophyses massive, heavily sclerotized, apophyses anteriores joined by heavily sclerotized band, apophyses anteriores and posteriores equal in length; ovipositor with few, outwardly directed teeth at apex.

The immature stages are unknown.

### *Obithome punctiferella* (Busck)

PL. 5, FIG. 59. TEXT FIGS. 4 *b*; 43*f*; 53*f* (McD. 7760).

*Mompha punctiferella* Busck, 1906, *Proc. U.S. Natl. Mus.*, **30**: 731.
Type-locality: Victoria, Texas. [USNM]

*Punctiferella* is superficially similar to *Afeda biloba* but can be separated by the presence of raised scales on the forewings and the hindwing lacking a closed cell. The few collected specimens of *punctiferella* vary in wing length from 3.8 to 5.0 mm and shade of gray or gray brown on the forewings. Male and female genitalia are as illustrated.

*Punctiferella* has been collected in coastal Texas from Richmond to Brownsville from 4 April to 24 June.

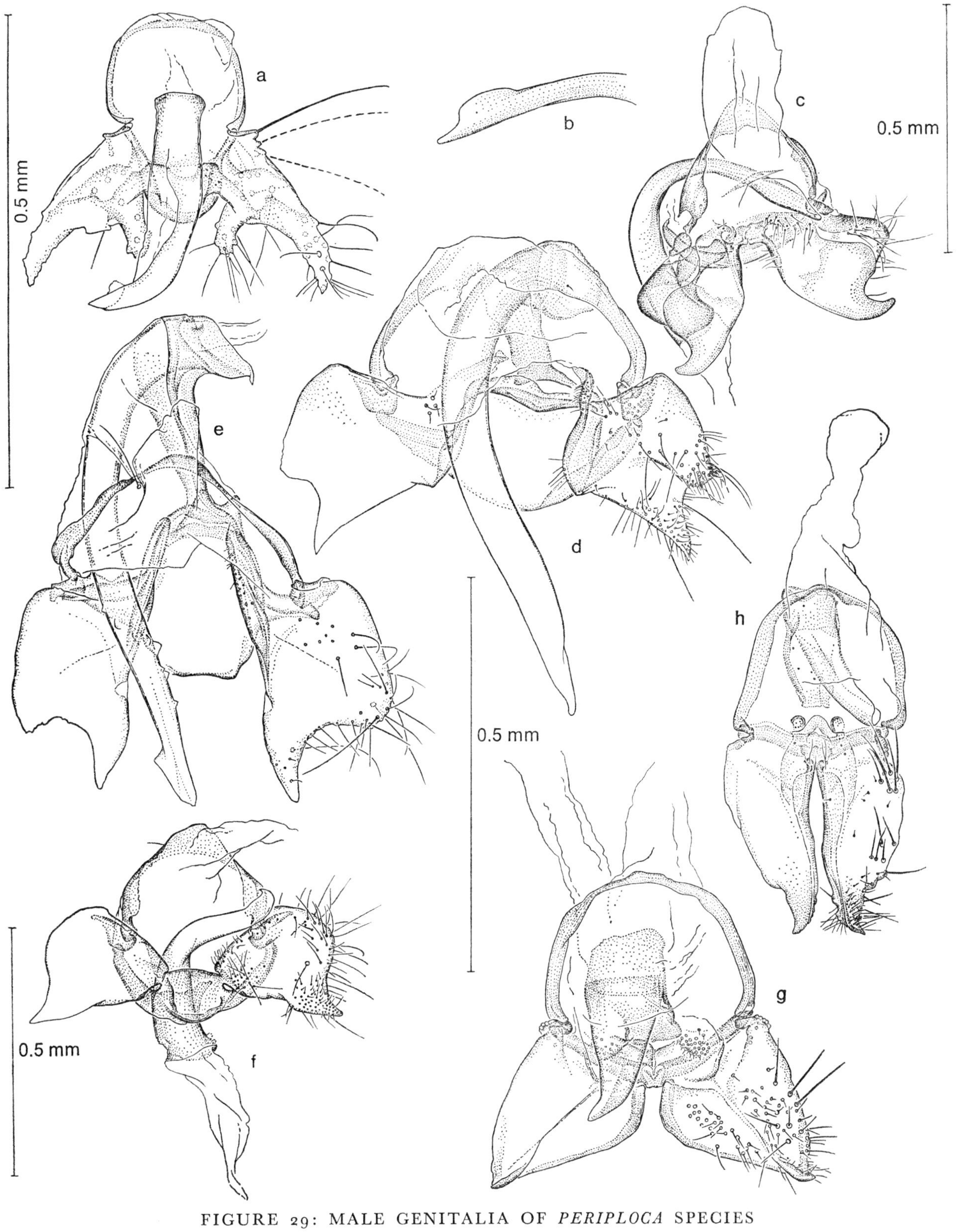


FIGURE 29: MALE GENITALIA OF *PERIPLOCA* SPECIES

*a. P. orichalcella.* *c. P. repanda.* *e. P. hostiata.* *g. P. ceanothiella.*
*b.* Apex of aedoeagus, *P. repanda.* *d. P. gleditschiaeella.* *f. P. intermedia.* *h. P. arsa.*

Different scales for *a*; *b, d, e, g, h*; and *c, f.*

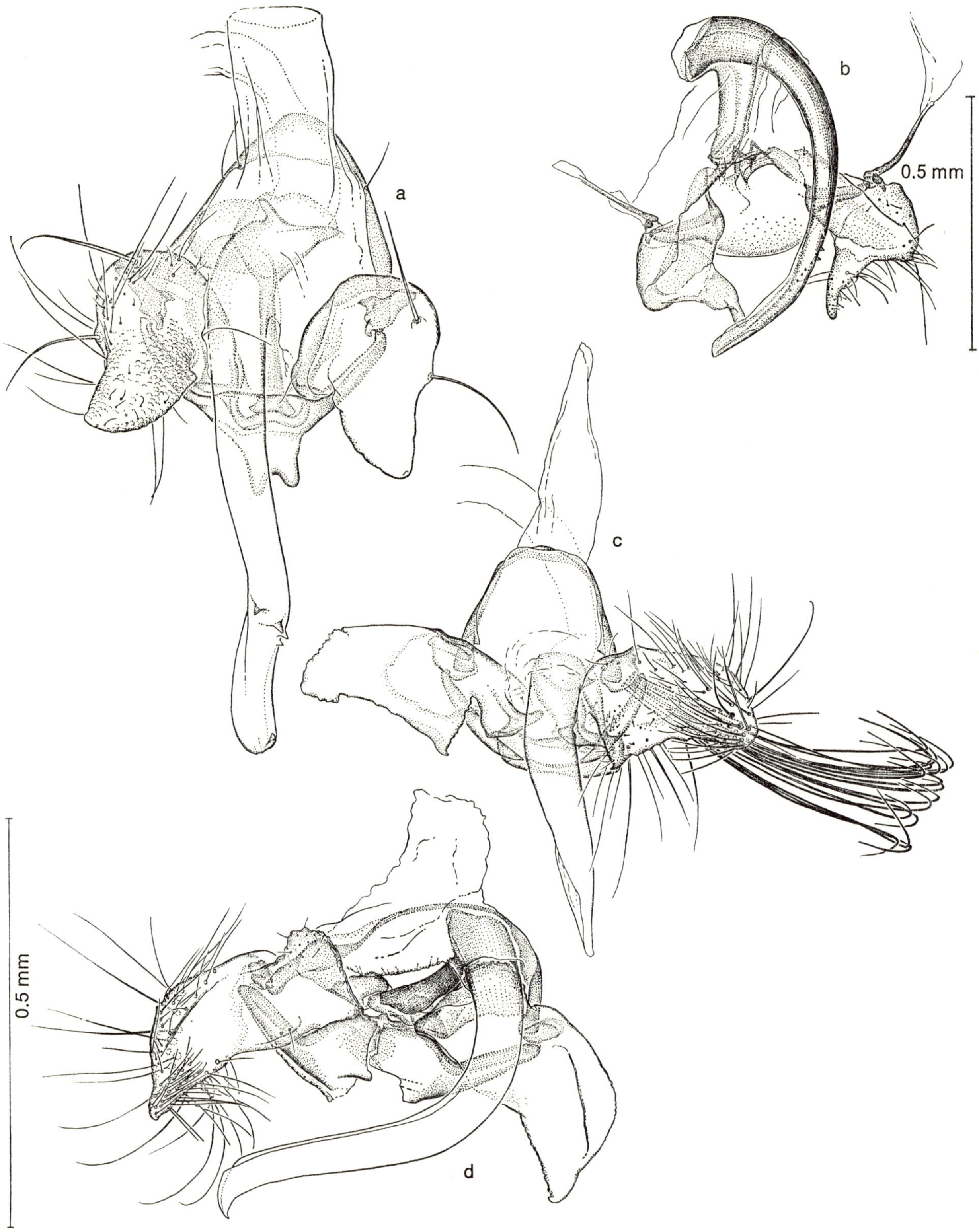


FIGURE 30: MALE GENITALIA OF *PERIPLOCA* SPECIES

*a. P. opinatrix.* *b. P. teres.* *c. P. atrata.* *d. P. hortatrix.*

All same scale except *b*.

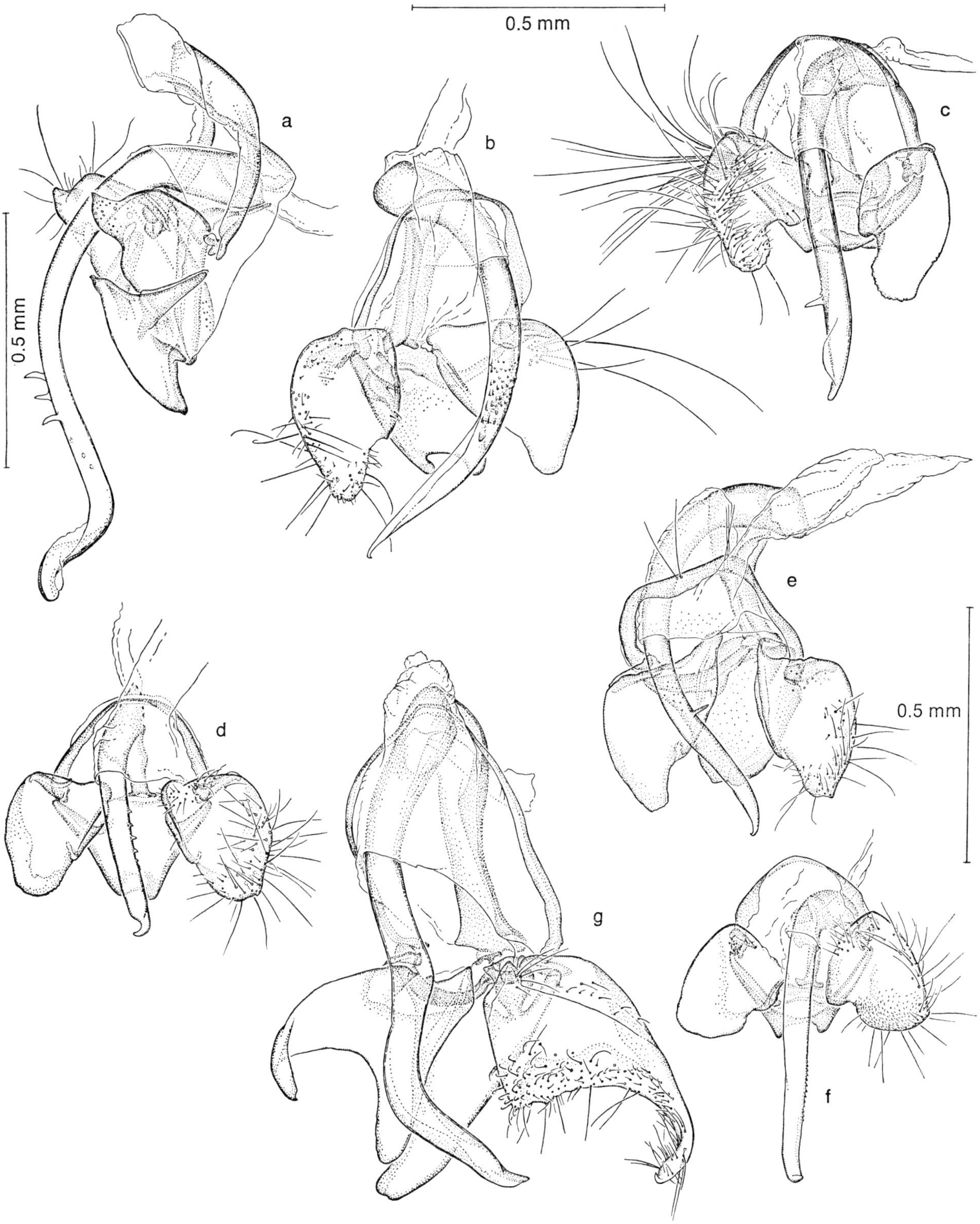


FIGURE 31: MALE GENITALIA OF *PERIPLOCA* SPECIES

*a. P. tridens.* *b. P. mimula.* *c. P. laeta.* *d. P. devia.* *e. P. nigra.* *f. P. cata.* *g. P. gulosa.*

All same scale.

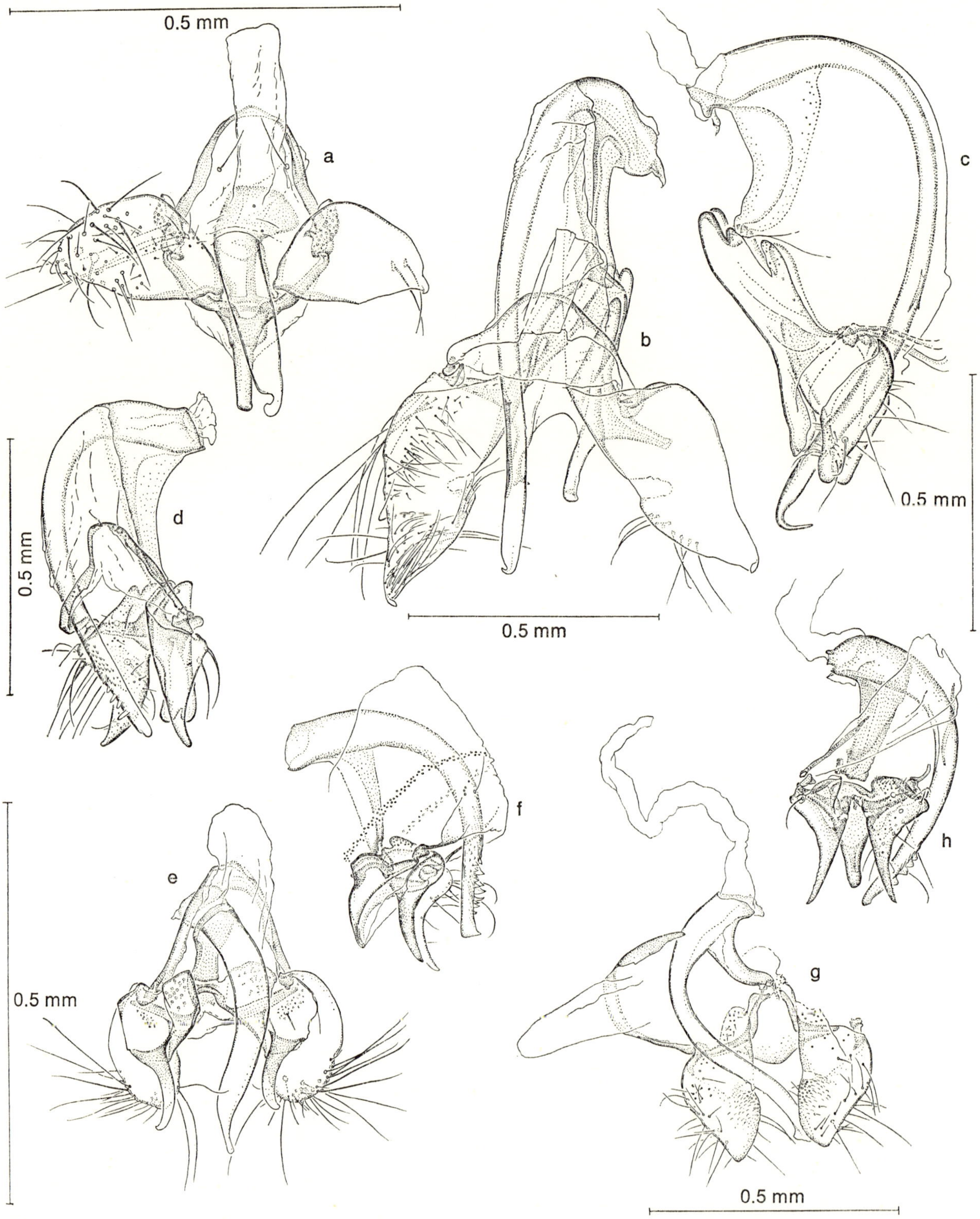


FIGURE 32: MALE GENITALIA OF *PERIPLOCA* SPECIES

*a. P. fessa.* *b. P. facula.* *c. P. juniperi.* *d. P. funebris.* *e. P. labes.* *f. P. dentella.* *g. P. dipapha.* *h. P. serrulata.*

All same scale except *a* and *e*.

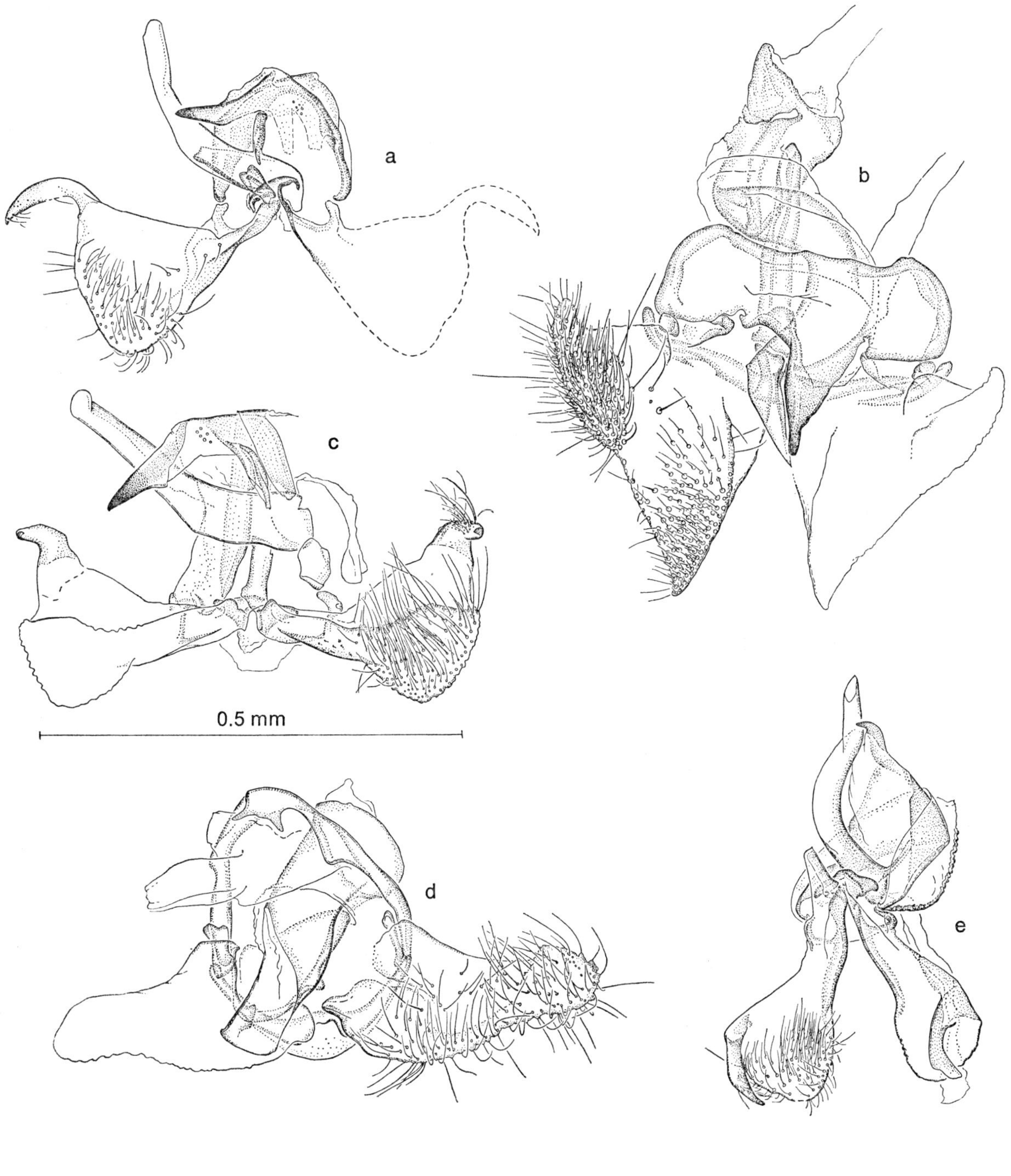


FIGURE 33: MALE GENITALIA OF CHRYSOPELEIINAE SPECIES

a. *Siskiwitia alticolans.* b. *Pristen corusca.* c. *Siskiwitia latebra.* d. *Synploca gumia.* e. *Siskiwitia falcata.*

All same scale.

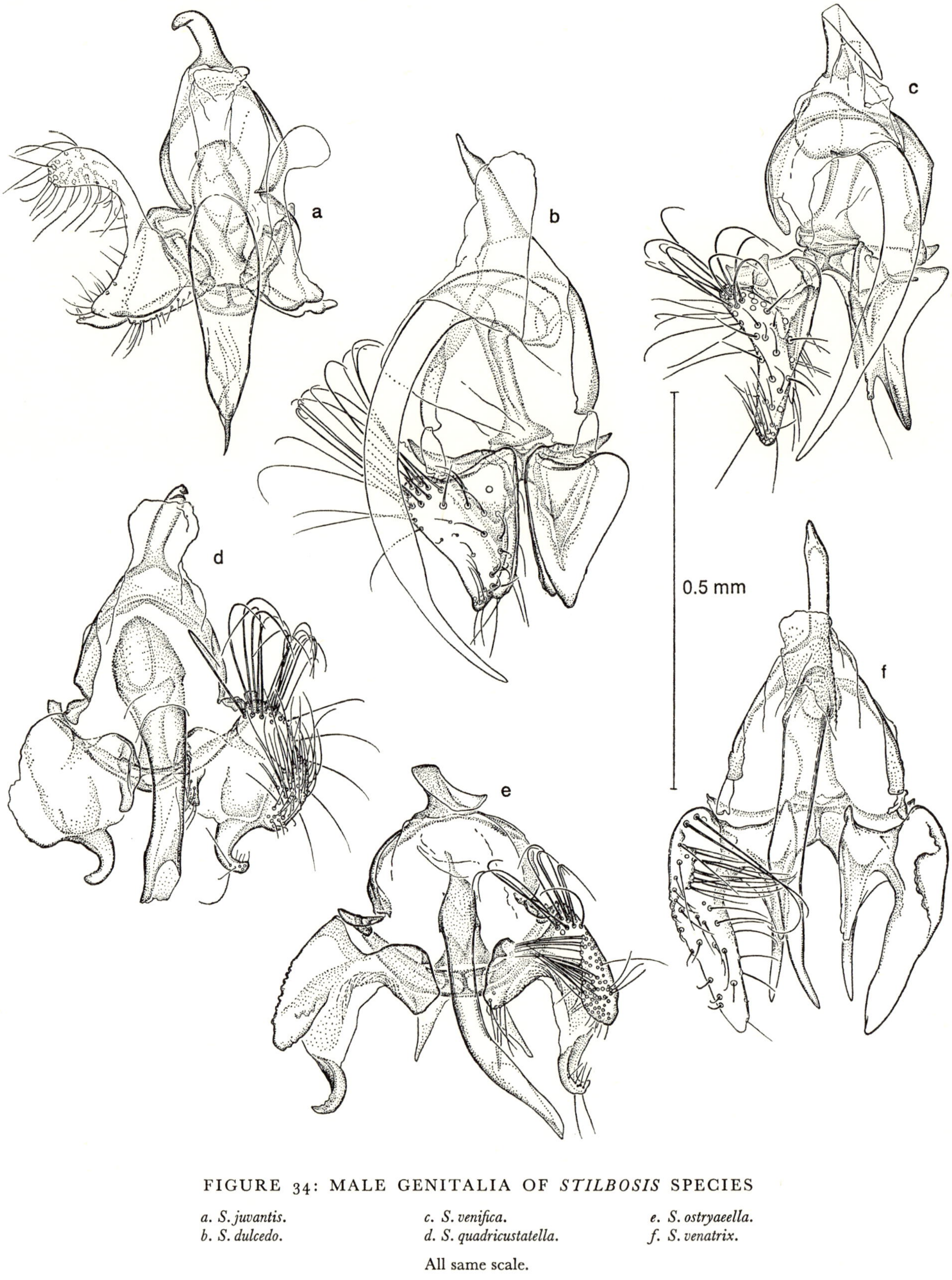


FIGURE 34: MALE GENITALIA OF *STILBOSIS* SPECIES

*a. S. juvantis.* *c. S. venifica.* *e. S. ostryaeella.*
*b. S. dulcedo.* *d. S. quadricustatella.* *f. S. venatrix.*

All same scale.

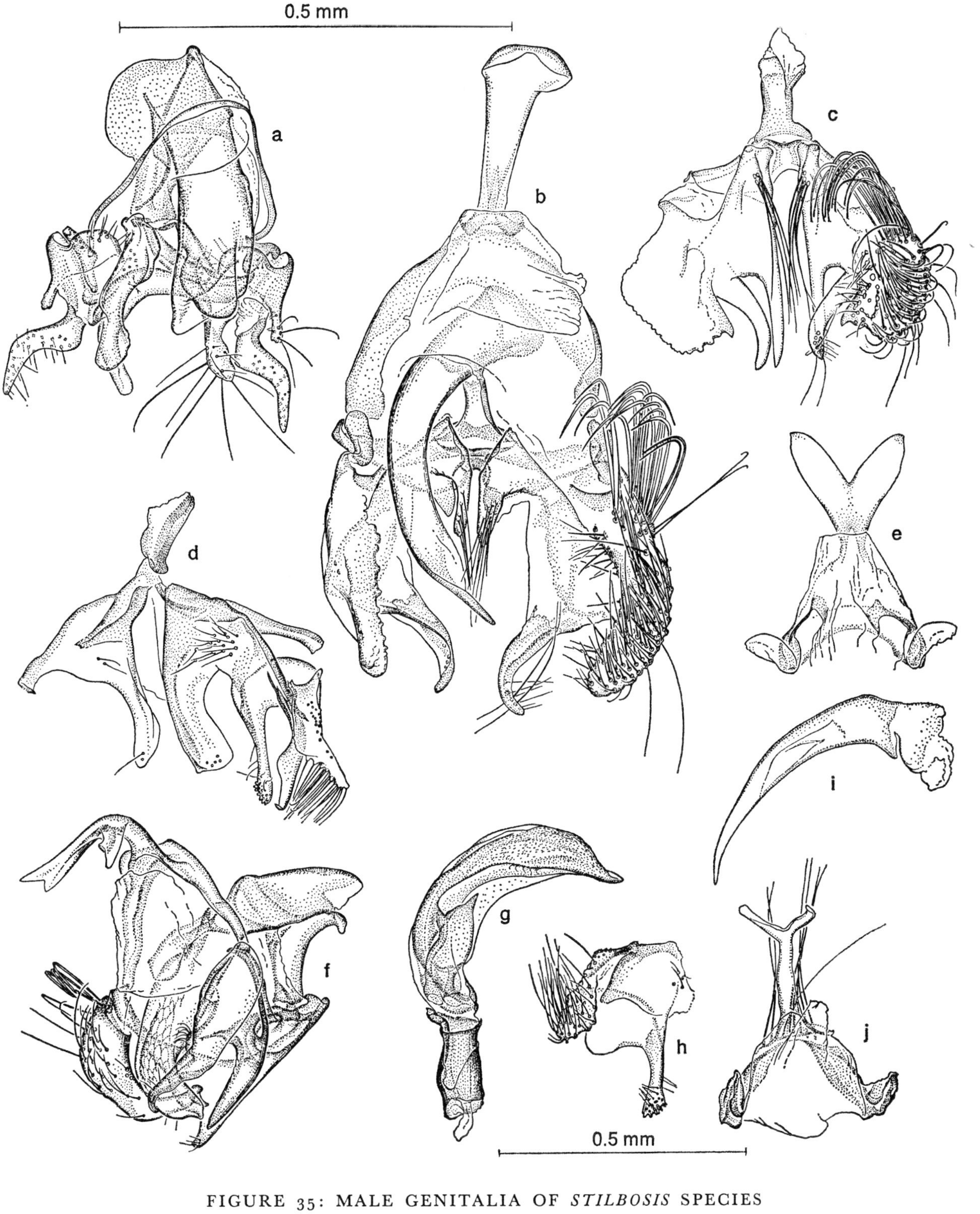


FIGURE 35: MALE GENITALIA OF *STILBOSIS* SPECIES

*a. S. risor.*
*b. S. stipator.*
*c.* Vinculum and valvae, *S. victor.*
*d.* Vinculum and right valva, *S. rhynchosiae.*
*e.* Tegumen and uncus, *S. victor.*
*f. S. sagana.*
*j.* Tegumen and uncus, *S. rhynchosiae.*
*g.* Aedoeagus, *S. rhynchosiae.*
*h.* Left valva, *S. rhynchosiae.*
*i.* Aedoeagus, *S. victor.*

Different scales for *a, b, c, e, f, i*; and *d, g, h, j*.

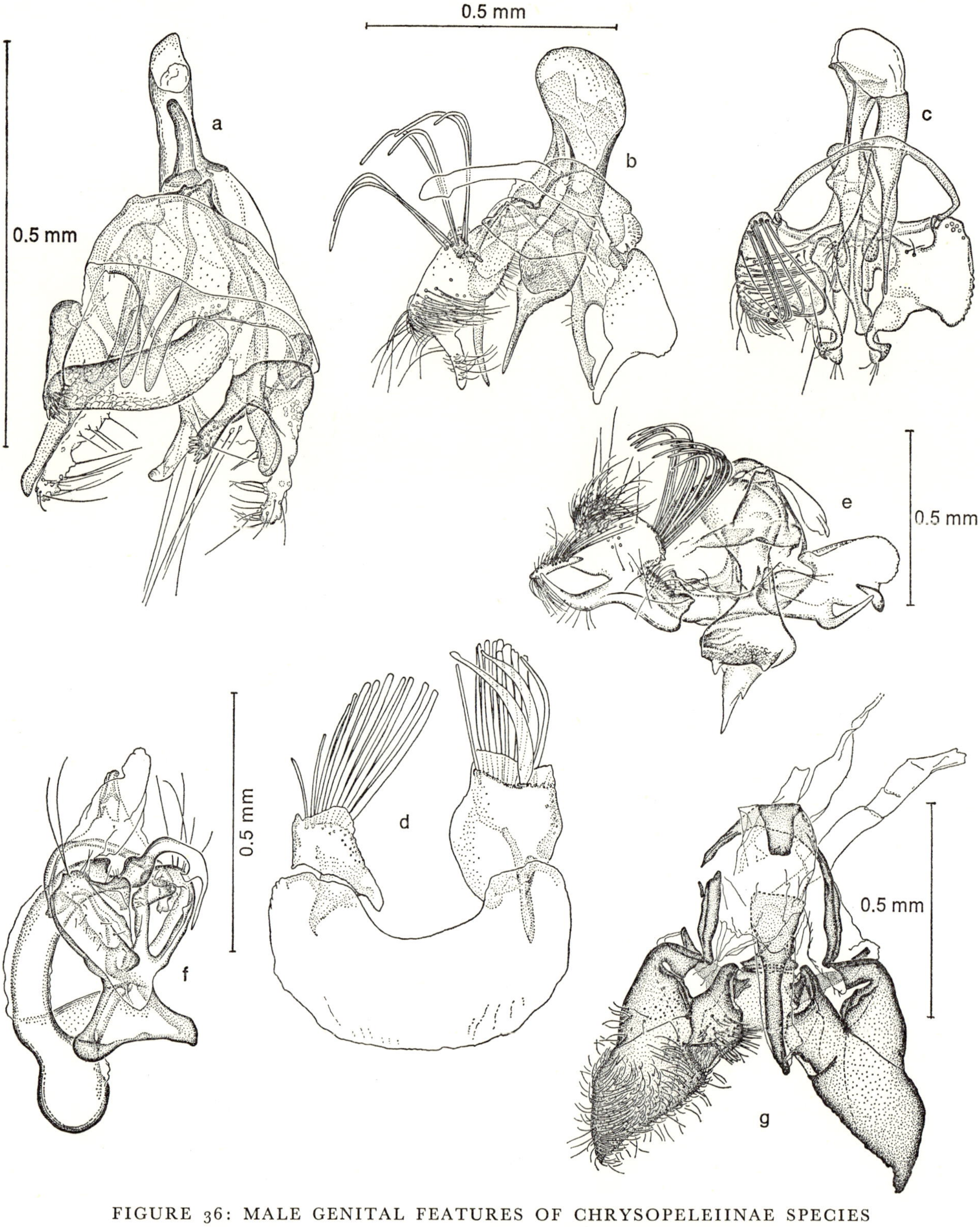


FIGURE 36: MALE GENITAL FEATURES OF CHRYSOPELEIINAE SPECIES

*a*. Genitalia, *Stilbosis extensa*.
*b*. Genitalia, *Stilbosis tesquella*.
*c*. Genitalia, *Stilbosis lonchocarpella*.
*d*. Seventh and eighth abdominal sterna, *Chrysopeleia purpuriella*.
*e*. Genitalia, *Stilbosis nubila*.
*f*. Genitalia, *Chrysopeleia purpuriella*.
*g*. Genitalia, *Walshia particornella*.

Different scales for *a*; *b*, *c*, *d*, *f*; *e*; and *g*.

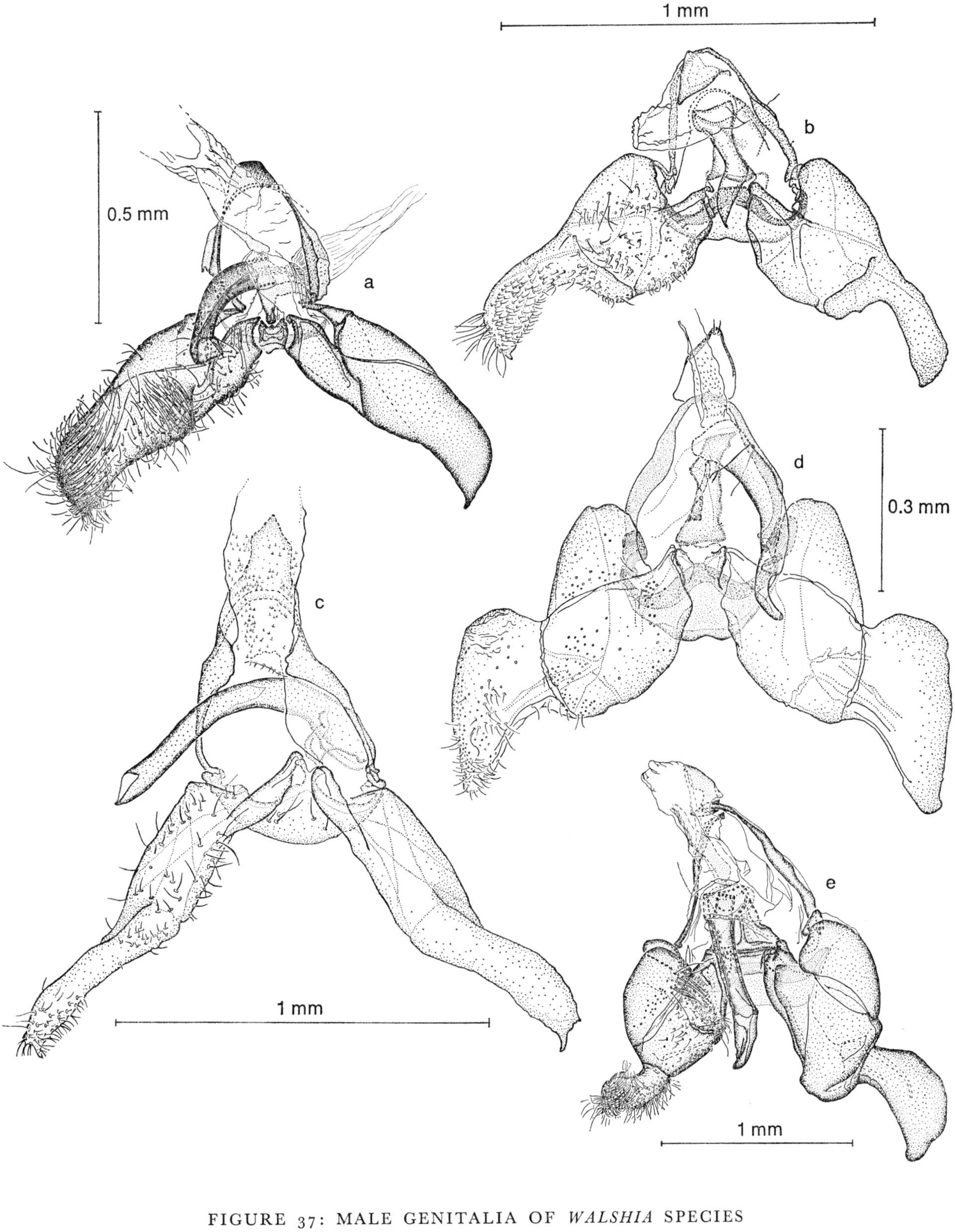


FIGURE 37: MALE GENITALIA OF *WALSHIA* SPECIES

*a. W. elegans.* *b. W. floridensis.* *c. W. miscecolorella.* *d. W. amorphella.* *e. W. dispar.*

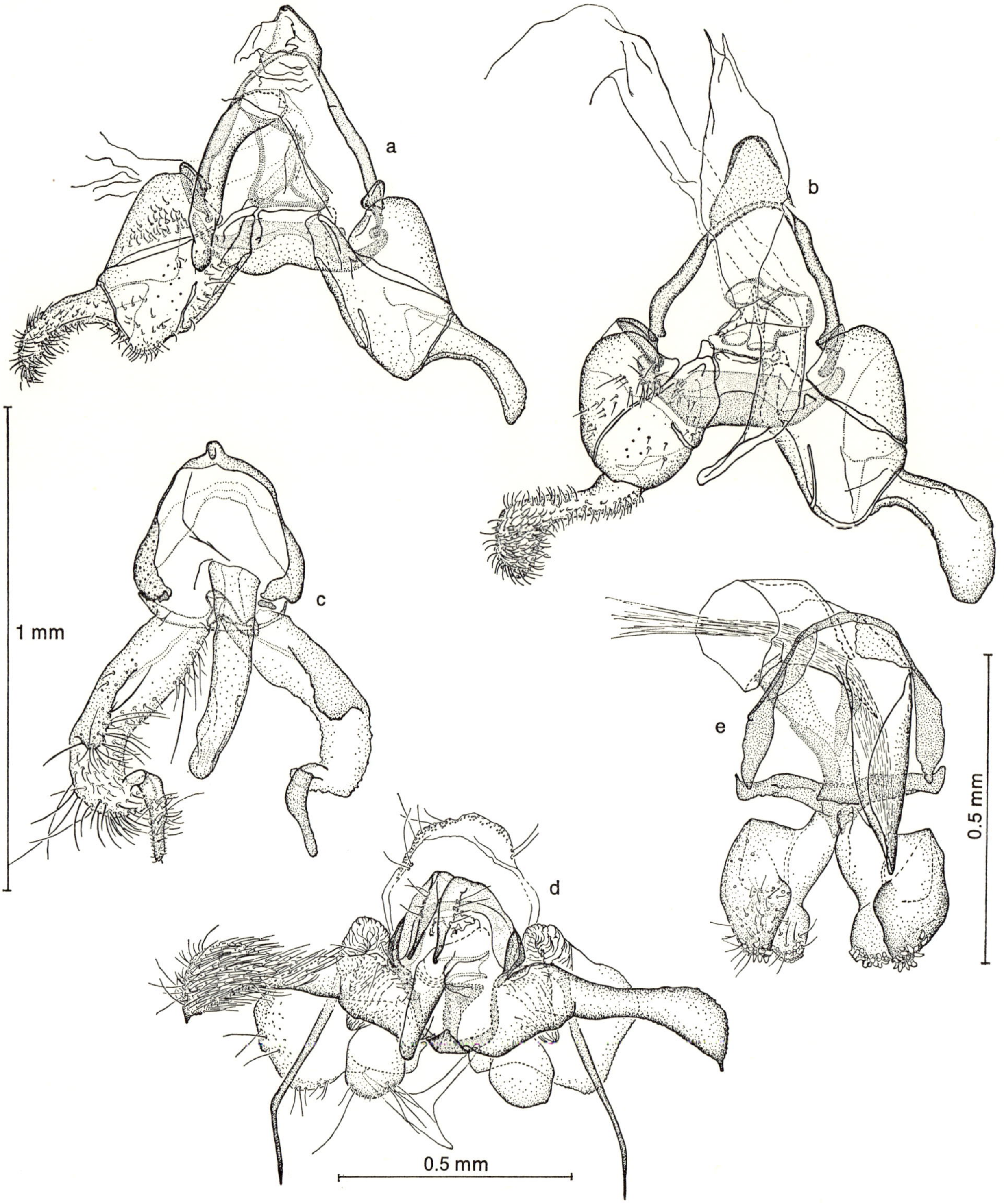


FIGURE 38: MALE GENITALIA OF CHRYSOPELEIINAE SPECIES

*a. Walshia exemplata.* *b. Walshia similis.* *c. Nepotula secura.* *d. Perimede erransella.* *e. Afeda biloba.*

Scales as shown. Not to scale: *a*, *b*.

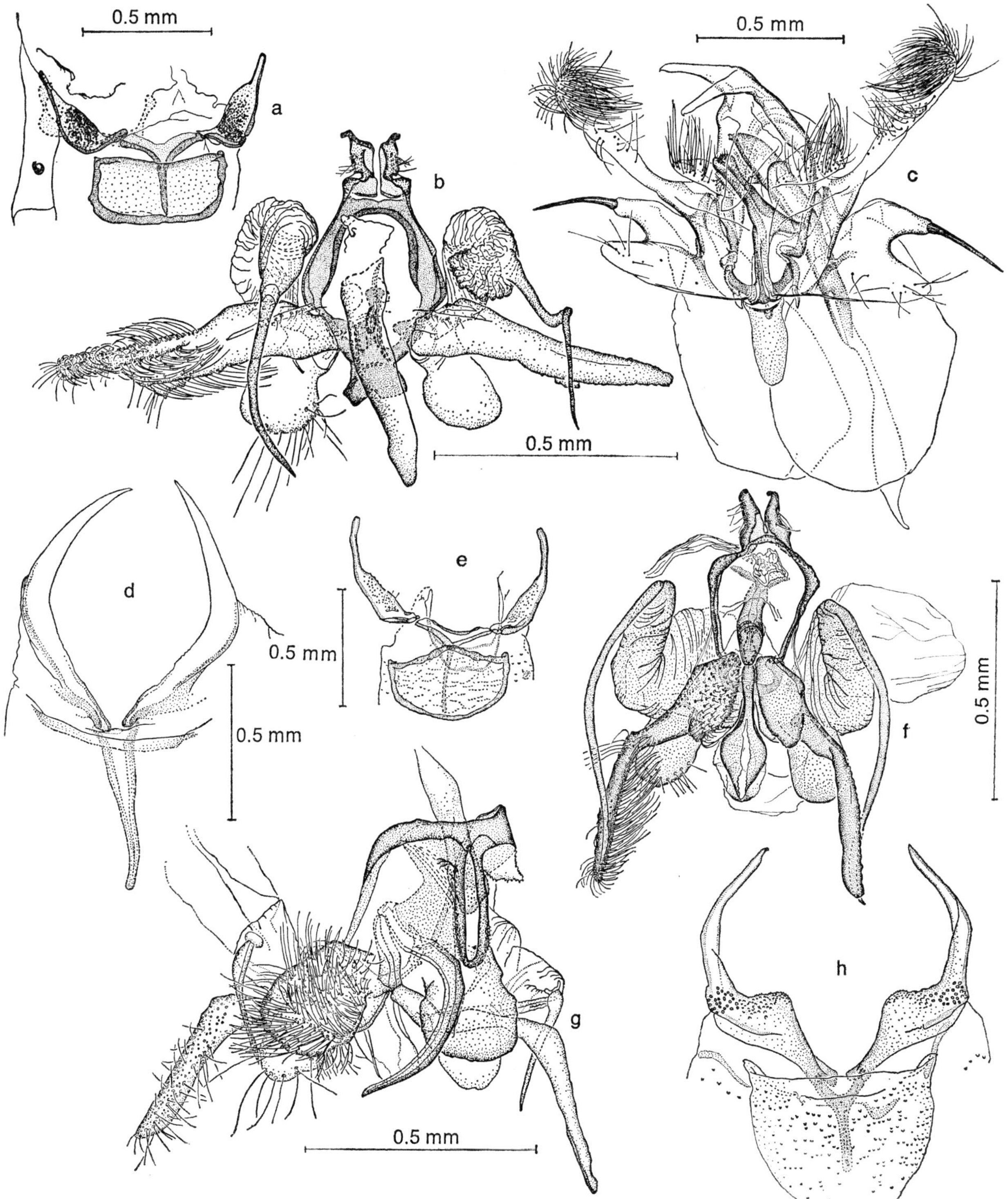


FIGURE 39: MALE GENITAL FEATURES OF *PERIMEDE* SPECIES

*a*. Seventh and eighth abdominal segments, *P. parilis*.
*b*. Genitalia, *P. battis*.
*c*. Genitalia, *P. grandis*.
*d*. Seventh and eighth abdominal sterna, *P. grandis*.
*e*. Seventh and eighth abdominal segments, *P. latris*.
*f*. Genitalia, *P. latris*.
*g*. Genitalia, *P. parilis*.
*h*. Seventh and eighth abdominal sterna, *P. battis*.

Scales as shown. Same scale: *b*, *h*.

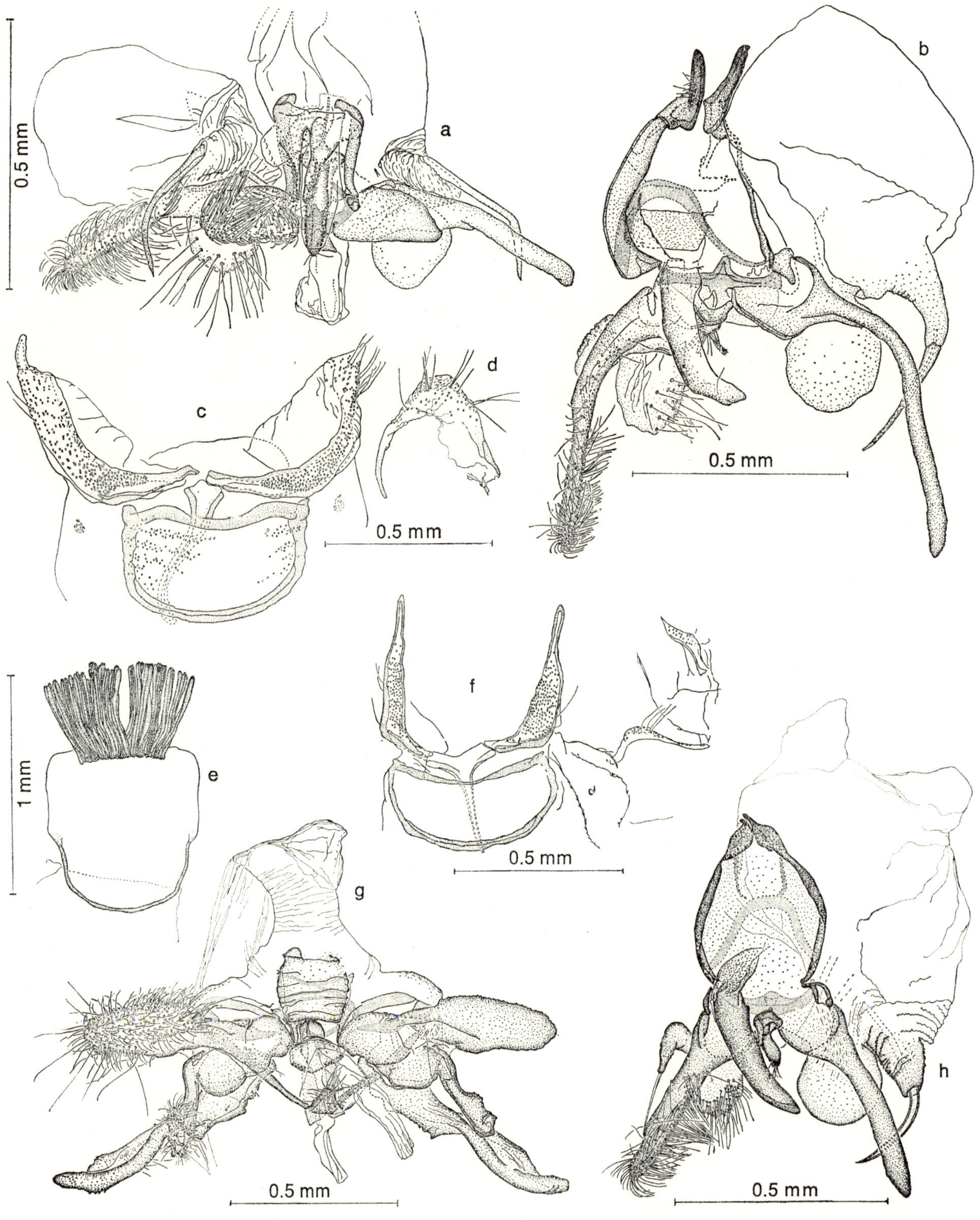


FIGURE 40: MALE GENITAL FEATURES OF *PERIMEDE* AND *SORHAGENIA* SPECIES

*a.* Genitalia, *P. circitor.*
*b.* Genitalia, *P. ricina.*
*c.* Seventh and eighth abdominal sterna, *P. ricina.*
*d.* Eighth abdominal tergum, *P. ricina.*
*e.* Eighth abdominal sternum, *S. nimbosa.*
*f.* Seventh and eighth abdominal segments, *P. circitor.*
*g.* Genitalia, *S. nimbosa.*
*h.* Genitalia, *P. falcata.*

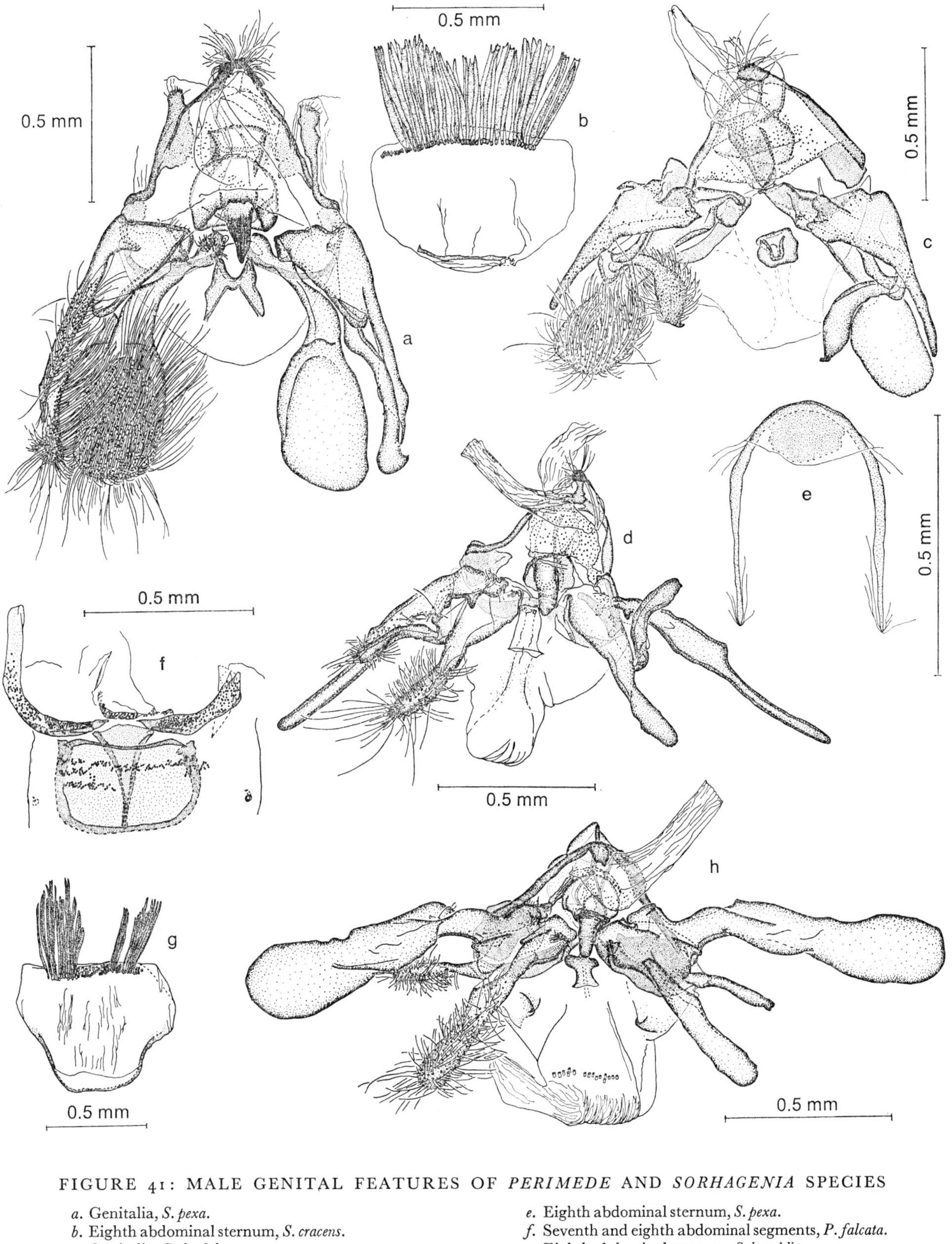


FIGURE 41: MALE GENITAL FEATURES OF *PERIMEDE* AND *SORHAGENIA* SPECIES

*a*. Genitalia, *S. pexa*.
*b*. Eighth abdominal sternum, *S. cracens*.
*c*. Genitalia, *S. daedala*.
*d*. Genitalia, *S. cracens*.
*e*. Eighth abdominal sternum, *S. pexa*.
*f*. Seventh and eighth abdominal segments, *P. falcata*.
*g*. Eighth abdominal sternum, *S. baucidis*.
*h*. Genitalia, *S. baucidis*.

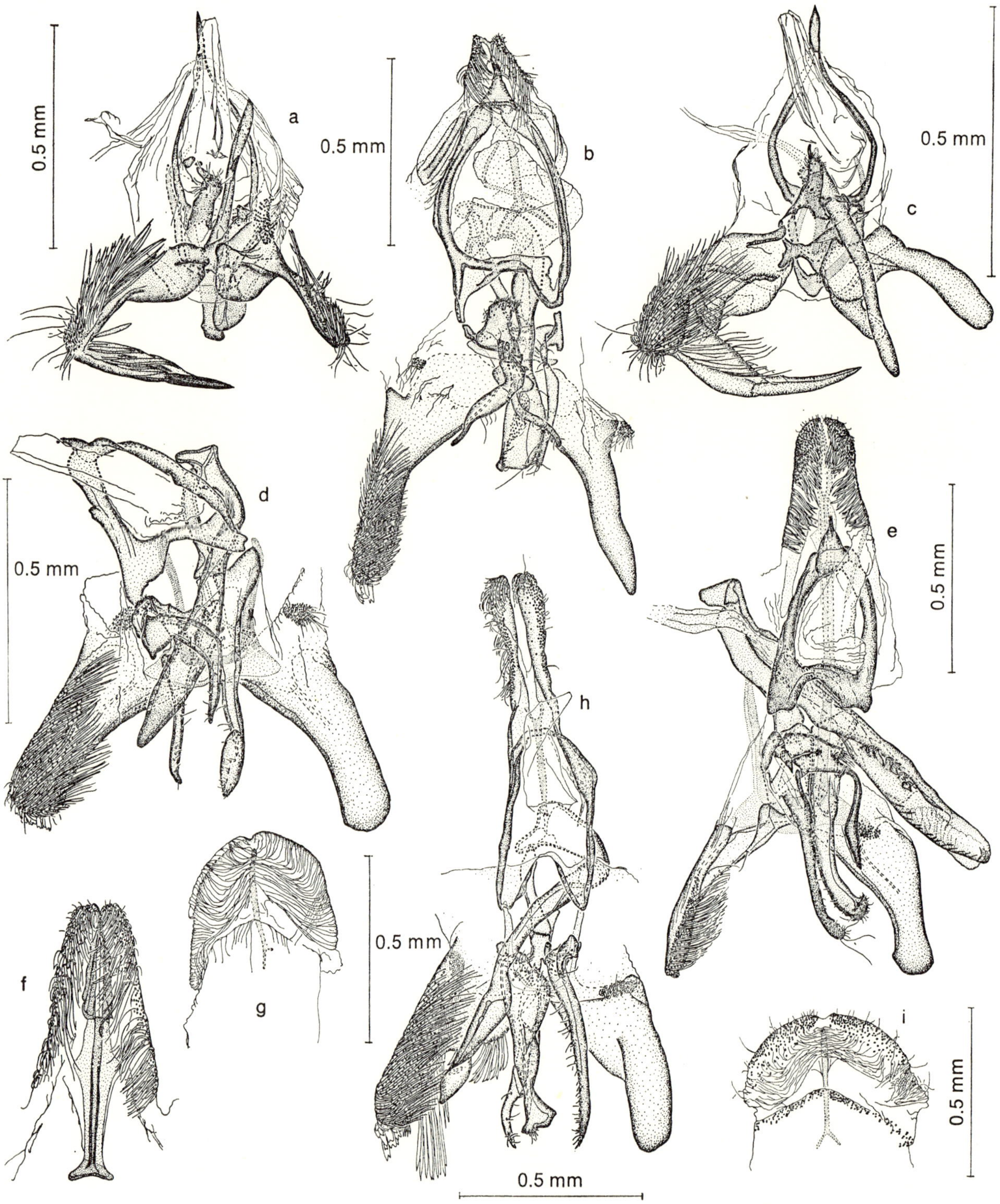


FIGURE 42: MALE GENITAL FEATURES OF *ITHOME* SPECIES

*a*. Genitalia, *I. concolorella*.
*b*. Eighth abdominal tergum and genitalia, *I. lassula*.
*c*. Genitalia, *I. curvipunctella*.
*d*. Genitalia, *I. edax*.
*e*. Eighth abdominal tergum and genitalia, *I. aquila*.
*f*. Eighth abdominal tergum, *I. edax*.
*g*. Eighth abdominal tergum, *I. curvipunctella*.
*h*. Eighth abdominal tergum and genitalia, *I. ferax*.
*i*. Eighth abdominal tergum, *I. concolorella*.

Scales as shown except *f* (not to scale).

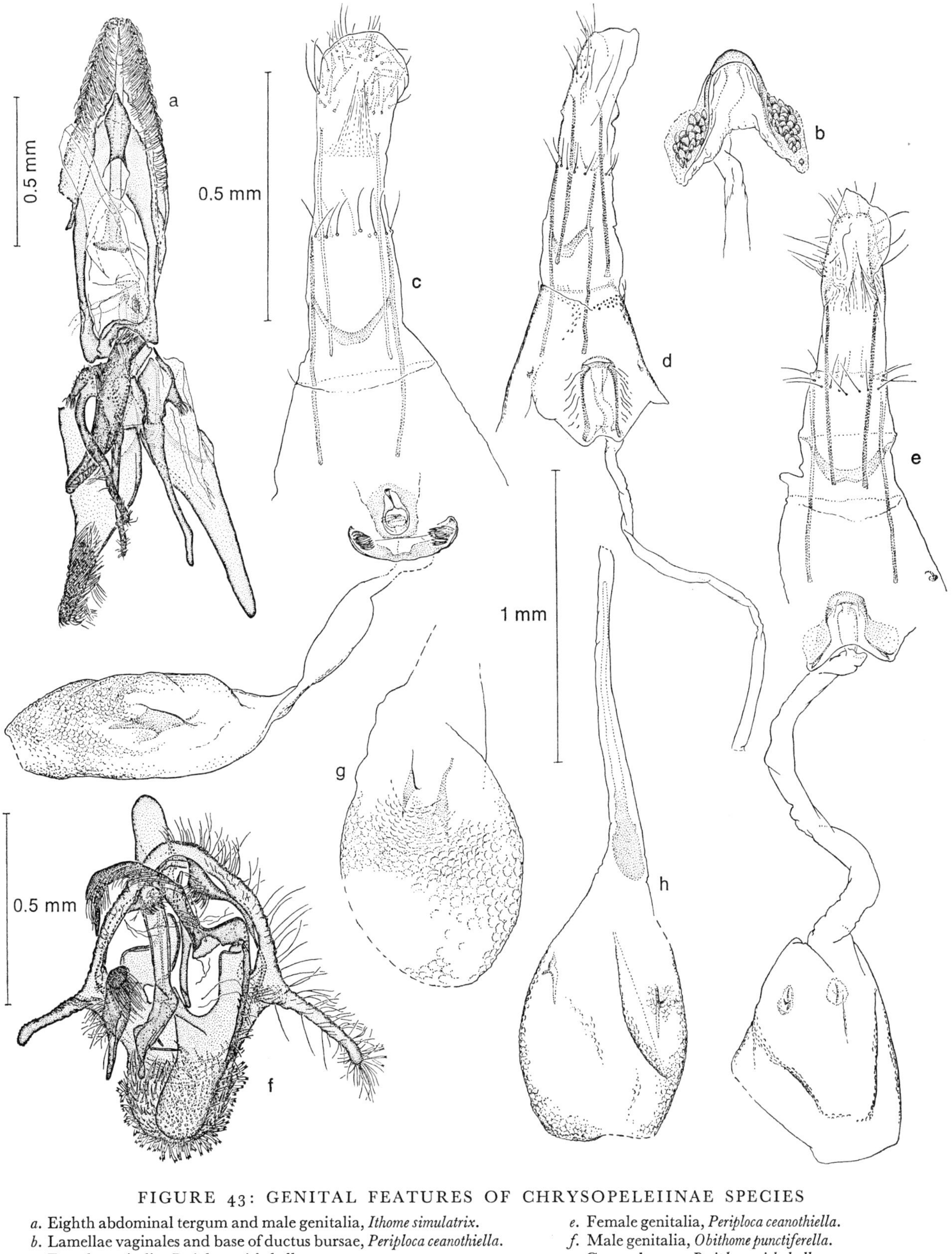


FIGURE 43: GENITAL FEATURES OF CHRYSOPELEIINAE SPECIES

*a.* Eighth abdominal tergum and male genitalia, *Ithome simulatrix*.
*b.* Lamellae vaginales and base of ductus bursae, *Periploca ceanothiella*.
*c.* Female genitalia, *Periploca orichalcella*.
*d.* Female genitalia, *Periploca arsa*.
*e.* Female genitalia, *Periploca ceanothiella*.
*f.* Male genitalia, *Obithome punctiferella*.
*g.* Corpus bursae, *Periploca orichalcella*.
*h.* Corpus bursae and ductus bursae, *Periploca arsa*.

Different scales for *a, b, e*; *c, g*; *d, h*; and *f*.

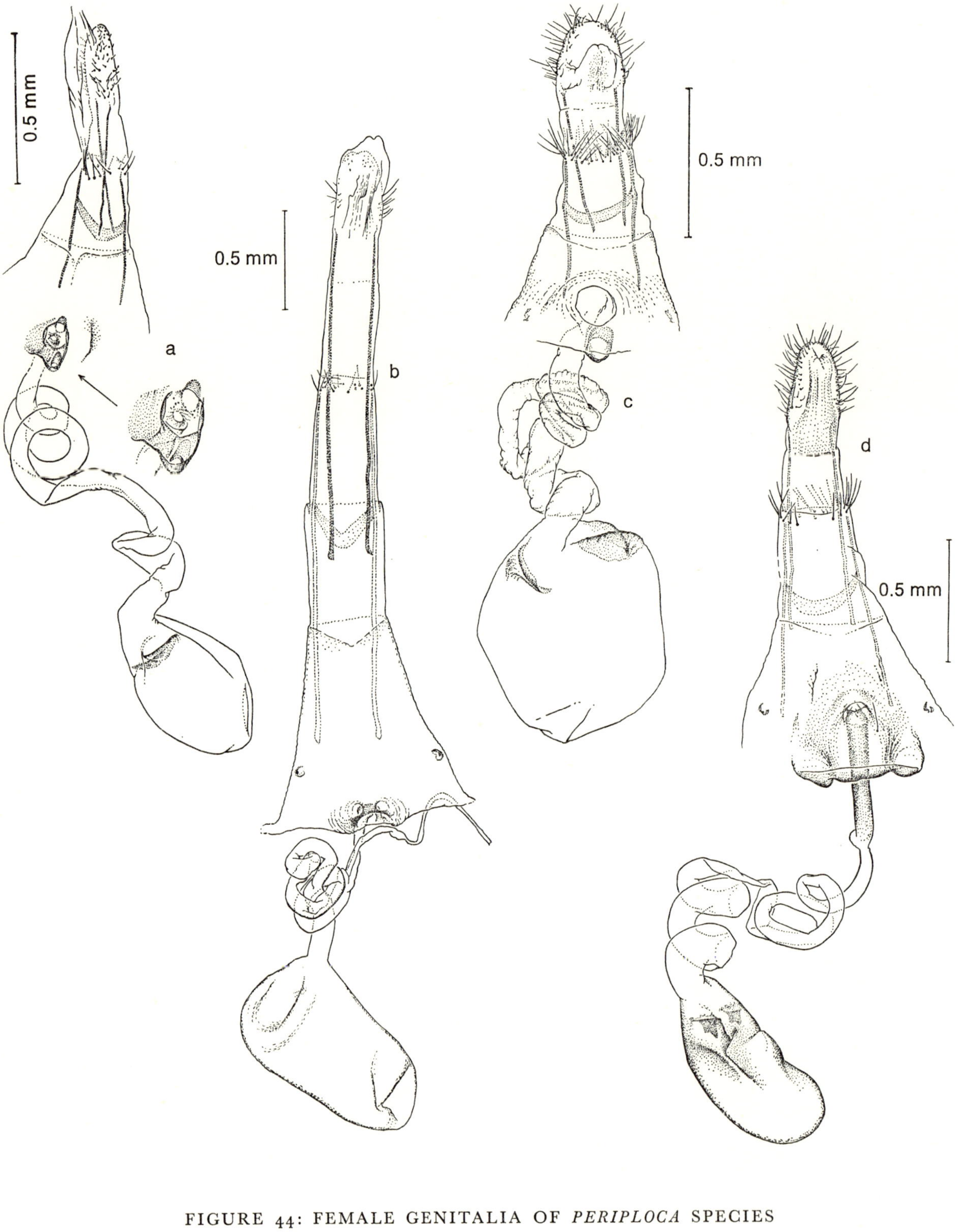


FIGURE 44: FEMALE GENITALIA OF *PERIPLOCA* SPECIES

*a. P. hortatrix.* *b. P. gleditschiaeella.* *c. P. opinatrix.* *d. P. mimula.*

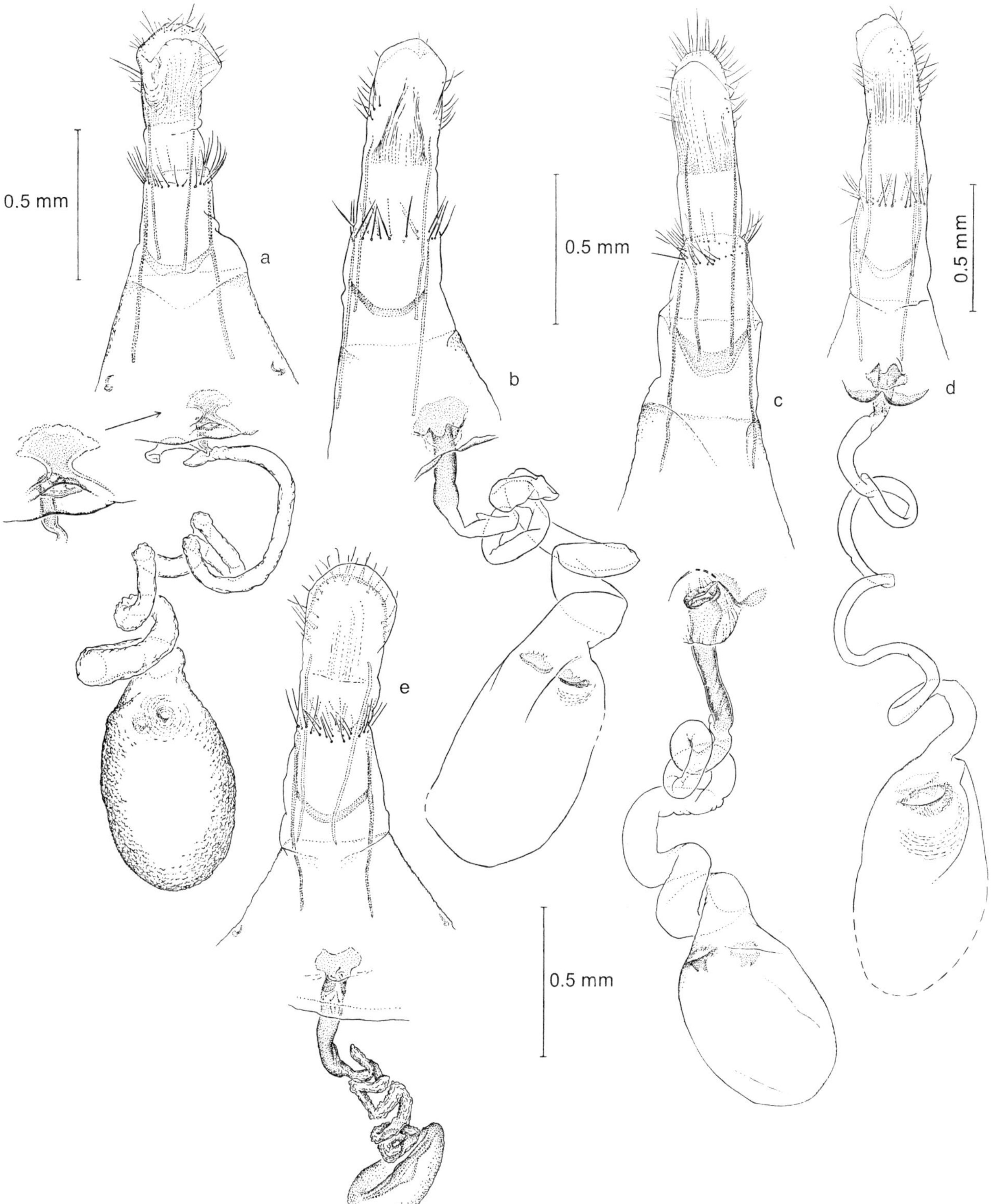


FIGURE 45: FEMALE GENITALIA OF *PERIPLOCA* SPECIES

*a. P. atrata.* *b. P. soror.* *c. P. laeta.* *d. P. devia.* *e. P. cata.*

All same scale except *d*.

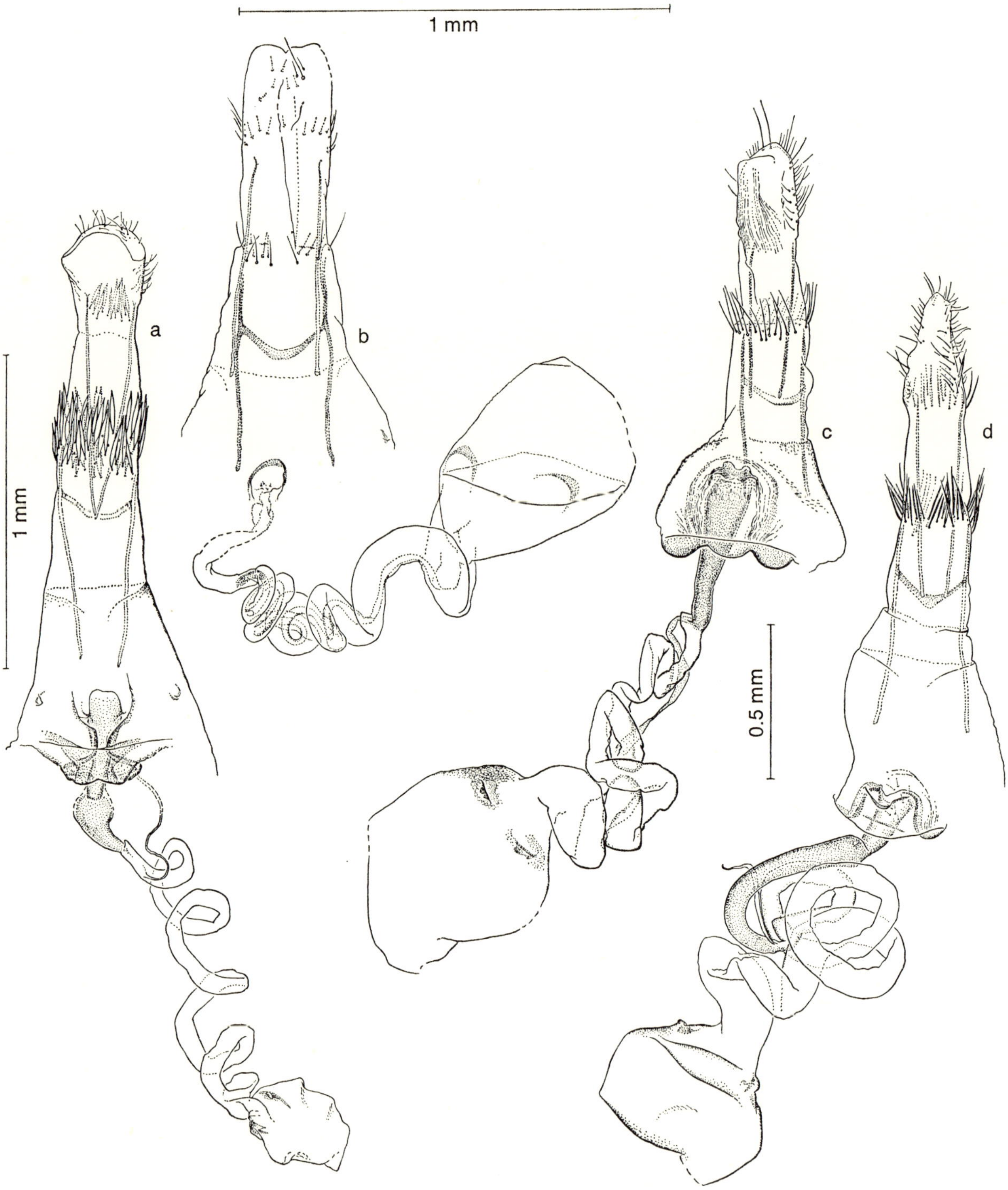


FIGURE 46: FEMALE GENITALIA OF *PERIPLOCA* SPECIES

*a. P. funebris.* *b. P. labes.* *c. P. nigra.* *d. P. juniperi.*

All same scale except *b*.

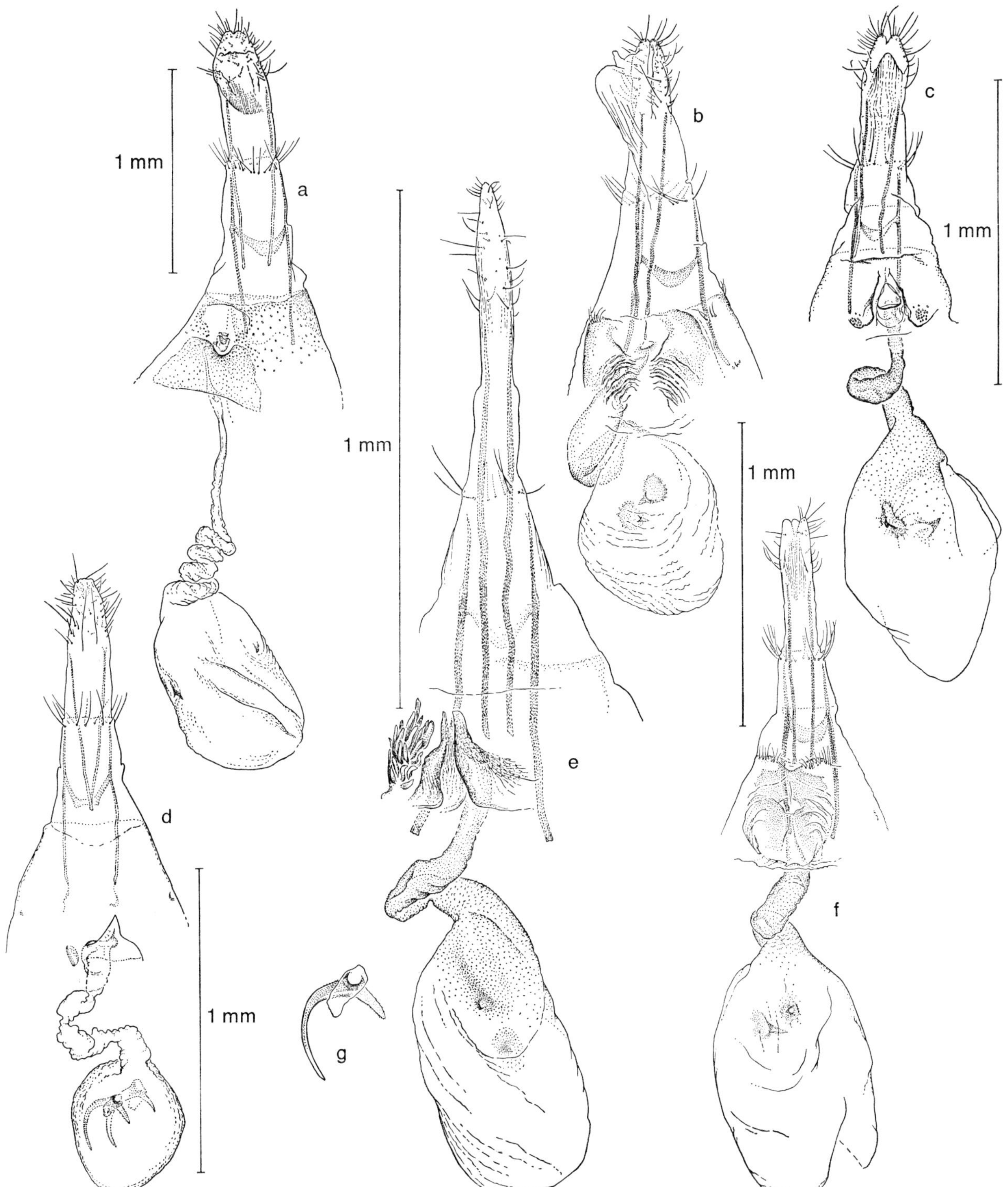


FIGURE 47: FEMALE GENITALIA OF CHRYSOPELEIINAE SPECIES

*a. Pristen corusca.*
*b. Stilbosis dulcedo.*
*c. Stilbosis quadricustatella.*
*d. Synploca gumia.*
*e. Stilbosis ostryaeella.*
*f. Stilbosis venifica.*
*g.* Signum, *Synploca gumia.*

Different scales for *a*; *b, c, d, f*; and *e*. Not to scale: *g*.

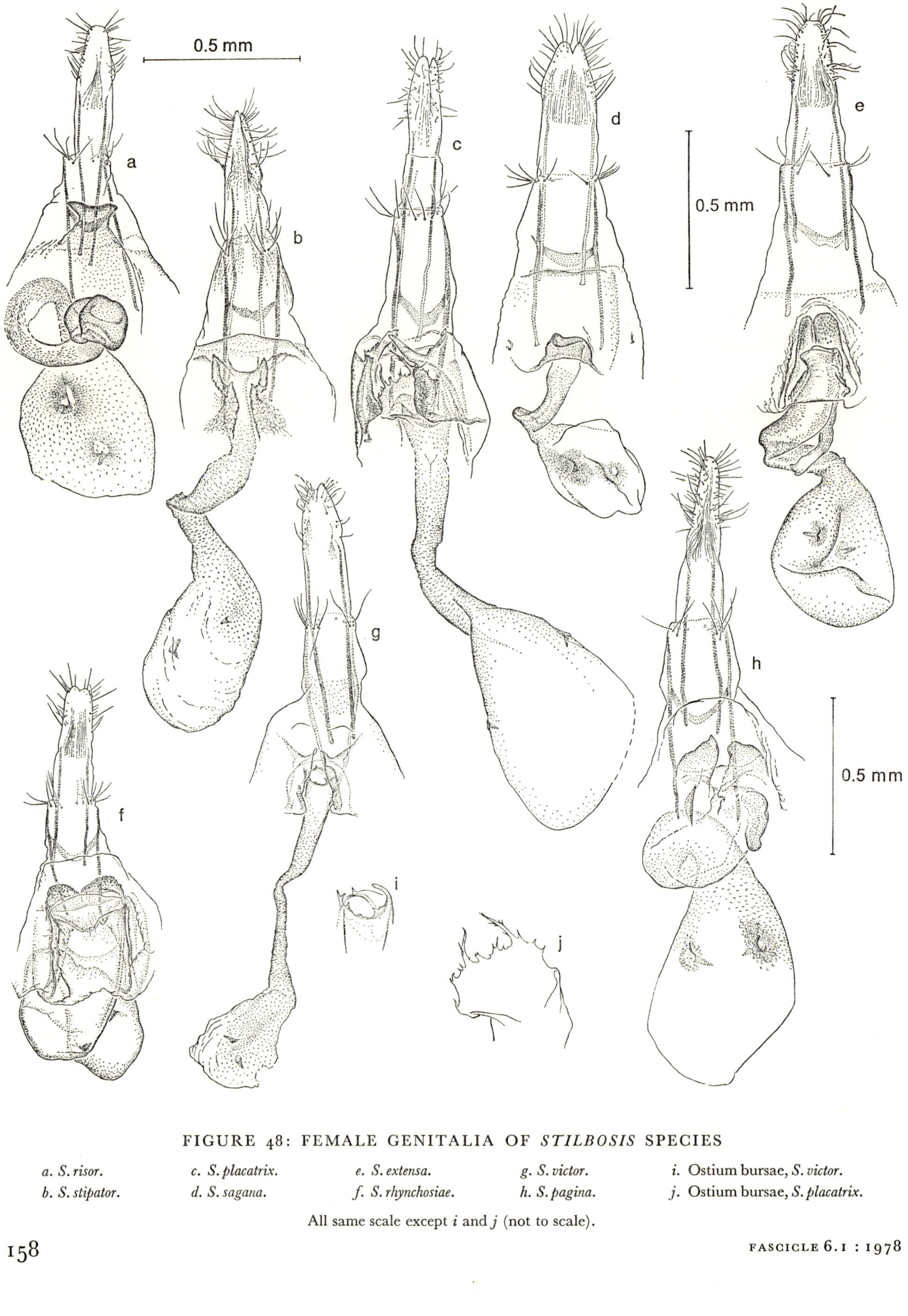


FIGURE 48: FEMALE GENITALIA OF *STILBOSIS* SPECIES

*a*. *S. risor*. *b*. *S. stipator*. *c*. *S. placatrix*. *d*. *S. sagana*. *e*. *S. extensa*. *f*. *S. rhynchosiae*. *g*. *S. victor*. *h*. *S. pagina*. *i*. Ostium bursae, *S. victor*. *j*. Ostium bursae, *S. placatrix*.

All same scale except *i* and *j* (not to scale).

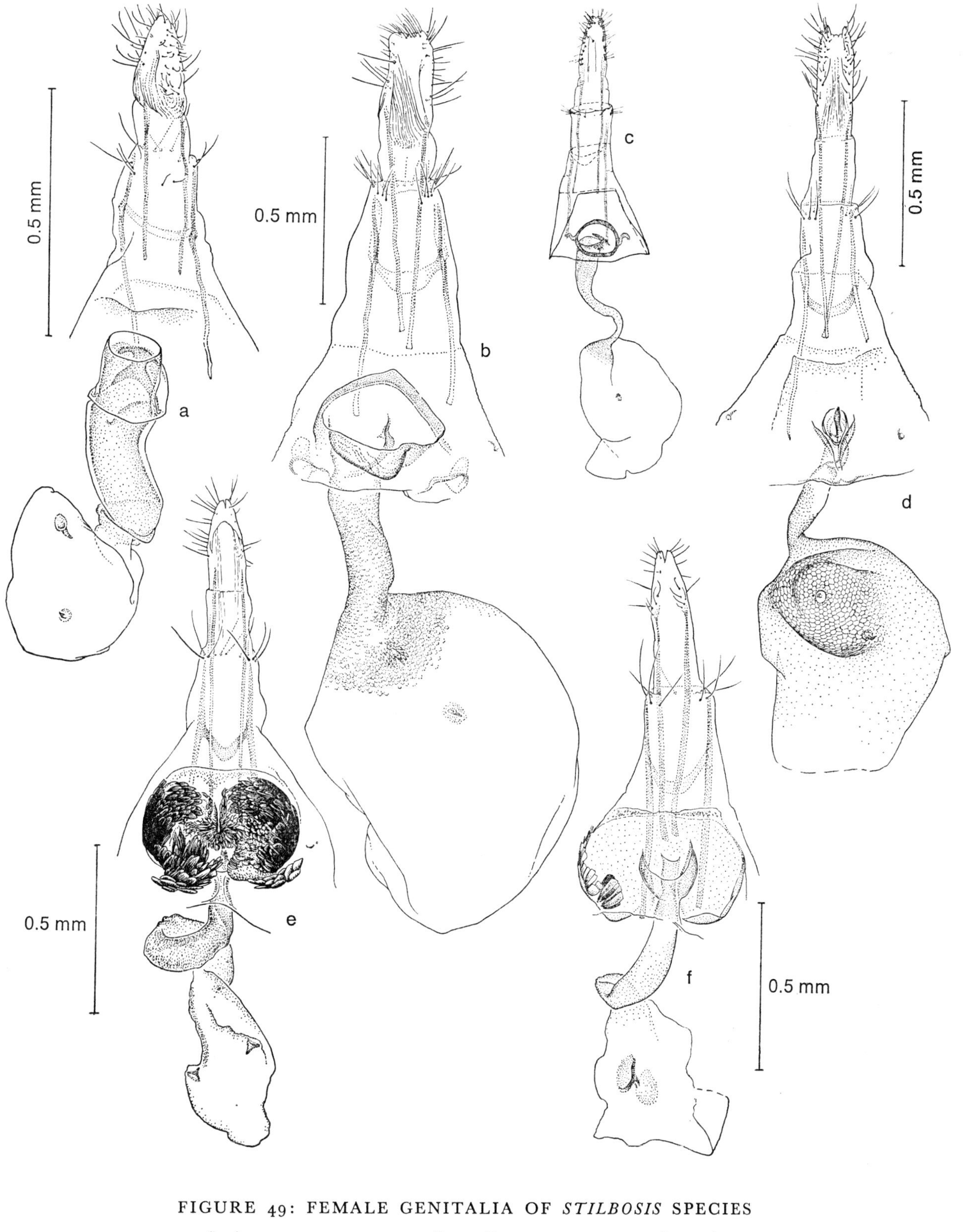


FIGURE 49: FEMALE GENITALIA OF *STILBOSIS* SPECIES

*a. S. scleroma.* *c. S. tesquella.* *e. S. ornatrix.*
*b. S. nubila.* *d. S. lonchocarpella.* *f. S. rotunda.*

Not to scale: *c*.

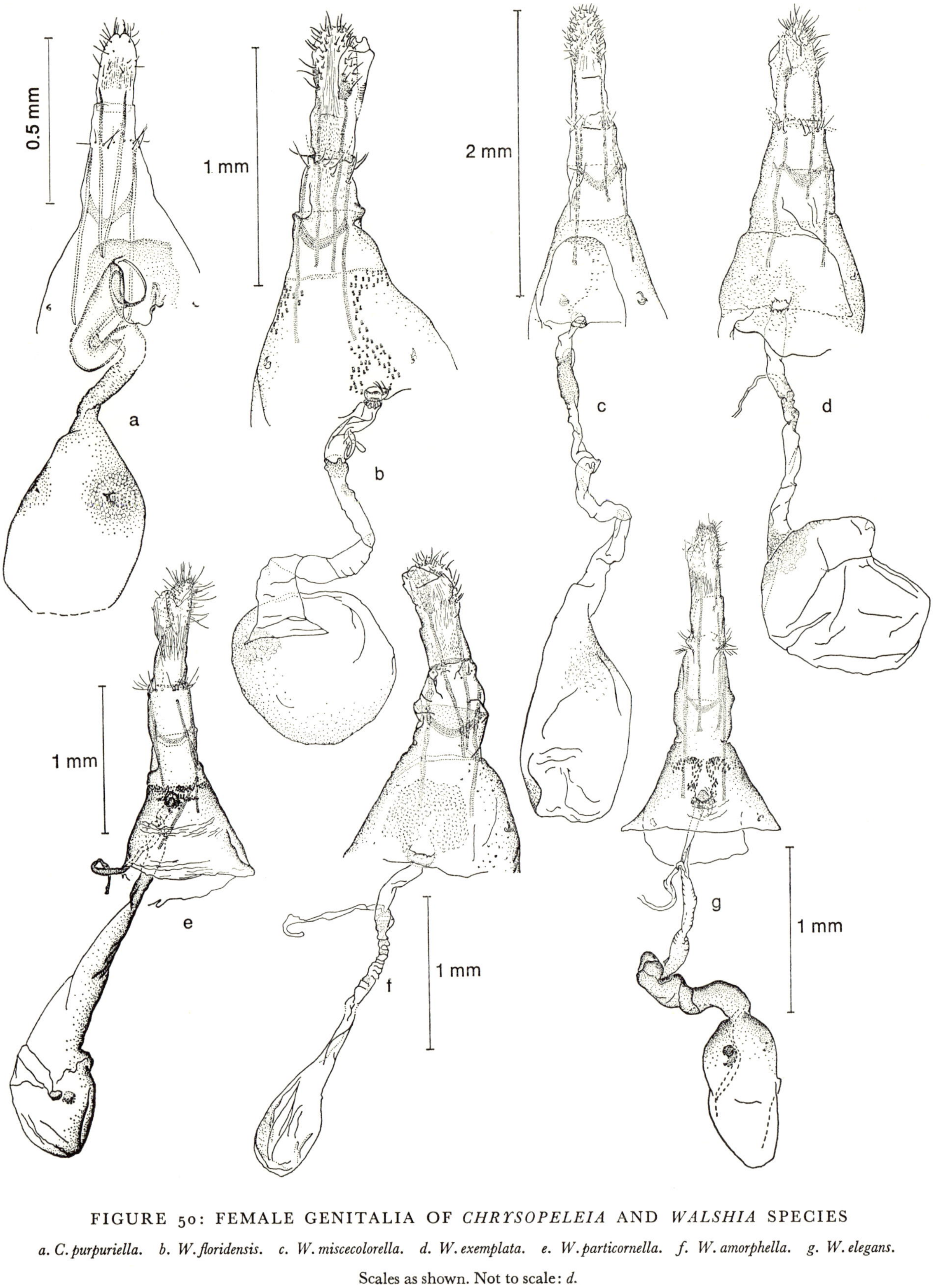


FIGURE 50: FEMALE GENITALIA OF *CHRYSOPELEIA* AND *WALSHIA* SPECIES

*a. C. purpuriella. b. W. floridensis. c. W. miscecolorella. d. W. exemplata. e. W. particornella. f. W. amorphella. g. W. elegans.*

Scales as shown. Not to scale: *d.*

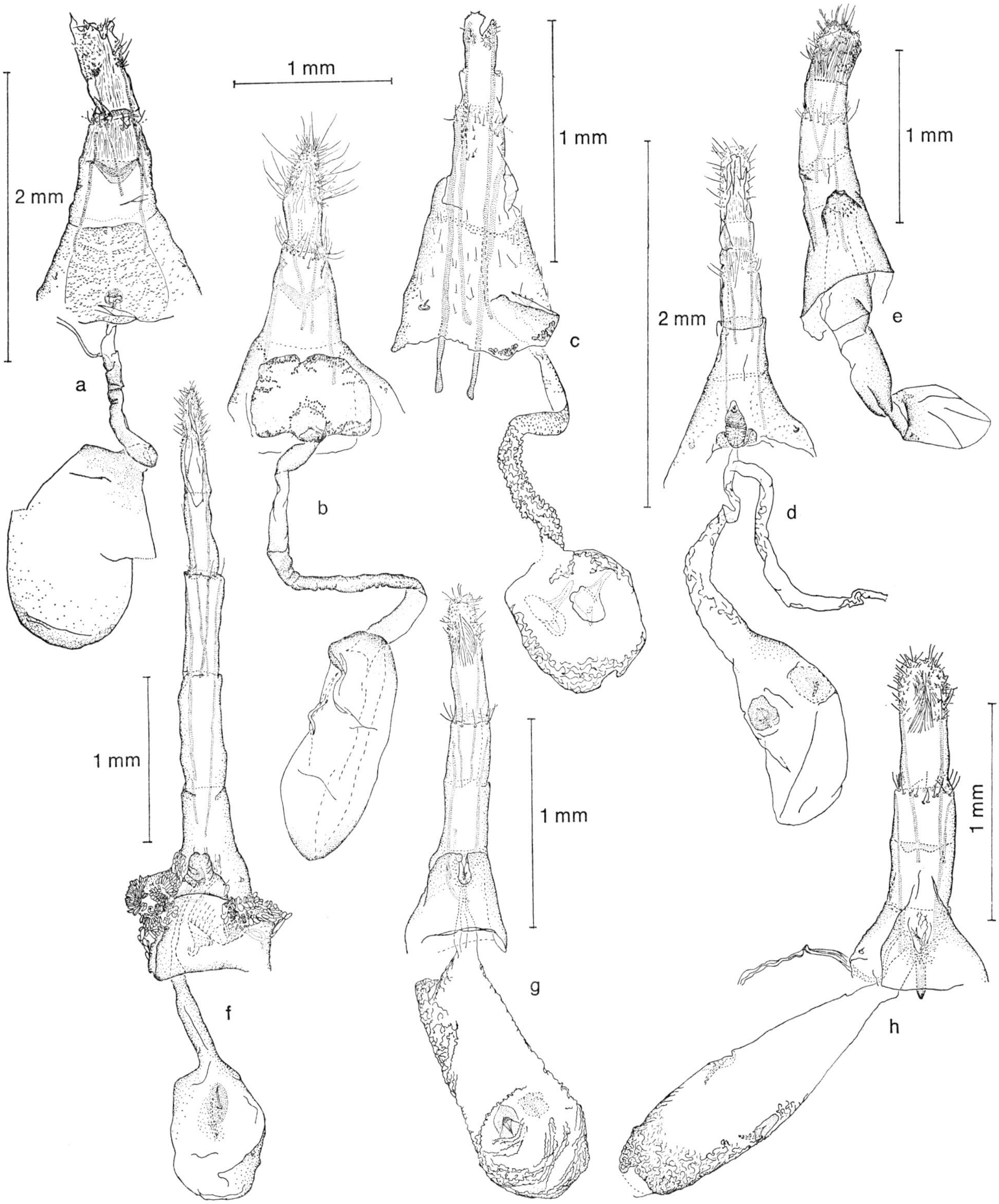


FIGURE 51: FEMALE GENITALIA OF CHRYSOPELEIINAE SPECIES

*a. Walshia similis.* *b. Walshia dispar.* *c. Afeda biloba.* *d. Perimede erransella.* *e. Perimede latris.* *f. Perimede battis.* *g. Perimede circitor.* *h. Perimede parilis.*

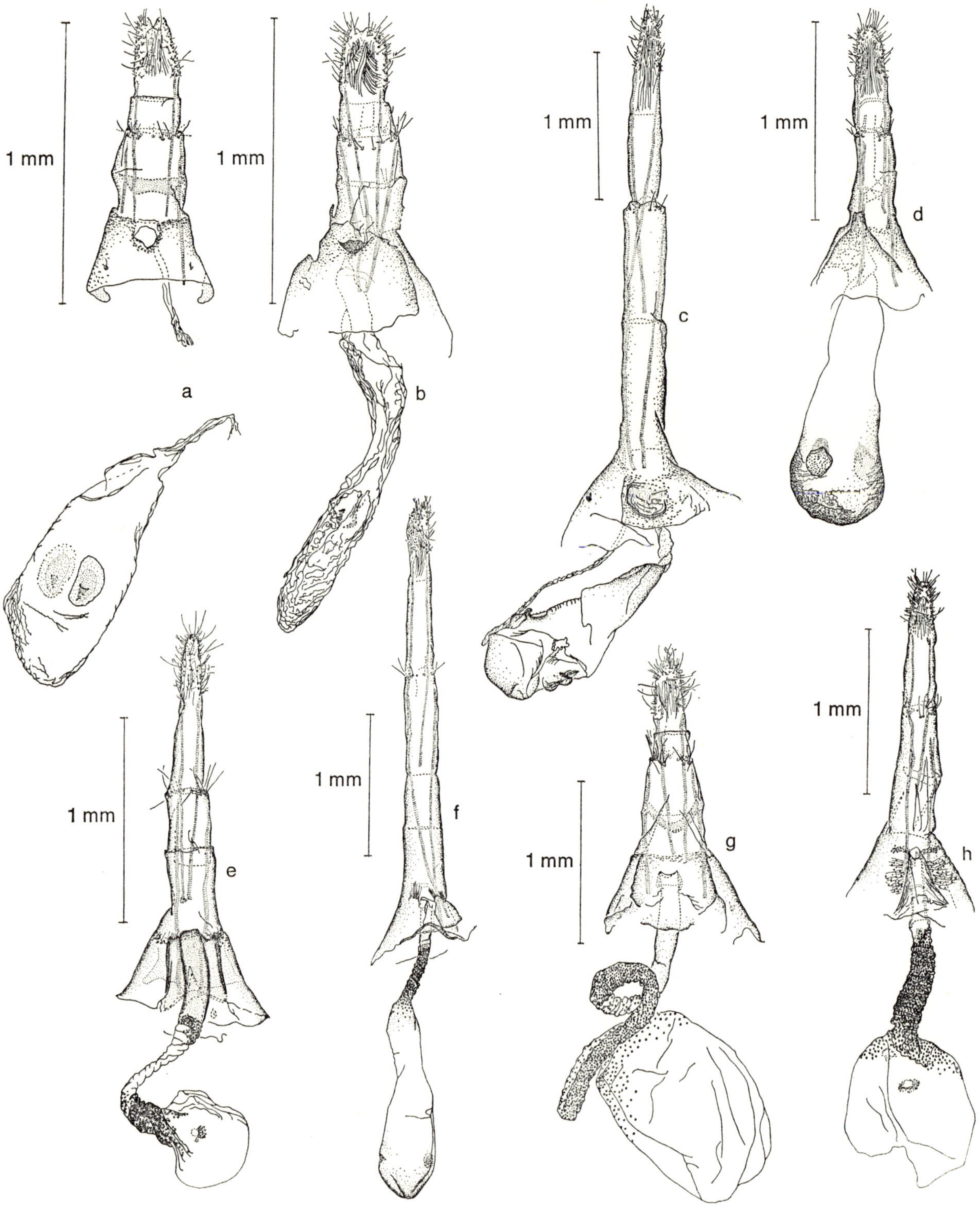


FIGURE 52: FEMALE GENITALIA OF *PERIMEDE* AND *SORHAGENIA* SPECIES

*a. P. erema.* *c. P. ricina.* *e. S. nimbosa.* *g. S. pexa.*
*b. P. maniola.* *d. P. falcata.* *f. S. daedala.* *h. S. baucidis.*

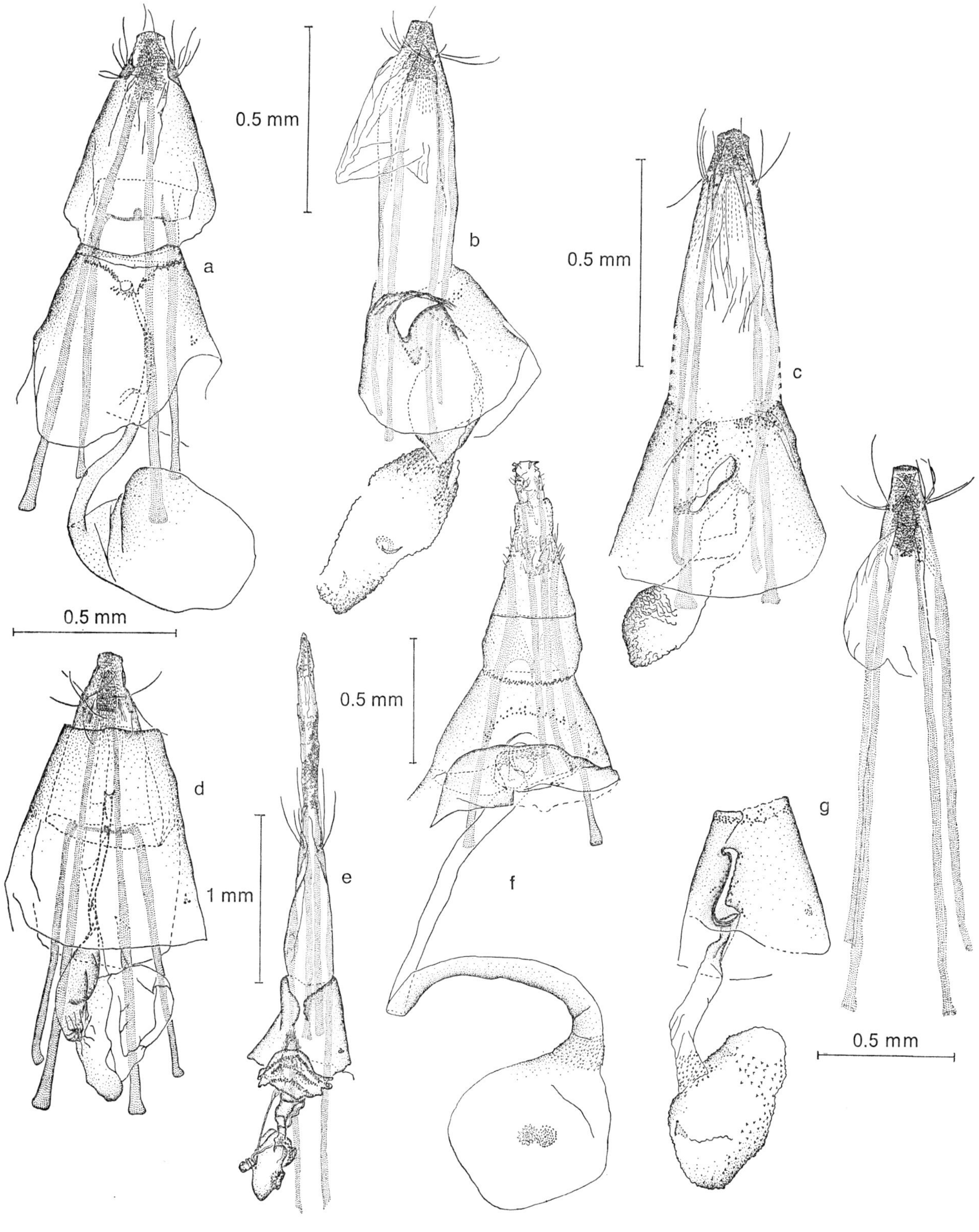


FIGURE 53: FEMALE GENITALIA OF *ITHOME* AND *OBITHOME* SPECIES

*a. I. curvipunctella.* *b. I. aquila.* *c. I. edax.* *d. I. concolorella.* *e. I. ferax.* *f. O. punctiferella.* *g. I. lassula.*

## LITERATURE

BOTTIMER, L. J. 1926. Notes on some Lepidoptera from eastern Texas. *Jour. Agric. Research*, **33**: 797–819, figs. 1–3.

BRADLEY, J. D. 1966. A comparative study of the coconut flat moth (*Agonoxena argaula* Meyr.) and its allies, including a new species (Lepidoptera, Agonoxenidae). *Bull. Ent. Research*, **56**: 453–472, figs. 1–38.

BRADLEY, J. D. 1972. *In* KLOET and HINKS. A check list of British insects, 2nd ed. (rev.), pt. 2: Lepidoptera. *Handbooks for the Identification of British Insects*, **11**, pt. 2: 1–40.

BRANDHORST, C. T. 1962. The microcommunity associated with the gall of *Walshia amorphella* (Lepidoptera: Cosmopterygidae [*sic*]) on *Amorpha fruticosa*. *Ann. Ent. Soc. Amer.*, **55**: 476–479, fig. 1.

BRAUN, A. F. 1915. New genera and species of Tineina. *Can. Ent.*, **47**: 188–197, figs. 18–23.

BRAUN, A. F. 1919. Notes on Cosmopterygidae [*sic*] with descriptions of new genera and species (Microlepidoptera). *Ent. News*, **30**: 260–264.

BRAUN, A. F. 1923. Microlepidoptera: notes and new species. *Trans. Amer. Ent. Soc.*, **49**: 115–127.

BRAUN, A. F. 1930. Notes on new species of Microlepidoptera from the Mineral Springs region of Adams County, Ohio. *Trans. Amer. Ent. Soc.*, **56**: 1–17.

BRAUN, A. F. 1948. Elachistidae of North America (Microlepidoptera). *Mem. Amer. Ent. Soc.*, **13**: 1–110+i, ii, figs. 1–137.

BUSCK, A. 1900 [1901]. New American Tineina. *Jour. New York Ent. Soc.*, **8**: 234–248, pl. IX.

BUSCK, A. 1903. Notes on Brackenridge Clemen's types of Tineina. *Proc. Ent. Soc. Washington*, **5**: 181–220.

BUSCK, A. 1906. A review of the American moths of the genus *Cosmopteryx* [sic] Hübner. *Proc. U.S. Natl. Mus.*, **30**: 707–713.

BUSCK, A. 1917. The pink bollworm, *Pectinophora gossypiella*. *Jour. Agric. Research*, **9** (10): 343–370, figs. 1–7, pl. 7–12.

BUSCK, A. 1934. Microlepidoptera of Cuba. *Ent. Americana*, **13** (n.s.): 151–203, pl. 30–36.

CHAMBERS, V. T. 1880. Notes upon some tineid larvae. *Psyche*, **3**: 63–68.

CHITTENDEN, F. H. 1916. The pink corn-worm: an insect destructive to corn in the crib. *U.S. Dept. Agric., Bull.*, **363**: 1–20, pl. 1–4, figs. 1–7.

CLAASSEN, P. W. 1921. *Typha* insects: their ecological relationships. *Cornell University Agric. Experiment Station, Mem.*, **47**: (457)–531, pl. 39–49.

CLARKE, J. F. G. 1941. Revision of the North American moths of the family Oecophoridae, with descriptions of new genera and species. *Proc. U.S. Natl. Mus.*, **90**: 33–286+i–viii, pl. 1–48.

CLARKE, J. F. G. 1942. A synopsis of the genus *Anoncia*, with one new species (Cosmopterygidae [*sic*]; Lepidoptera). *Can. Ent.*, **74**: 17 19, figs. 1–3.

CLARKE, J. F. G. 1955–69. *Catalogue of the Type Specimens of Microlepidoptera in the British Museum (Natural History) Described by Edward Meyrick*. Vols. 1, 2, 4–7. Trustees of the British Museum. London.

CLEMENS, B. 1872. *In* H. T. STAINTON [editor]. *The Tineina of North America*, xv+282 pages. John Van Voorst. London. (Reprints of Clemens' papers from various journals.)

COMMON, I. F. B. 1970. *In* I. M. MACKERRAS. *The Insects of Australia*, 765–866, figs. 36.1–36.52. Victoria.

COSENS, A. 1908. A new lepidopterous gall-producer. *Can. Ent.*, **40**: 107–108.

DIAKONOFF, A. 1954. Microlepidoptera of New Guinea, part IV. *Verh. der Koninklijke Nederlandse Akademie van Wetenschappen, AFD. Natuurkunde*, **50** (1): 1–191, figs. 552–719.

DIAKONOFF, A. 1967. Microlepidoptera of the Philippine Islands. *U.S. Natl. Mus., Bull.*, **257**: 1–484, figs. 1–846.

DIETZ, W. G. 1900. On *Pigritia* Clem. *Trans. Amer. Ent. Soc.*, **27**: 100–120, pl. 5, 6.

DIETZ, W. G. 1910. Revision of the Blastobasidae of North America. *Trans. Amer. Ent. Soc.*, **36**: 1–72, pl. 1–4.

DUCKWORTH, W. D. 1964. North American Stenomidae (Lepidoptera: Gelechioidea). *Proc. U.S. Natl. Mus.*, **116**: 23–71, pl. 1–4, maps 1–12, figs. 1–45.

DUCKWORTH, W. D. 1973. The Old World Stenomidae: a preliminary survey of the fauna, notes on relationships, and revision of the genus *Eriogenes* (Lepidoptera: Gelechioidea). *Smithsonian Contrib. Zool.*, **147**: 1–21, figs. 1–9.

DYAR, H. G. 1902 [1903]. A list of North American Lepidoptera. *U.S. Natl. Mus., Bull.*, **52**: xix+1–723.

FLETCHER, T. B. 1933. Life-histories of Indian Microlepidoptera (second series). *Imperial Council of Agric. Research, Scientific Monograph*, **4**: 1–83, pl. 1–77.

FORBES, W. T. M. 1923. The Lepidoptera of New York and Neighboring States. *Cornell University Agric. Experiment Station, Mem.*, **68**: 1–729, figs. 1–439.

FORBES, W. T. M. 1931. Supplementary report of the Heterocera or moths of Porto Rico. *Jour. Dept. Agric. Porto Rico*, **4**: 339–394, pl. 42–47.

FRACKER, S. B. 1915. The classification of lepidopterous larvae. *Illinois Biological Mon.*, **2** (1): 1–169, pl. 1–10.

GOZMÁNY, L. A. 1957. Notes on the generic group *Symmoca* Hbn. (Lep. Gelechiidae). *Annales Historico-Naturales Musei Nationalis Hungarici*, (n.s.), **8**: 325–346, figs. 1–9.

GOZMÁNY, L. A. 1962. On the genus *Paradoris* Meyr., and some notes on symmocid taxa (Lepidoptera: Gelechiidae). *Acta Zoologica Academiae Scientiarum Hungaricae*, **8**: 39–65, figs. 1–19.

GOZMÁNY, L. A. 1963. The family Symmocidae and the description of new taxa mainly from the Near East (Lepidoptera). *Acta Zoologica Academiae Scientiarum Hungaricae*, **9**: 67–134, figs. 1–71.

GOZMÁNY, L. A. 1967. The family Holcopogonidae fam. nov. (Lepidoptera) and its constituent taxa. *Acta Zoologica Academiae Scientiarum Hungaricae*, **13**: 271–278, figs. 1–5.

HEINEMANN, H. V., and M. F. WOCKE. 1870, [1876] 1877. *Die Schmetterlinge Deutschlands und der Schweiz*, **2**: 825+102 pages. Schwetschke und Sohn. Braunschweig.

HEINRICH, C. 1921. Some Lepidoptera likely to be confused with the pink bollworm. *Jour. Agric. Research*, **20**: 807–836, pl. 93–109.

HINTON, H. E. 1946. On the homology and nomenclature of the setae of the lepidopterous larvae, with some notes on the phylogeny of the Lepidoptera. *Trans. Ent. Soc. London*, **97** (1): 1–37, figs. 1–24.

HODGES, R. W. 1961. A review of the genus *Walshia* Clemens with descriptions of new species (Lepidoptera: Gelechioidea). *Bull. Brooklyn Ent. Soc.*, **56**: 66–80, figs. 1–18.

HODGES, R. W. 1962. The genus *Ithome* in North America north of Mexico (Walshiidae). *Jour. Lep. Soc.*, **15**: 81–90, figs. 1–12.

HODGES, R. W. 1962. A revision of the Cosmopterigidae of America north of Mexico, with a definition of the Momphidae and Walshiidae (Lepidoptera: Gelechioidea). *Ent. Americana*, **42** (n.s.): 1–166, figs. 1–199.

HODGES, R. W. 1962. A review of the genus *Periploca* with descriptions of nine new species (Lepidoptera: Gelechioidea). *Pan-Pacific Ent.*, **38**: 83–97, figs. 1–24.

HODGES, R. W. 1962. The genus *Perimede* Chambers in North America north of Mexico. *Proc. Ent. Soc. Washington*, **64**: 145–154, figs. 1–14.

HODGES, R. W. 1964. A review of the North American moths of the family Walshiidae (Lepidoptera: Gelechioidea). *Proc. U.S. Natl. Mus.*, **115**: 289–330, figs. 1–66.

HODGES, R. W. 1969. Nearctic Walshiidae, notes and new taxa (Lepidoptera: Gelechioidea). *Smithsonian Contrib. Zool.*, **18**: 1–30, figs. 1–46.

HODGES, R. W. 1974. *In* R. B. DOMINICK *et al.* Gelechioidea: Oecophoridae (in part). *The Moths of America North of Mexico*, Fasc. **6.2**: x, 1–142, figs. 1–32, pl. A, 1–7.

JANSE, A. J. T. 1949–1963. *The Moths of South Africa*, vols. 5, 6, Gelechiidae. Transvaal Museum. Pretoria.

KASY, F. 1968. Die Walshiidae-Gattung *Calycobathra* (Lepidoptera, Gelechioidea). *Ann. Naturhistorisches Mus. Wien*, **72**: 177–195, figs. 1–12.

KASY, F. 1968. Ergebnisse der zoologischen Nubien-Expedition 1962. Teil XXXV. Lepidoptera: *Walshiidae*. *Ann. Naturhistorisches Mus. Wien*, **72**: 497–525, figs. 1–26.

KASY, F. 1969. Vorläufige Revision der Gattung *Ascalenia* Wocke (Lepidoptera, Walshiidae). *Ann. Naturhistorisches Mus. Wien*, **73**: 339–375, figs. 1–27, pl. 1.

KASY, F. 1970. Beitrag zur Kenntnis der Gattungen *Bifascia* Amsel, *Bifascioides* Kasy, und *Gisilia* Kasy (Lepidoptera, Walshiidae). *Ann. Naturhistorisches Mus. Wien*, **74**: 187–194, figs. 1–6.

KASY, F. 1973. Beitrag zur Kenntnis der Familie Stathmopodidae Meyrick (Lepidoptera, Gelechioidea). *Tijds. voor Ent.*, **116**: 227–299, figs. 1–98.

KEIFER, H. H. 1935. California microlepidoptera VII. *Mo. Bull. Dept. Agric. California*, **24**: 195–218, pl. I–V.

KOEHLER, C. S., and M. TAUBER. 1964. *Periploca nigra*, a major cause of dieback on ornamental juniper in California. *Jour. Econ. Ent.*, **57**: 563–566, figs. 1–3.

KUZNETSOV, N. J. 1915. Description of *Parametriotes theae*, gen. n., sp. n. (Lepidoptera, Tineidae), a new enemy of the tea bush in Transcaucasia. *Rev. Russe d'Entomologie* (*Russkoe Entomologischeskoe Obozrenie*), **15**: 627–651, pl. 5–9.

LEMARCHAND, S. 1947. Les *Tineina* Gelechiidae. *Rev. Française de Lép.*, **11**: 145–163, figs. 1–10.

MACKAY, M. R. 1972. Larval sketches of some microlepidoptera, chiefly North American. *Mem. Ent. Soc. Canada*, **88**: 1–83, figs. 1–54.

MALICKY, H., and R. SOBHIAN. 1971. Untersuchungen über Bionomie, Präimaginalstadien und Verbreitung der europäischen *Sorhagenia*-Arten. *Beiträge Ent.*, **21**: 179–190, figs. 1–5.

MANGLITZ, G. R., H. J. GORZ, and H. J. STEVENS, JR. 1971. Biology of the sweetclover root borer. *Jour. Econ. Ent.*, **64**: 1154–1158, figs. 1, 2, tables 1, 2.

MCDUNNOUGH, J. 1939. Check list of the Lepidoptera of Canada and the United States of America. Part II. Microlepidoptera. *Mem. So. California Acad. Sci.*, **2** (1): 1–171.

MEYRICK, E. 1915. *Exotic Microlepidoptera*, **1** (11): 321–352.

MEYRICK, E. 1922. *In* M. P. WYTSMAN. Lepidoptera Heterocera Fam. Oecophoridae. *Genera Insectorum*, **180**: 1–224, pl. 1–6.

MEYRICK, E. 1925. *In* M. P. WYTSMAN. Lepidoptera Heterocera Fam. Gelechiadae. *Genera Insectorum*, **184**: 1–290, pl. 1–5.

MOSHER, E. 1916. A classification of the Lepidoptera based on characters of the pupa. *Bull. Illinois State Laboratory Nat. Hist.*, **12**: 17–259, pl. 19–27.

NAMBA, R. 1956. Descriptions of the immature states and notes on the biology of *Ithome concolorella* (Chambers) (Lepidoptera: Cosmopterygidae [*sic*]), a pest of kiawe in the Hawaiian Islands. *Proc. Hawaiian Ent. Soc.*, **16**: 95–100, fig. 1, tables 1, 2.

OPLER, P. A. 1974. Biology, ecology, and host specificity of Microlepidoptera associated with *Quercus agrifolia* (Fagaceae). *University of California Publications in Ent.*, **75**: 1–83, pl. 1–7.

PIERCE, F. N., and J. W. METCALFE. 1935. *The Genitalia of the Tineid Families of the Lepidoptera of the British Islands.* xxii+116 pages, pl. 1–68. F. N. Pierce. Oundle.

POWELL, J. A. 1963. Observations on larval and pupal habits of the juniper cone moth, *Periploca atrata* Hodges (Lepidoptera: Gelechioidea). *Pan-Pacific Ent.*, **39**: 177–181, figs. 1–3.

POWELL, J. A. 1973. A systematic monograph of New World ethmiid moths (Lepidoptera: Gelechioidea). *Smithsonian Contrib. Zool.*, **120**: iv+1–302, figs. 1–294, pl. 1–22.

RIEDL, T. 1969. Matériaux pour la connaissance de Momphidae paléarctiques (Lepidoptera). *Polskie Pismo Ent.*, **39**: 635–919, figs. 1–414.

RILEY, C. V. 1870. *Second Annual Report on the Noxious, Beneficial and other Insects, of the State of Missouri.* 135 pages. Horace Wilcox. Jefferson City.

SATTLER, K. 1967. *In* H. G. AMSEL *et al.* Ethmiidae. *Microlepidoptera Palaearctica*, **2**: xvi, 1–185, pl. 1–106.

SATTLER, K. 1973. A catalogue of the family-group and genus-group names of the Gelechiidae, Holcopogonidae, Lecithoceridae and Symmocidae (Lepidoptera). *Bull. British Museum (Nat. Hist.) Ent.*, **28**: 155–282.

TOLL, S. 1964. *Klucze do Oznaczania Owadów Polski, 27 (Lepidoptera), part 35 (Oecophoridae)*: 1–174, figs. 1–755. [English translation. 1975. U.S. Dept. Commerce, Natl. Technical Information Service, Springfield, Virginia 22161.]

WALSINGHAM. 1907. *In* D. SHARP, *Fauna Hawaiiensis*, **1** (4): 469–759, pl. 10–25.

COLOR PLATES

# Gelechioidea

PLATE 1

# Gelechioidea

## COSMOPTERIGIDAE

figs. 1–64

TWICE NATURAL SIZE 2:1

1. *Scythris* species, ♂. Hart Prairie, 8500′, 10 mi NNW Flagstaff, Coconino Co., Arizona, 30 June 1961, Ronald W. Hodges (USNM).

2. *Coleophora discostriata* (Wlsm.), ♀. Vail Lake Road, 6500′, $9\frac{1}{2}$ mi SE Flagstaff, Coconino Co., Arizona, 11 July 1961, Ronald W. Hodges (USNM).

3. *Gerdana caritella* Bsk., ♂. Devil's Den St. Pk., Wash. Co., Arkansas, 10 July 1966, R. W. Hodges (USNM).

4. *Valentinia glandulella* (Riley), ♂. Devil's Den St. Pk., Wash. Co., Arkansas, 19 May 1966, R. W. Hodges (USNM).

5. *Blastobasis* species, ♀. Florida, Lake Placid, Archbold Bio. Sta., 1–7 May 1964, R. W. Hodges (USNM).

6. *Holcocera lepidophaga* Clarke, ♂. Hartford, Arkansas, Sebastian Co., 8 June 1966, R. W. Hodges (USNM).

7. *Holcocera gigantella* (Cham.), ♂. Arizona, Santa Cruz Co., Madera Canyon, Bog Springs Campground, 5100′, Santa Rita Mts., 10–26 July 1964, D. R. Davis (USNM).

8. *Menesta melanella* Murt., ♂. Devil's Den St. Pk., Wash. Co., Arkansas, 14 July 1966, R. W. Hodges (USNM).

9. *Antaeotricha schlaegeri* (Zell.), ♀. Madera Canyon, 4880′, Santa Rita Mts., Arizona, 14 Oct. 1959, R. W. Hodges (USNM).

10. *Ethmia monticola* (Wlsm.), ♂. Paratype of *emmeli* Powell. Fort Valley, 7350′, 7½ mi NW Flagstaff, Coconino Co., Arizona, 22 June 1961, Ronald W. Hodges (USNM).

11. *Setiostoma fernaldella* Riley, ♀. Oak Ck. Cyn., Coconino Co., Arizona, em. 3 July 1968, P. A. Opler (USNM).

12. *Ethmia macelhosiella* Bsk., ♂. Maryland, Plummers I., 18 Oct. 1964, J. F. G. Clarke (USNM).

13. *Ethmia albitogata* (Wlsm.), ♂. San Bruno Mts., San Mateo Co., California, 7 Feb. 1964, J. Powell (USNM).

14. *Ethmia albistrigella* (Wlsm.), ♂. Washington, Blue Mts., 5700′, Godman Springs, 23 June 1970, J. F. G. Clarke (USNM).

15. *Ethmia zelleriella* (Cham.), ♂. Six Mile Creek, Ithaca, New York, 29 May 1959, R. W. Hodges (USNM).

16. *Pyramidobela angelarum* Keif., ♀. Paratype. Los Angeles, California, em. 24 July 1935 (USNM).

17. *Limnaecia phragmitella* Staint., ♀. Michigan, Ocqueoc L., 25–26 July 1970, R. W. & E. R. Hodges (USNM). (p. 48).

18. *Triclonella xuthocelis* Hodges, ♂. Paratype. Madera Canyon, 4880′, Santa Rita Mts., Arizona, 11 Sept. 1959, R. W. Hodges (USNM). (p. 54).

19. *Martyrhilda canella* (Bsk.), ♀. Six Mile Creek, Ithaca, New York, 25 June 1947, J. G. Franclemont (JGF).

20. *Anoncia furvicosta* Hodges, ♂. Paratype. Madera Canyon, 4880′, Santa Rita Mts., Arizona, 23 July 1959, R. W. Hodges (USNM). (p. 66).

21. *Semioscopis inornata* Wlsm., ♂. Six Mile Creek, Ithaca, New York, 7 May 1961, R. W. Hodges (USNM).

22. *Agonopterix robiniella* (Pack.), ♀. Six Mile Creek, Ithaca, New York, 23 July 1960, R. W. Hodges (USNM).

23. *Depressaria alienella* Bsk., ♂. Bar Harbor, Maine, 24 Aug. 1935, A. E. Brower (USNM).

24. *Psilocorsis cryptolechiella* (Cham.), ♀. Devil's Den St. Pk., Wash. Co., Arkansas, 4 June 1966, R. W. Hodges (USNM).

25. *Polix coloradella* (Wlsm.), ♂. South Dakota, Hardy Work Center, T3N, R1E, S30, 21 July 1965, R. W. Hodges (USNM).

26. *Carcina quercana* (F.), ♀. Munich [Germany], "1/70", Hartmann (BMNH).

27. *Martyringa latipennis* (Wlsm.), ♂. Six Mile Creek, Ithaca, New York, 6 July 1954, J. G. Franclemont (JGF).

28. *Pleurota albastrigulella* (Kft.), ♀. Mint Canyon Summit, L. A. Co., California, 3 May 1945 (USNM).

29. *Mompha definitella* (Zell.), ♀. South Dakota, Stovehole Park, T2N, R1E, S2,3, 4–6 July 1965, R. W. Hodges (USNM).

30. *Mompha edithella* B. & Bsk., ♂. West Fork, 6500′, 16 mi SW Flagstaff, Coconino Co., Arizona, 10 Aug. 1961, Ronald W. Hodges (USNM).

31. *Homaledra heptathalama* Bsk., ♂. Florida, Parker Is., Highlands Co., 26–29 May 1964, R. W. Hodges (USNM).

32. *Batrachedra praeangusta* (Haw.), ♀. Hart Prairie, 8500′, 10 mi NNW Flagstaff, Coconino Co., Arizona, 24 July 1961, Ronald W. Hodges (USNM).

33. *Blastodacna bicristatella* (Cham.), ♀. Putnam Co., Illinois, 6 June 1944, M. O. Glenn (USNM).

34. *Glyphipteryx linneella* (Cl.), ♂. Ithaca, New York, 28 July 1960, R. W. Hodges (USNM).

35. *Glyphidocera septentrionella* Bsk., ♂. Hart Prairie, 8500′, 10 mi NNW Flagstaff, Coconino Co., Arizona, 14 July 1961, Ronald W. Hodges (USNM).

36. *Oegoconia quadripuncta* (Haw.), ♂. Adelphi, Maryland, 25 June 1971, R. W. Hodges (USNM).

37. *Metzneria lappella* (L.), ♀. McLean Bogs Reserve, Tompkins Co., New York, 20 July, 1957, J. G. Franclemont (JGF).

38. *Isophrictis magnella* (Bsk.), ♀. Wyoming, 6 mi NW Newcastle, 18 July 1965, R. W. Hodges (USNM).

39. *Aristotelia argentifera* Bsk., ♂. Wyoming, 6 mi NW Newcastle, 18 July 1965, R. W. Hodges (USNM).

40. "*Aristotelia*" *placidella* (Zell.), ♀. West Fork, 6500′, 16 mi SW Flagstaff, Coconino Co., Arizona, 13 July 1961, Ronald W. Hodges (USNM).

41. *Coleotechnites milleri* (Bsk.), ♂. California, 14 mi S Lee Vining, Mono Co., em. 11 July 1964, R. E. Stevens (USNM).

42. *Coleotechnites atrupictella* (Dietz), ♂. Highlands, 3865′, Macon Co., North Carolina, 16 July 1958, R. W. Hodges (USNM).

43. *Telphusa longifasciella* (Clem.), ♀. Six Mile Creek, Ithaca, New York, 7 May 1961, R. W. Hodges (USNM).

44. *Gelechia lynceella* Zell., ♀. South Dakota, Hardy Work Center, T3N, R1E, S30, 21 July 1965, R. W. Hodges (USNM).

45. *Lita variabilis* (Bsk.), ♂. Hart Prairie, 8500′, 10 mi NNW Flagstaff, Coconino Co., Arizona, 11 Aug. 1961, Ronald W. Hodges (USNM).

46. *Neodactylota basilica* Hodges, ♂. Paratype. Peña Blanca Canyon, Santa Cruz Co., Arizona, 11 Aug. 1959, R. W. Hodges (USNM).

47. *Chionodes lugubrella* (F.), ♂. South Dakota, Hardy Work Center, T3N, R1E, S30, 11 July 1965, R. W. Hodges (USNM).

48. *Chionodes bicostomaculella* (Cham.), ♂. Florida, Lake Placid, Archbold Bio. Sta., 1–7 May 1964, R. W. Hodges (USNM).

49. *Filatima albicostella* Clarke, ♂. Fort Valley, 7350′, 7½ mi NW Flagstaff, Coconino Co., Arizona, 16 July 1961, Ronald W. Hodges (USNM).

50. *Aroga websteri* (Clarke), ♂. Klamath Co., Oregon, 10 mi SE Klamath Falls, em. 29 July 1964, E. A. Dickason (USNM).

51. *Polyhymno luteostrigella* Cham., ♂. Madera Canyon, 4880′, Santa Rita Mts., Arizona, 4 Aug. 1959, R. W. Hodges (USNM).

52. *Anacampsis niveopulvella* (Cham.), ♂. Hart Prairie, 8500′, 10 mi NNW Flagstaff, Coconino Co., Arizona, 14 July 1961, Ronald W. Hodges (USNM).

53. *Strobisia proserpinella* F. & B., ♂. Arkansas, Cameron Bluff, 9 June 1964, J. F. G. and Thelma M. Clarke (USNM).

54. *Prostomeus brunneus* Bsk., ♂. Florida, Parker Is., Highlands Co., 4–7 June 1964, R. W. Hodges (USNM).

55. *Dichomeris ligulella* Hbn., ♂. Devil's Den St. Pk., Wash. Co., Arkansas, 5 July 1966, R. W. Hodges (USNM).

56. *Dichomeris marginella* (F.), ♀. Lansing, Michigan, 20 June 1961, R. W. Hodges (USNM).

57. *Brachmia fernaldella* (Bsk.), ♂. South Dakota, Hardy Work Center, T3N, R1E, S30, 9 July 1965, R. W. Hodges (USNM).

58. *Pectinophora gossypiella* (Saund.), ♀. Presidio, Texas, April 1939 (USNM).

59. *Deoclona yuccasella* Bsk., ♀. Victorville, California, em. 28 Aug. 1939, K. Saunders (USNM).

60. *Anarsia lineatella* Zell., ♂. [Europe] (USNM).

61. *Sitotroga cerealella* (Olivier), ♂. Columbia, Missouri, em. 9 June 1969, R. W. Hodges (USNM).

62. *Periploca mimula* Hodges, ♂. Wyoming, 6 mi NW Newcastle, 18 July 1965, R. W. Hodges (USNM). (p. 107).

63. *Walshia miscecolorella* (Cham.), ♂. Hart Prairie, 8500′, 10 mi NNW Flagstaff, Coconino Co., Arizona, 30 June 1961, Ronald W. Hodges (USNM). (p. 124).

64. *Stilbosis nubila* Hodges, ♂. West Fork, 6500′, 16 mi SW Flagstaff, Coconino Co., Arizona, 8 July 1961, Ronald W. Hodges (USNM). (p. 122).

1
2
3
4
5
6
7
8
9
10
11
12
13
14
15
16
17
18
19
20
21
22
23
24
25
26
27
28
29
30
31
32
33
34
35
36
37
38
39
40
41
42
43
44
45
46
47
48
49
50
51
52
53
54
55
56
57
58
59
60
61
62
63
64

PLATE 2

# Gelechioidea

## COSMOPTERIGIDAE

figs. 1–43

FOUR TIMES NATURAL SIZE 4:1

1. *Scythris eboracensis* (Zell.), ♀. Falls Church, Virginia, Fairfax Co., em. 13 July 1961, J. F. G. Clarke (USNM).
2. *Coleophora spissicornis* (Haw.), ♂. South Dakota, Hardy Work Center, T3N, R1E, S30, 11 July 1965, R. W. Hodges (USNM).
3. *Coleophora bidentella* McD., ♂. Paratype. Hemmingford, Quebec, 28 June 1925, G. H. Hammond (USNM).
4. *Epigritia* species, ♀. Devil's Den St. Pk., Wash. Co., Arkansas, 9 July 1966, R. W. Hodges (USNM).
5. *Dryope* species, ♂. Devil's Den St. Pk., Wash. Co., Arkansas, 18 June 1966, R. W. Hodges (USNM).
6. *Decantha boreasella* (Cham.), ♀. Devil's Den St. Pk., Wash. Co., Arkansas, 22 June 1966, R. W. Hodges (USNM).
7. *Coelopoeta glutinosi* Wlsm., ♂. Los Angeles Co., California, em. 18 June 1938, C. Dammers (USNM).
8. *Hemiprosopa albella* (Cham.), ♀. Wyoming, 6 mi NW Newcastle, 23 June 1965, R. W. Hodges (USNM).
9. *Elachista tanyopa* Meyr., ♀. Lakehurst, New Jersey, 31 May 1962, R. W. Hodges (USNM).
10. *Elachista* species, ♀. South Dakota, Hardy Work Center, T3N, R1E, S30, 21 July 1965, R. W. Hodges (USNM).
11. *Elachista cuculata* Braun, ♀. Devil's Den St. Pk., Wash. Co., Arkansas, 4 June 1966, R. W. Hodges (USNM).
12. *Cosmiotes illectella* Clem., ♂. West Fork, 6500′, 16 mi SW Flagstaff, Coconino Co., Arizona, 13 July 1961, Ronald W. Hodges (USNM).
13. *Dicranoctetes* species, ♂. Florida, Fisheating Cr., Palmdale, 7–10 May 1964, R. W. Hodges (USNM).
14. *Mompha bottimeri* (Bsk.), ♀. Florida, Fisheating Cr., Palmdale, 7–10 May 1964, R. W. Hodges (USNM).
15. *Aetia bipunctella* Cham., ♂. Devil's Den St. Pk., Wash. Co., Arkansas, 2 June 1966, R. W. Hodges (USNM).
16. *Ymeldia janae* Hodges, ♂. Florida, Lake Placid, Archbold Bio. Sta., 1–7 May 1964, R. W. Hodges (USNM).
17. *Stereomita andropogonis* Braun, ♀. Putnam Co., Illinois, 12 Aug. 1963, M. O. Glenn (USNM).
18. *Helice constrictella* (Zell.), ♂. Devil's Den St. Pk., Wash. Co., Arkansas, 9 July 1966, R. W. Hodges (USNM).
19. *Ithome concolorella* (Cham.), ♂. Devil's Den St. Pk., Wash. Co., Arkansas, 30 June 1966, R. W. Hodges (USNM). (p. 135).
20. *Cosmopterix nitens* Wlsm., ♂. Cotype. Texas (USNM). (p. 24).
21. *Cosmopterix molybdina* Hodges, ♂. Holotype. Bar Harbor, Maine, 13 July 1936, A. E. Brower (CU). (p. 25).
22. *Cosmopterix pulchrimella* Cham., ♀. Devil's Den St. Pk., Wash. Co., Arkansas, 22 July 1966, R. W. Hodges (USNM). (p. 25).
23. *Cosmopterix pulchrimella* Cham., ♂. Baton Rouge, E. Baton Rouge Par., Louisiana, 13 April 1970, G. Strickland (USNM). (p. 25).
24. *Cosmopterix bendidia* Hodges, ♂. Paratype. Madera Canyon, 4880′, Santa Rita Mts., Arizona, 11 July 1959, R. W. Hodges (USNM). (p. 26).
25. *Cosmopterix attenuatella* (Wlk.), ♂. Florida, Lake Placid, Archbold Bio. Sta., 16–22 May 1964, R. W. Hodges (USNM). (p. 26).
26. *Cosmopterix attenuatella* (Wlk.), ♀. Florida, Fisheating Cr., Palmdale, 7–10 May 1964, R. W. Hodges (USNM). (p. 26).
27. *Cosmopterix clandestinella* Bsk., ♀. Paralectotype. Twining, Maryland, em. 20 July 1899, Aug. Busck (USNM). (p. 27).
28. *Cosmopterix montisella* Cham., ♂. West Fork, 6500′, 16 mi SW Flagstaff, Coconino Co., Arizona, 4 July 1961, Ronald W. Hodges (USNM). (p. 27).
29. *Cosmopterix montisella* Cham., ♂. West Fork, 6500′, 16 mi SW Flagstaff, Coconino Co., Arizona, 13 Aug. 1961, Ronald W. Hodges (USNM). (p. 27).
30. *Cosmopterix montisella* Cham., ♂. Vail Lake Rd., 6500′, 9½ mi SE Flagstaff, Coconino Co., Arizona, 18 July 1961, Ronald W. Hodges (USNM). (p. 27).
31. *Cosmopterix magophila* Meyr., ♂. Washington Co., Arkansas, 25 Aug. 1966, R. L. Brown (USNM). (p. 28).
32. *Cosmopterix gracilens* Hodges, ♂. Holotype. Palmerlee, Arizona (USNM). (p. 28).
33. *Cosmopterix dapifera* Hodges, ♂. Paratype. Madera Canyon, 4880′, Santa Rita Mts., Arizona, 5 Sept. 1959, R. W. Hodges (USNM). (p. 29).
34. *Cosmopterix dapifera* Hodges, ♂. Florida, Lake Placid, Archbold Bio. Sta., 16–22 May 1964, R. W. Hodges (USNM). (p. 29).
35. *Cosmopterix delicatella* Wlsm., ♂. Florida, Parker Is., Highlands Co., 4–7 June 1964, R. W. Hodges (USNM). (p. 29).
36. *Cosmopterix delicatella* Wlsm., ♂. Florida, Lake Placid, Archbold Bio. Sta., 8–15 May 1964, R. W. Hodges (USNM). (p. 29)
37. *Cosmopterix dicacula* Hodges, ♂. Holotype. Gainesville, Florida, 7 July 1927, J. Speed Rogers (CU). (p. 30).
38. *Cosmopterix lespedezae* Wlsm., ♂. Holotype. No locality [Dallas, Texas] (USNM). (p. 30).
39. *Cosmopterix opulenta* Braun, ♀. Madera Canyon, 4400′, Santa Rita Mts., Arizona, 12 Oct. 1959, R. W. Hodges (USNM). (p. 31).
40. *Cosmopterix opulenta* Braun, ♂. Vacaville, 2 mi SE, California, em. 30 July 1968, P. A. Opler (USNM). (p. 31).
41. *Cosmopterix quadrilineella* Cham., ♂. Madera Canyon, 4400′, Santa Rita Mts., Arizona, 10 Oct. 1959, R. W. Hodges (USNM). (p. 32).
42. *Cosmopterix quadrilineella* Cham., ♀. Madera Canyon, 4880′, Santa Rita Mts., Arizona, 22 Aug. 1959, R. W. Hodges (USNM). (p. 32).
43. *Cosmopterix minutella* Beutenmüller, ♂. Lake Placid, Florida, Archbold Bio. Sta., 29 March 1959, R. W. Hodges (USNM). (p. 33).

1
2
3
4
5
6
7
8
9
10
11
12
13
14
15
16
17
18
19
20
21
22
23
24
25
26
27
28
29
30
31
32
33
34
35
36
37
38
39
40
41
42
43

PLATE 3

# Gelechioidea

## COSMOPTERIGIDAE

figs. 1–43

FOUR TIMES NATURAL SIZE 4:1

1. *Cosmopterix abdita* Hodges, ♂. Florida, Fisheating Cr., Palmdale, 7–10 May 1964, R. W. Hodges (USNM). (p. 33).
2. *Cosmopterix abdita* Hodges, ♀. Florida, Fisheating Cr., Palmdale, 7–10 May 1964, R. W. Hodges (USNM). (p. 33).
3. *Cosmopterix inopis* Hodges, ♂. Holotype. Madera Canyon, 4880′, Santa Rita Mts., Arizona, 23 Sept. 1959, R. W. Hodges (CU). (p. 33).
4. *Cosmopterix chalybaeella* Wlsm., ♂. Holotype. SW Texas, 1884, Morrison (BMNH). (p. 34).
5. *Cosmopterix gemmiferella* Clem., ♀. Devil's Den St. Pk., Wash. Co., Arkansas, 20 June 1966, R. W. Hodges (USNM). (p. 34).
6. *Cosmopterix bacata* Hodges, ♂. Florida, Lake Placid, Archbold Bio. Sta., 8–15 May 1964, R. W. Hodges (USNM). (p. 35).
7. *Cosmopterix damnosa* Hodges, ♂. Holotype. Lake Placid, Florida, Archbold Bio. Sta., 27 March 1959, R. W. Hodges (CU). (p. 35).
8. *Cosmopterix clemensella* Staint., ♀. Falls Church, Virginia, Holmes Run, 19 June 1961, W. W. Wirth (USNM). (p. 35).
9. *Cosmopterix clemensella* Staint., ♂. Slaterville Springs, New York, 6 July 1960, R. W. Hodges (USNM). (p. 35).
10. *Cosmopterix scirpicola* Hodges, ♀. Wyoming, 6 mi NW Newcastle, 18 July 1965, R. W. Hodges (USNM). (p. 36).
11. *Cosmopterix scirpicola* Hodges, ♂. Florida, Lake Placid, Archbold Bio. Sta., 16–22 May 1964, R. W. Hodges (USNM). (p. 36).
12. *Cosmopterix ebriola* Hodges, ♀. Holotype. Lake Placid, Florida, Archbold Bio. Sta., 30 March 1959, R. W. Hodges (CU). (p. 37).
13. *Cosmopterix fernaldella* Wlsm., ♀. Michigan, Ocqueoc L., 25–26 July 1970, R. W. & E. R. Hodges (USNM). (p. 37).
14. *Cosmopterix floridanella* Beutenmüller, ♀. Florida, Lake Placid, Archbold Bio. Sta., 1–7 May 1964, R. W. Hodges (USNM). (p. 37).
15. *Cosmopterix facunda* Hodges, ♀. Holotype. Brownsville, Texas, 27 March 1928, F. H. Benjamin (USNM). (p. 38).
16. *Tanygona lignicolorella* Braun, ♀. Devil's Den St. Pk., Wash. Co., Arkansas, 13 July 1966, R. W. Hodges (USNM). (p. 39).
17. *Pebops ipomoeae* (Bsk.), ♂. Key Largo Key, Monroe Co., Florida, 22 Nov. 1965, Mrs. Spencer Kemp (USNM). (p. 38).
18. *Eralea albalineella* (Cham.), ♀. Arizona, Santa Cruz Co., Bog Springs Campground, Madera Canyon, 5100′, Santa Rita Mts., 10–26 July 1964, D. R. Davis (USNM). (p. 40).
19. *Eralea albalineella* (Cham.), ♂. Holotype of *Eralea striata* Hodges. Siesta Key, Sarasota Co., Florida, 13 May 1960, C. P. Kimball (CU). (p. 40).
20. *Eralea abludius* Hodges, ♂. Paratype. Florida, Lake Placid, Archbold Bio. Sta., 16–22 May 1964, R. W. Hodges (USNM). (p. 40).
21. *Melanocinclis lineigera* Hodges, ♂. Florida, Lake Placid, Archbold Bio. Sta., 1–7 May 1964, R. W. Hodges (USNM). (p. 42).
22. *Melanocinclis nigrilineella* (Cham.), ♂. Madera Canyon, 4880′, Santa Rita Mts., Arizona, 30 June 1959, R. W. Hodges (USNM). (p. 42).
23. *Melanocinclis sparsa* Hodges, ♂. Holotype. Devil's Den St. Pk., Wash. Co., Arkansas, 6 July 1966, R. W. Hodges (USNM). (p. 42).
24. *Melanocinclis sparsa* Hodges, ♀. Paratype. Devil's Den St. Pk., Wash. Co., Arkansas, 22 May 1966, R. W. Hodges (USNM). (p. 42).
25. *Melanocinclis gnoma* Hodges, ♂. Holotype. Florida, Lake Placid, Archbold Bio. Sta., 8–15 May 1964, R. W. Hodges (USNM). (p. 43).
26. *Melanocinclis vibex* Hodges, ♂. Holotype. Florida, Lake Placid, Archbold Bio. Sta., 16–22 May 1964, R. W. Hodges (USNM). (p. 43).
27. *Melanocinclis vibex* Hodges, ♀. Paratype. Florida, Lake Placid, Archbold Bio. Sta., 16–22 May 1964, R. W. Hodges (USNM). (p. 43).
28. *Stagmatophora sexnotella* (Cham.), ♂. Florida, Lake Placid, Archbold Bio. Sta., 16–22 May 1964, R. W. Hodges (USNM). (p. 45).
29. *Pyroderces rileyi* (Wlsm.), ♀. Beeville, Texas, em. 25 April 1896 (USNM). (p. 47).
30. *Pyroderces rileyi* (Wlsm.), ♀. Washington, D.C., Aug. Busck (USNM). (p. 47).
31. *Pyroderces badia* (Hodges), ♀. Florida, Lake Placid, Archbold Bio. Sta., 8–15 May 1964, R. W. Hodges (USNM). (p. 47).
32. *Pyroderces badia* (Hodges), ♂. Florida, Lake Placid, Archbold Bio Sta., 8–15 May 1964, R. W. Hodges (USNM). (p. 47).
33. *Pyroderces badia* (Hodges), ♀. Florida, Lake Placid, Archbold Bio Sta., 16–22 May 1964, R. W. Hodges (USNM). (p. 47).
34. *Pyroderces albistrigella* (Mösch.), ♀. Florida, Parker Is., Highlands Co., 26–29 May 1964, R. W. Hodges (USNM). (p. 48).
35. *Teladoma helianthi* Bsk., ♂. Oklahoma City, Oklahoma, 10 Aug 1955, D. R. Davis (USNM). (p. 50).
36. *Teladoma helianthi* Bsk., ♂. Devil's Den St. Pk., Wash. Co. Arkansas, 3 July 1966, R. W. Hodges (USNM). (p. 50).
37. *Teladoma astigmatica* (Meyr.), ♂. Paralectotype. Fort Davis, 5000′, Texas, Oct. 1927 (USNM). (p. 50).
38. *Teladoma murina* Hodges, ♂. Paratype. Peña Blanca Canyon Santa Cruz Co., Arizona, 26 Aug. 1959, R. W. Hodges (USNM) (p. 51).
39. *Teladoma nebula* Hodges, ♂. Holotype. Peña Blanca Canyon Santa Cruz Co., Arizona, 26 Aug. 1959, R. W. Hodges (CU). (p. 52).
40. *Teladoma exigua* Hodges, ♂. Holotype. Madera Canyon, 4880′ Santa Rita Mts., Arizona, 11 July 1959, R. W. Hodges (CU) (p. 52).
41. *Teladoma incana* Hodges, ♂. Devil's Den St. Pk., Wash. Co. Arkansas, 28 May 1966, R. W. Hodges (USNM). (p. 51).
42. *Teladoma incana* Hodges, ♀. Devil's Den St. Pk., Wash. Co. Arkansas, 7 July 1966, R. W. Hodges (USNM). (p. 51).
43. *Teladoma habra* Hodges, ♂. Holotype. Devil's Den St. Pk., Wash Co., Arkansas, 22 May 1966, R. W. Hodges (USNM). (p. 51).

1
2
3
4
5
6
7
8
9
10
11
12
13
14
15
16
17
18
19
20
21
22
23
24
25
26
27
28
29
30
31
32
33
34
35
36
37
38
39
40
41
42
43

# PLATE 4

# Gelechioidea

## COSMOPTERIGIDAE

figs. 1–65

TWICE NATURAL SIZE 2:1

1. *Stagmatophora wyattella* B. & Bsk., ♀. Putnam Co., Illinois, 30 July 1959, M. O. Glenn (USNM). (p. 45).
2. *Stagmatophora iridella* (Bsk.), ♀. Berkeley, California, 2 April 1959, D. D. Linsdale (USNM). (p. 45).
3. *Stagmatophora iridella* (Bsk.), ♀. San Diego, California, 30 Aug. 1921 (USNM). (p. 45).
4. *Stagmatophora enchrysa* (Hodges), ♂. Holotype. Cajon V., S. Bernardino Co., California, em. 30 June 1938, C. Dammers (USNM). (p. 46).
5. *Limnaecia phragmitella* Staint., ♀. Michigan, Ocqueoc L., 25–26 June 1970, R. W. & E. R. Hodges (USNM). (p. 48).
6. *Limnaecia phragmitella* Staint., ♀. Michigan, Ocqueoc L., 25–26 June 1970, R. W. & E. R. Hodges (USNM). (p. 48).
7. *Triclonella pergandeella* Bsk., ♂. Holotype. AB 14, on *Desmodium*, issued 1 July 1899 (USNM). (p. 54).
8. *Triclonella xuthocelis* Hodges, ♂. Paratype. Madera Canyon, 4880′, Santa Rita Mts., Arizona, 11 Sept. 1959, R. W. Hodges (USNM). (p. 54).

9. *Triclonella xuthocelis* Hodges, ♂. Paratype. Madera Canyon, 4880′, Santa Rita Mts., Arizona, 4 July 1959, R. W. Hodges (USNM). (p. 54).
10. *Triclonella antidectis* (Meyr.), ♀. Madera Canyon, 4880′, Santa Rita Mts., Arizona, 14 July 1959, R. W. Hodges (USNM). (p. 54).
11. *Triclonella determinatella* (Zell.), ♂. Devil's Den St. Pk., Wash. Co., Arkansas, 28 June 1966, R. W. Hodges (USNM). (p. 54).
12. *Triclonella bicoloripennis* Hodges, ♂. Baton Rouge, E. Baton Rouge Par., Louisiana, 27 April 1970, G. Strickland (USNM). (p. 55).
13. *Triclonella bicoloripennis* Hodges, ♀. Baton Rouge, E. Baton Rouge Par., Louisiana, 21 May 1970, G. Strickland (USNM). (p. 55).
14. *Anoncia conia* (Wlsm.), ♀. La Mesa, California, em. 3 Aug. 1950, V. O. Miller (USNM). (p. 59).
15. *Anoncia conia* (Wlsm.), ♀. Paratype of *marinensis* Keif. Phoenix L., Marin Co., California, em. 27 June 1934, Keifer (USNM). (p. 59).
16. *Anoncia conia* (Wlsm.), ♀. Paratype of *marinensis* Keif. Phoenix L., Marin Co., California, em. 29 June 1927, H. H. Keifer (USNM). (p. 59).
17. *Anoncia sphacelina* (Keif.), ♂. Paratype. Shingle Spr., El Dorado Co., California, em. 30 April 1934, Keifer (USNM). (p. 60).
18. *Anoncia brunneipes* Hodges, ♂. Holotype. San Diego, California, 20 June 1911, W. S. Wright (USNM). (p. 59).
19. *Anoncia glacialis* Hodges, ♂. Holotype. Riverside, California, 8 July 1933, Keifer (CU). (p. 61).
20. *Anoncia bitoqua* Hodges, ♂. Holotype. Surprise Cyn., Inyo Co., Panamint Mts., California, 24 April 1957, J. Powell (UCB). (p. 61).
21. *Anoncia fasciata* (Wlsm.), ♂. Paratype. Head of Noyo R., Mendocino Co., California, 8–11 June 1871, Wlsm. (USNM). (p. 60).
22. *Anoncia mosa* Hodges, ♂. California, S. L. O. Co., 2 mi W Paso Robles, 25 April 1968, Powell & Chemsak (UCB). (p. 71).
23. *Anoncia porriginosa* Hodges, ♂. Holotype. Claremont, California, Metz (USNM). (p. 62).
24. *Anoncia episcia* (Wlsm.), ♀. Pozo, S. L. O. Co., California, 1 May 1962, R. L. Langston (USNM). (p. 69).
25. *Anoncia alboligula* Hodges, ♂. Holotype. Olancha, Inva [*sic*] Co., California, 24–30 June (USNM). (p. 60).
26. *Anoncia nebritis* Hodges, ♀. Paratype. Madera Canyon, 4880′, Santa Rita Mts., Arizona, 25 July 1959, R. W. Hodges (USNM). (p. 65).
27. *Anoncia diveni* (Heinr.), ♀. Zamorana, Honduras, N. L. H. Krause (USNM). (p. 66).
28. *Anoncia episcia* (Wlsm.), ♂. Paratype. Dry Creek, Sonoma Co., California, 20–21 May 1871, Wlsm. (USNM). (p. 69).
29. *Anoncia loexya* Hodges, ♂. Arizona, Pine, Gila Co., 4 June 1968, Powell & Opler (USNM). (p. 68).
30. *Anoncia episcia* (Wlsm.), ♂. Pozo, S. L. O. Co., California, 1 May 1962, J. Powell (USNM). (p. 69).
31. *Anoncia episcia* (Wlsm.), ♂. Pozo, S. L. O. Co., California, 27 April 1962, J. Powell (USNM). (p. 69).
32. *Anoncia orites* (Wlsm.), ♀. Holotype. Mt. Shasta, Siskiyou Co., California, 2 Aug.–1 Sept. 1871, Wlsm. (BMNH). (p. 67).
33. *Anoncia callida* Hodges, ♂. West Fork, 6500′, 16 mi SW Flagstaff, Coconino Co., Arizona, 10 Aug. 1961, Ronald W. Hodges (USNM). (p. 64).
34. *Anoncia piperata* Hodges, ♂. Holotype. Smoky Val., Tulare Co., California, 6200′, 8 June 1944 (USNM). (p. 71).
35. *Anoncia leucoritis* (Meyr.), ♂. Holotype of *mentzeliae* Clarke. Snake River, Washington, Whitman Co., opp. Clarkston, J. F. G. Clarke (USNM). (p. 66).
36. *Anoncia longa* (Meyr.), ♂. Paradise, Cochise Co., Arizona, 8–15 May (USNM). (p. 66).
37. *Anoncia longa* (Meyr.), ♂. Walnut Canyon, 6500′, $6\frac{1}{3}$ mi E by S Flagstaff, Coconino Co., Arizona, 4 Aug. 1965, J. G. Franclemont (JGF). (p. 66).
38. *Anoncia longa* (Meyr.), ♀. Silver Creek Wash, 4880′, 0.7 mi W Portal, Cochise Co., Arizona, 23 Aug. 1966, J. G. Franclemont (JGF). (p. 66).
39. *Anoncia furvicosta* Hodges, ♂. Paratype. Madera Canyon, 4880′, Santa Rita Mts., Arizona, 11 July 1959, R. W. Hodges (USNM). (p. 66).
40. *Anoncia furvicosta* Hodges, ♂. Paratype. Madera Canyon, 4880′, Santa Rita Mts., Arizona, 23 July 1959, R. W. Hodges (USNM). (p. 66).
41. *Anoncia nocticola* Hodges, ♀. Holotype. Madera Canyon, 4880′, Santa Rita Mts., Arizona, 3 Aug. 1959, R. W. Hodges (CU). (p. 65).
42. *Anoncia aciculata* (Meyr.), ♂. Clark Co., Nevada, 16–23 April (USNM). (p. 63).
43. *Antequera acertella* (Bsk.), ♂. Paratype. San Diego, California, 20 Nov. 1911, W. S. Wright (USNM). (p. 17).
44. *Antequera acertella* (Bsk.), ♂. Piru Creek, NE base Alamo Mt., Ventura Co., California, 6 May 1959, G. I. Stage (UCB). (p. 17).
45. *Periploca ceanothiella* (Cosens), ♀. Glendale, California, em. 16 May 1934, Tower (USNM). (p. 103).
46. *Periploca intermedia* Hodges, ♂. Holotype. Devil's Den St. Pk., Wash. Co., Arkansas, 11 July 1966, R. W. Hodges (USNM). (p. 104).
47. *Periploca gleditschiaeella* (Cham.), ♂. Kentucky, em. Feb. 1903, Aug. Busck (USNM). (p. 104).
48. *Periploca atrata* Hodges, ♀. Paratype. Madera Canyon, 5600′, Santa Rita Mts., Arizona, 1 Aug. 1959, R. W. Hodges (USNM). (p. 106).
49. *Pristen corusca* (Hodges), ♂. Arizona, Santa Cruz Co., Madera Canyon, 5100′, Santa Rita Mts., 10–26 July 1964, D. R. Davis (USNM). (p. 113).
50. *Walshia particornella* (Bsk.), ♀. Devil's Den St. Pk., Wash. Co., Arkansas, 13 July 1966, R. W. Hodges (USNM). (p. 124).
51. *Walshia amorphella* Clem., ♀. Putnam Co., Illinois, 25 June 1961, M. O. Glenn (USNM). (p. 125).
52. *Walshia amorphella* Clem., ♂. Putnam Co., Illinois, em. 25 June 1961, M. O. Glenn (USNM). (p. 125).
53. *Walshia miscecolorella* (Cham.), ♂. Argus Mts., California, 12 May 1944, C. Henne (USNM). (p. 124).
54. *Walshia miscecolorella* (Cham.), ♀. Hart Prairie, 8500′, 10 mi NNW Flagstaff, Coconino Co., Arizona, 30 June 1961, Ronald W. Hodges (USNM). (p. 124).
55. *Walshia miscecolorella* (Cham.), ♀. Lincoln, Nebraska, em. 16 March 1967, G. R. Manglitz (USNM). (p. 124).
56. *Walshia miscecolorella* (Cham.), ♀. Madera Canyon, 4880′, Santa Rita Mts., Arizona, 7 July 1959, R. W. Hodges (USNM). (p. 124).
57. *Walshia miscecolorella* (Cham.), ♂. Devil's Den St. Pk., Wash. Co., Arkansas, 5 June 1966, R. W. Hodges (USNM). (p. 124).
58. *Walshia miscecolorella* (Cham.), ♂. Florida, Lake Placid, Archbold Bio. Sta., 1–7 May 1964, R. W. Hodges (USNM). (p. 124).
59. *Walshia miscecolorella* (Cham.), ♂. Fort Valley, 7350′, $7\frac{1}{2}$ mi NW Flagstaff, Coconino Co., Arizona, 24 July 1961, Ronald W. Hodges (USNM). (p. 124).
60. *Walshia dispar* Hodges, ♂. Paratype. Six Mile Creek, Ithaca, New York, 11 Aug. 1956, J. G. Franclemont (USNM). (p. 126).
61. *Walshia similis* Hodges, ♀. Paratype. Cherry Hill Recrtn. Area—Rte. 107, 2000′, Oconee Co., South Carolina, 20 Aug. 1958, R. W. Hodges (USNM). (p. 126).
62. *Walshia exemplata* Hodges, ♂. Holotype. Plummer's [*sic*] I., Maryland, July 1902, Aug. Busck (LACM). (p. 126).
63. *Nepotula secura* Hodges, ♂. Holotype. Pensacola, Florida, 28 Oct. 1961, Shirley Hills (USNM). (p. 126).
64. *Sorhagenia pexa* Hodges, ♀. Paratype. Devil's Den St. Pk., Wash. Co., Arkansas, 13 July 1966, R. W. Hodges (USNM). (p. 134).
65. *Perimede ricina* Hodges, ♀. Falls Church, Virginia, 9 July 1962, O. S. Flint (USNM). (p. 131).

1
2
3
4
5
6
7
8
9
10
11
12
13
14
15
16
17
18
19
20
21
22
23
24
25
26
27
28
29
30
31
32
33
34
35
36
37
38
39
40
41
42
43
44
45
46
47
48
49
50
51
52
53
54
55
56
57
58
59
60
61
62
63
64
65

PLATE 5

# Gelechioidea

## COSMOPTERIGIDAE

figs. 1–59

FOUR TIMES NATURAL SIZE 4:1

1. *Periploca orichalcella* (Clem.), ♀. Oroville, California, 7 March 1928, H. H. Keifer (USNM). (p. 103).
2. *Periploca hostiata* Hodges, ♂. Holotype. Washington, Yakima Co., 2.5 mi W Ft. Simcoe, 31 July 1962, J. F. G. Clarke (USNM). (p. 105).
3. *Periploca hortatrix* Hodges, ♀. Holotype. Devil's Den St. Pk., Wash. Co., Arkansas, 20 July 1966, R. W. Hodges (USNM). (p. 106).
4. *Periploca opinatrix* Hodges, ♂. Holotype. Wyoming, 6 mi NW Newcastle, 13 July 1965, R. W. Hodges (USNM). (p. 106).
5. *Periploca mimula* Hodges, ♂. Wyoming, 6 mi NW Newcastle, 18 July 1965, R. W. Hodges (USNM). (p. 107).
6. *Periploca laeta* Hodges, ♂. Paratype. Madera Canyon, 4880′, Santa Rita Mts., Arizona, 22 July 1959, R.W. Hodges (USNM). (p. 107).
7. *Periploca cata* Hodges, ♀. Paratype. Putnam Co., Illinois, em. 30 May 1960, M. O. Glenn (USNM). (p. 107).
8. *Periploca cata* Hodges, ♀. Paratype. Putnam Co., Illinois, em. 9 May 1960, M. O. Glenn (USNM). (p. 107).
9. *Periploca devia* Hodges, ♂. Holotype. Fort Valley, 7350′, 7½ mi NW Flagstaff, Coconino Co., Arizona, 22 June 1961, Ronald W. Hodges (USNM). (p. 108).
10. *Periploca nigra* Hodges, ♂. Holotype. Sacramento, Sac. Co., California, em. 14 Feb. 1961, A. Forbes (USNM). (p. 108).
11. *Periploca nigra* Hodges, ♂. Devil's Den St. Pk., Wash. Co., Arkansas, 30 May 1966, R. W. Hodges (USNM). (p. 108).
12. *Periploca fessa* Hodges, ♂. Wyoming, 6 mi NW Newcastle, 13 July 1965, R. W. Hodges (USNM). (p. 109).
13. *Periploca gulosa* Hodges, ♂. Paratype. Madera Canyon, 4880′, Santa Rita Mts., Arizona, 20 Aug. 1959, R. W. Hodges (USNM). (p. 109).
14. *Periploca juniperi* Hodges, ♀. Holotype. Wyoming, 6 mi NW Newcastle, em. 3 Aug. 1965, R. W. Hodges (USNM). (p. 109).
15. *Periploca funebris* Hodges, ♂. Paratype. Madera Canyon, 4880′, Santa Rita Mts., Arizona, 9 July 1959, R. W. Hodges (USNM). (p. 110).
16. *Periploca dipapha* Hodges, ♂. Holotype. Madera Canyon, 4880′, Santa Rita Mts., Arizona, 27 July 1959, R. W. Hodges (CU). (p. 111).
17. *Periploca labes* Hodges, ♀. Holotype. Vail Lake Rd., 6500′, 9½ mi SE Flagstaff, Coconino Co., Arizona, 11 July 1961, R. W. Hodges (USNM). (p. 111).
18. *Synploca gumia* Hodges, ♀. Paratype. Madera Canyon, 4880′, Santa Rita Mts., Arizona, 2 July 1959, R. W. Hodges (USNM).(p. 114).
19. *Stilbosis juvantis* (Hodges), ♂. Paratype. Madera Canyon, 4880′, Santa Rita Mts., Arizona, 2 July 1959, R. W. Hodges (USNM). (p. 117).
20. *Stilbosis dulcedo* (Hodges), ♀. 2 mi E Antioch, Contra Costa Co., California, em. 16 May 1968, P. A. Opler (USNM). (p. 117).
21. *Stilbosis venifica* (Hodges), ♂. Putnam Co., Illinois, 3 June 1962, M. O. Glenn (USNM). (p. 117).
22. *Stilbosis venifica* (Hodges), ♂. Devil's Den St. Pk., Wash. Co., Arkansas, 12 June 1966, R. W. Hodges (USNM). (p. 117).
23. *Stilbosis venatrix* (Hodges), ♂. Paratype. Madera Canyon, 4880′, Santa Rita Mts., Arizona, 11 July 1959, R. W. Hodges (USNM). (p. 117).
24. *Stilbosis ostryaeella* (Cham.), ♂. Devil's Den St. Pk., Wash. Co., Arkansas, 6 July 1966, R. W. Hodges (USNM). (p. 117).
25. *Stilbosis quadricustatella* (Cham.), ♂. Florida, Lake Placid, Archbold Bio. Sta., 16–22 May 1964, R. W. Hodges (USNM). (p. 118).
26. *Stilbosis quadricustatella* (Cham.), ♂. Devil's Den St. Pk., Wash. Co., Arkansas, 5 July 1966, R. W. Hodges (USNM). (p. 118).
27. *Stilbosis stipator* (Hodges), ♂. Holotype. Vail Lake Rd., 6500′, 9½ mi SE Flagstaff, Coconino Co., Arizona, 11 July 1961, Ronald W. Hodges (USNM). (p. 118).
28. *Stilbosis risor* (Hodges), ♀. Madera Canyon, 4880′, Santa Rita Mts., Arizona, 19 Sept. 1959, R. W. Hodges (USNM). (p. 118).
29. *Stilbosis victor* (Hodges), ♂. Lakehurst, New Jersey, 31 May 1962, R. W. Hodges (USNM). (p. 118).
30. *Stilbosis placatrix* (Hodges), ♀. Holotype. Devil's Den St. Pk., Wash. Co., Arkansas, 20 June 1966, R. W. Hodges (USNM). (p. 118).
31. *Stilbosis sagana* (Hodges), ♂. Paratype. Madera Canyon, 4880′, Santa Rita Mts., Arizona, 28 Sept. 1959, R. W. Hodges (USNM). (p. 118).
32. *Stilbosis rhynchosiae* (Hodges), ♀. Paratype. Chickasaw Forest, Chester Co., Tennessee, em. 1 Aug. 1938, A. F. Braun (USNM). (p. 119).
33. *Stilbosis extensa* (Braun), ♀. Antioch, Contra Costa Co., California, 15 May 1958, J. Powell (USNM). (p. 119).
34. *Stilbosis nubila* Hodges, ♂. West Fork, 16 mi SW Flagstaff, Coconino Co., Arizona, 8 July 1961, Ronald W. Hodges (USNM). (p. 122).
35. *Stilbosis tesquella* Clem., ♀. Highlands, 3865′, Macon Co., North Carolina, 18 July 1958, R. W. Hodges (USNM). (p. 121).
36. *Stilbosis tesquella* Clem., ♀. D.C., em. 6 Aug. 1899 (USNM). (p. 121).
37. *Stilbosis nubila* Hodges, ♂. Paratype. Madera Canyon, 4880′, Santa Rita Mts., Arizona, 22 Aug. 1959, R. W. Hodges (USNM). (p. 122).
38. *Stilbosis lonchocarpella* Bsk., ♀. Key Largo, Monroe Co., Florida, 13 Feb. 1967, Mrs. Spencer Kemp (USNM). (p. 122).
39. *Chrysopeleia purpuriella* Cham., ♀. Devil's Den St. Pk., Wash. Co., Arkansas, 3 July 1966, R. W. Hodges (USNM). (p. 122).
40. *Sorhagenia nimbosa* (Braun), ♀. Wiemar, Placer Co., California, 7 July 1932, Fourness (USNM). (p. 132).
41. *Sorhagenia daedala* Hodges, ♂. Holotype. Mt. Shasta City, Siskiyou Co., California, 4 July 1958, J. Powell (USNM). (p. 133).
42. *Sorhagenia baucidis* Hodges, ♀. Paratype. Putnam Co., Illinois, 2 July 1966, M. O. Glenn (USNM). (p. 133).
43. *Siskiwitia alticolans* Hodges, ♂. Holotype. Arizona, Santa Cruz Co., Madera Canyon, Bog Springs Campground, 5100′, Santa Rita Mts., 10–26 July 1964, D. R. Davis (USNM). (p. 112).
44. *Siskiwitia latebra* Hodges, ♂. Devil's Den St. Pk., Wash. Co., Arkansas, 11 July 1966, R. W. Hodges (USNM). (p. 112).
45. *Perimede erransella* Cham., ♂. Devil's Den St. Pk., Wash. Co., Arkansas, 15 June 1966, R. W. Hodges (USNM). (p. 129).
46. *Perimede falcata* Braun, ♀. Florida, Parker Is., Highlands Co., 4–7 June 1964, R. W. Hodges (USNM). (p. 131).
47. *Perimede erema* Hodges, ♀. Holotype. Devil's Den St. Pk., Wash. Co., Arkansas, 27 June 1966, R. W. Hodges (USNM). (p. 131).
48. *Perimede parilis* Hodges, ♀. Holotype. Devil's Den St. Pk., Wash. Co., Arkansas, 10 July 1966, R. W. Hodges (USNM). (p. 130).
49. *Perimede circitor* Hodges, ♀. Holotype. Devil's Den St. Pk., Wash. Co., Arkansas, 18 July 1966, R. W. Hodges (USNM). (p. 130).
50. *Perimede maniola* Hodges, ♀. Holotype. Putnam Co., Illinois, 19 Aug. 1965, M. O. Glenn (USNM). (p. 131).
51. *Perimede latris* Hodges, ♂. Paratype. Madera Canyon, 4880′, Santa Rita Mts., Arizona, 18 Aug. 1959, R. W. Hodges (USNM). (p. 130).
52. *Perimede battis* Hodges, ♀. Paratype. Peña Blanca Canyon, Santa Cruz Co., Arizona, 8 Aug. 1959, R. W. Hodges (USNM). (p. 129).
53. *Ithome concolorella* (Cham.), ♂. Devil's Den St. Pk., Wash. Co., Arkansas, 30 June 1966, R. W. Hodges (USNM). (p. 135).
54. *Ithome curvipunctella* (Wlsm.), ♂. Siesta Key, Sarasota Co., Florida, 31 May 1957, C. P. Kimball (USNM). (p. 135).
55. *Ithome edax* Hodges, ♀. Paratype. Brownsville, [Texas], em. 17 June 1944 (USNM). (p. 136).
56. *Ithome lassula* Hodges, ♀. Paratype. Key West, Florida, em. 16 March 1945 (USNM). (p. 137).
57. *Ithome ferax* Hodges, ♂. Paratype. Siesta Key, Sarasota Co., Florida, 29 Feb. 1960, C. P. Kimball (USNM). (p. 136).
58. *Ithome aquila* Hodges, ♂. Holotype. Tavernier Plantation Key, Florida, em. 5 Jan. 1971, J. Powell (UCB). (p. 136).
59. *Obithome punctiferella* (Bsk.), ♂. Holotype. Victoria, Texas, "24-4", H. S. Barber (USNM). (p. 138).

1
2
3
4
5
6
7
8
9
10
11
12
13
14
15
16
17
18
19
20
21
22
23
24
25
26
27
28
29
30
31
32
33
34
35
36
37
38
39
40
41
42
43
44
45
46
47
48
49
50
51
52
53
54
55
56
57
58
59

PLATE 6

# Gelechioidea

## COSMOPTERIGIDAE

figs. 1–34

TWICE NATURAL SIZE 2:1

1. *Cosmopterix sinelinea* Hodges, ♂. Holotype. McClellanville, South Carolina, The Wedge, 26 Aug. 1971, R. W. Hodges (USNM). (p. 24).
2. *Cosmopterix chisosensis* Hodges, ♂. Holotype. Texas, Brewster Co., Chisos Mts., Green Gulch, 5500′, 6 June 1973, R. W. Hodges (USNM). (p. 31).
3. *Teladoma tonia* Hodges, ♂. Holotype. Madera Canyon, 5600′, Santa Rita Mts., Arizona, 1 Aug. 1959, R. W. Hodges (CU). (p. 50).
4. *Anoncia slales* Hodges, ♀. Holotype. Mt. Hamilton, Santa Clara Co., California, 11 July 1967, W. C. Gagné (UCB). (p. 61).
5. *Anoncia flegax* Hodges, ♂. Holotype. Mt. Shasta City, Siskiyou Co., California, 3 July 1958, J. Powell (UCB). (p. 62).
6. *Anoncia mones* Hodges, ♂. Holotype. Texas, Culberson Co., Guadalupe Mts., Cherry Cn., 5096′, 22 May 1973, R. W. Hodges (USNM). (p. 63).
7. *Anoncia naclia* Hodges, ♂. Holotype. Baboquivari Mts., Pima Co., Arizona, 1–15 July 1921, O. C. Poling (USNM). (p. 64).
8. *Anoncia psepsa* Hodges, ♀. Holotype. White Mts., 7000′, Arizona, 5–15 July 1925, O. C. Poling (USNM). (p. 65).
9. *Anoncia fregeis* Hodges, ♀. Holotype. Carrizo Plains, Kern Co., California, 20 Aug. 1935, G. D. Hanna (USNM). (p. 67).
10. *Anoncia noscres* Hodges, ♀. Holotype. Kern Co., California, Piute Mts., 4350′, T29S, R33E, Sec. 23, 1–3 June 1973, J. P. & K. E. Donahue (LACM). (p. 68).
11. *Anoncia venis* Hodges, ♀. Holotype. Arizona, Yuma Co., Quartzite, 4 April 1973, J. H. Baker (USNM). (p. 70).
12. *Anoncia smogops* Hodges, ♂. Holotype. Baboquivari Mts., Pima Co., Arizona, 1–15 May 1924, O. C. Poling (USNM). (p. 72).
13. *Anoncia smogops* Hodges, ♀. Paratype. Olancha, Inyo Co., California, 8–15 June (USNM). (p. 72).
14. *Euclemensia bassettella* (Clem.), ♂. Devil's Den St. Pk., Wash. Co., Arkansas, 12 July 1966, R. W. Hodges (USNM). (p. 18).
15. *Euclemensia schwarziella* Bsk., ♀. Nogales, Arizona, 8 July 1945 (USNM). (p. 19).
16. *Periploca arsa* Hodges, ♂. Paratype. Key Largo Key, Monroe Co., Florida, 1 Oct. 1971, Mrs. S. Kemp (USNM). (p. 103).
17. *Periploca repanda* Hodges, ♂. Holotype. Texas, Brewster Co., Chisos Mts., Nugent Mt., 4 June 1973, R. W. Hodges (USNM). (p. 104).
18. *Periploca teres* Hodges, ♂. Holotype. Rustler Park, 8500′, Chiricahua Mts., Arizona, 20 Aug. 1972, J. Powell (UCB). (p. 105).
19. *Anoncia psentia* Hodges, ♀. Holotype. 14 mi NE Mojave, Kern Co., California, 13 April 1962, C. MacNeill, D. Rentz & R. Brown (CAS). (p. 73).
20. *Periploca tridens* Hodges, ♂. Holotype. Twain Harte, Tuolumne Co., California, 20 June 1969, M. Lundgren (UCB). (p. 105).
21. *Periploca soror* Hodges, ♀. Holotype. Texas, Brewster Co., Chisos Mts., Green Gulch, 5500′, 6 June 1973, R. W. Hodges (USNM). (p. 108).
22. *Periploca facula* Hodges, ♂. Paratype. Madera Canyon, 4400′, Santa Rita Mts., Arizona, 10 Oct. 1959, R. W. Hodges (USNM). (p. 109).
23. *Periploca serrulata* Hodges. ♂. Holotype. Johnson's Pass, 5 mi W Clover, Tooele Co., Utah, 13 Aug. 1973, J. Powell (UCB). (p. 110).
24. *Siskiwitia falcata* Hodges, ♂. Holotype. Lake Placid, Archbold Bio. Sta., 16–22 May 1964, R. W. Hodges (USNM). (p. 112).
25. *Stilbosis pagina* Hodges, ♀. Holotype. Madera Canyon, 4880′, Santa Rita Mts., Arizona, 24 July 1959, R. W. Hodges (CU). (p. 119).
26. *Stilbosis ornatrix* Hodges, ♀. Holotype. Madera Canyon, 4880′, Santa Rita Mts., Arizona, 20 July 1959, R. W. Hodges (CU). (p. 120).
27. *Stilbosis scleroma* Hodges, ♀. Holotype. Madera Canyon, 4880′, Santa Rita Mts., Arizona, 3 Aug. 1959, R. W. Hodges (CU). (p. 120).
28. *Stilbosis rotunda* Hodges, ♀. Holotype. Madera Canyon, 4880′, Santa Rita Mts., Arizona, 9 July 1959, R. W. Hodges (CU). (p. 121).
29. *Walshia elegans* Hodges, ♂. Holotype. Fluker, Tangipahoa Par., Louisiana, 18 Sept. 1971, G. Strickland (USNM). (p. 124).
30. *Walshia floridensis* Hodges, ♂. Holotype. Lake Alfred, Florida, 31 Dec. 1929, L. Bottimer (USNM). (p. 125).
31. *Afeda biloba* Hodges, ♀. Holotype. Tavernier, Monroe Co., Florida, 6 Oct. 1955, J. N. Todd (USNM). (p. 127).
32. *Perimede grandis* Hodges, ♂. Holotype. Texas, Culberson Co., Guadalupe Mts., McKittrick Cn., 5000′, 23 May 1973, R. W. Hodges (USNM). (p. 130).
33. *Sorhagenia cracens* Hodges, ♂. Holotype. Spirit Lake, Idaho, 16–23 July (USNM). (p. 133).
34. *Ithome simulatrix* Hodges, ♂. Holotype. Everglade, Florida, 8–15 April (USNM). (p. 137).

1
2
3
4
5
6
7
8
9
10
11
12
13
14
15
16
17
18
19
20
21
22
23
24
25
26
27
28
29
30
31
32
33
34

# NOTES

1. ABBREVIATIONS FOR COLLECTORS AND COLLECTIONS

| | |
|---|---|
| AB | André Blanchard |
| ABK | Alexander B. Klots |
| AEB | A. E. Brower |
| AFB | Annette F. Braun |
| AMNH | American Museum of Natural History |
| ANSP | Academy of Natural Sciences, Philadelphia |
| BH | Bernard Heineman |
| BM | Bryant Mather |
| BMNH | British Museum (Natural History) |
| CAS | California Academy of Sciences, San Francisco |
| CM | Carnegie Museum |
| CNC | Canadian National Collection |
| CPK | C. P. Kimball |
| CU | Cornell University |
| DCF | D. C. Ferguson |
| FMNH | Field Museum of Natural History |
| GS | Gayle Strickland |
| HUMB | Museum Alexander Humboldt, Berlin University |
| INHS | Illinois Natural History Survey |
| JGF | J. G. Franclemont |
| JN | John Newman |
| JRH | J. R. Heitzman |
| KWP | K. W. Philip |
| LACM | Los Angeles County Museum of Natural History |
| LEM | Lyman Entomological Museum |
| MCZ | Museum of Comparative Zoology, Harvard University |
| MNSA | Museu Argentino de Ciencias Naturales "Bernardino Rivadavia", Buenos Aires |
| MOG | Murray O. Glenn |
| MSU | Michigan State University |
| NHMV | Naturhistorisches Museum, Vienna |
| NSM | Nova Scotia Museum |
| PMBC | Provincial Museum of British Columbia |
| RHL | Ronald H. Leuschner |
| ROK | R. O. Kendall |
| ROM | Royal Ontario Museum, University of Toronto |
| SAH | Sidney A. Hessel |
| UA | University of Alberta |
| UBC | University of British Columbia |
| UCB | University of California, Berkeley |
| UCD | University of California, Davis |
| UM | University of Michigan |
| UMO | University Museum, Oxford |
| USNM | United States National Museum |
| WEM | William E. Miller |
| WPC | Wedge Plantation Collection (R. B. Dominick) |
| YPM | Peabody Museum of Natural History, Yale University |
| ZSBS | Zoologische Sammlung des Bayerischen Staates, Munich |

2. COMMON NAMES

The use of an asterisk* in the text denotes a name listed in *Common Names of Insects Approved by The Entomological Society of America* (1970).

French-language common names have been taken from Auclair, J. L., *et al.*, 1964, *French Names of Insects of Canada*, 3rd edition, published for the Quebec Society for the Protection of Plants by Department of Agriculture and Colonization, Quebec. The abbreviation "m." after a name indicates that it is masculine, "f." that it is feminine.

3. CITATIONS OF AUTHORITIES

Authors' names without parentheses indicate that the specific name is associated with the genus in which it was described.

Authors' names in parentheses indicate that the specific name has been transferred from the genus in which it was described to another genus.

4. WING LENGTHS

Wing length is the measurement in millimeters from the base to the apex of the forewing.

5. LOCATION OF TYPE-SPECIMEN

The current location of the type-specimen is given by the appropriate abbreviation in square brackets immediately following the type-locality.

# INDEX TO ANIMAL NAMES

Principal entries are given in bold face
Plate references are given as (3:20)

Generic names cited only in combination with specific names, whether in synonymy or text, are not given in the index. Look for such entries under the specific name. For example, *Cosmopterix scirpicola* will be found under *scirpicola*, but not under *Cosmopterix*.

## INDEX

# INDEX TO PLANT NAMES

INDEX

MASSACHUSETTS AUDUBON SOCIETY
SOUTH GREAT ROAD
LINCOLN, MASSACHUSETTS 01773